"This v[illegible] tion—in these ve[illegible]!"

Rudy Rucker
Author of *Mind Tools*
and *The Fourth Dimension*

"The adventure of mathematics outstrips the exploration of the stars and of atoms. Kasner and Newman revive the adventure for a new generation of readers who like to think. Here we appreciate amazing achievements in the science of pure thought, its fundamental importance as a model of the manifest."

A.K. Dewdney
"Computer Recreations" columnist,
Scientific American;
Author of *The Planiverse*

"MATHEMATICS AND THE IMAGINATION is an exciting and instructive book which teaches you mathematics and delights the imagination."

Raymond Smullyan
Author of *Forever Undecided: A Puzzle Guide to Gödel, The Lady or the Tiger? & Other Logic Puzzles*

"Those who enjoy fooling around in the infinitely surprising bypaths of mathematics will find this book a handy guide."

The New Yorker

"The reader is actually taken to the outposts of mathematics instead of being kept in endless training in the rear."

The New York Times

"This book, which lives up to its title in every way, might well have been merely terrifying, whereas it proves to be both charming and exciting. Even minds normally predisposed to shy at precise definitions and the logical derivation of consequences therefrom can hardly fail to be attracted to so engaging an exposition of what is to be found in the mathematician's chamber of horrors."

The Saturday Review of Literature

MATHEMATICS AND THE IMAGINATION

"A brilliant and entirely charming book about the subject from which most of us run away in terror...."
The New Republic

T E M P U S™

Edward Kasner
James R. Newman

PUBLISHED BY
Tempus Books of Microsoft Press
A Division of Microsoft Corporation
16011 N.E. 36th Way, Box 97017
Redmond, Washington 98073-9717

Library of Congress Cataloging in Publication Data
Kasner, Edward, 1878-1955.
Mathematics and the imagination / Edward Kasner, James Newman.
p. cm.
Reprint. Originally published: New York : Simon and Schuster, 1940.
Bibliography: p.
Includes index.
ISBN 1-55615-104-7 (pbk.) : $8.95
1. Mathematics. 2. Mathematical recreations. I. Newman, James
Roy, 1907-1966. II. Title.
QA93.K3 1989 87-19705
510--dc19 CIP
Rev.

Printed and bound in the United States of America.

1 2 3 4 5 6 7 8 9 HCHC 3 2 1 0 9

Distributed to the book trade in the United States by Harper & Row.

Distributed to the book trade in Canada by General Publishing Company, Ltd.

To R.G.
without whose selfless help
and understanding there would
have been no book.

ACKNOWLEDGMENT

We are indebted to many books, too many to enumerate. Some of them are listed in the selected bibliography.

And we wish to acknowledge particularly the services of Mr. Don Mittleman of Columbia University, whose help in preparation of the manuscript has been generous and invaluable.

Table of Contents

Introduction

I. NEW NAMES FOR OLD 3

Easy words for hard ideas . . . Transcendental . . . Non-simple curve . . . Simple curve . . . Simple group . . . Bolsheviks and giraffes . . . Turbines . . . Turns and slides . . . Circles and cycles . . . Patho-circles . . . Clocks . . . Hexagons and parhexagons . . . Radicals, hyperradicals, and ultraradicals (nonpolitical) . . . New numbers for the nursery . . . Googol and googolplex . . . Miracle of the rising book . . . The mathescope.

II. BEYOND THE GOOGOL 27

Counting—the language of number . . . Counting, matching, and "Going to Jerusalem" . . . Cardinal numbers . . . Cosmic chess and googols . . . The sand reckoner . . . Mathematical induction . . . The infinite and its progeny . . . Zeno . . . Puzzles and quarrels . . . Bolzano . . . Galileo's puzzle . . . Cantor . . . Measuring the measuring rod . . . The whole is no greater than some of its parts . . . The first transfinite—Aleph$_0$ *. . . Arithmetic for morons . . . Common sense hits a snag . . . Cardinality of the continuum . . . Extravagances of a mathematical madman . . . The tortoise unmasked . . . Motionless motion . . . Private life of a number . . . The house that Cantor built.*

III. π, i, e (PIE) 65

Chinamen and chandeliers . . . Twilight of common sense . . . π, i, e *. . . Squaring the circle and its cousins . . . Mathematical impossibility . . . Silk purse, sow's ear, ruler and compass . . . Rigor mortis . . . Algebraic equations and tran-*

scendental numbers . . . Galois and Greek epidemics . . . Cube duplicators and angle trisectors . . . Biography of π *. . . Infancy: Archimedes, the Bible, the Egyptians . . . Adolescence: Vieta, Van Ceulen . . . Maturity: Wallis, Newton, Leibniz . . . Old Age: Dase, Richter, Shanks . . . Victim of schizophrenia . . . Boon to insurance companies . . .* (e) *. . . Logarithms or tricks of the trade . . . Mr. Briggs is surprised . . . Mr. Napier explains . . . Biography of* e*; or* e*, the banker's boon . . . Pituitary gland of mathematics: the exponential function . . .* (i) *. . . Humpty Dumpty, Doctor of Semantics . . . Imaginary numbers . . . The* $\sqrt{-1}$*, or "Where am I?" . . . Biography of* i*, the self-made amphibian . . . Omar Khayyám, Cardan, Bombelli, and Gauss . . .* i *and Soviet Russia . . . Program music of mathematics . . . Breakfast in bed; or, How to become a great mathematician . . . Analytic geometry . . . Geometric representation of* i *. . . Complex plane . . . A famous formula, faith, and humility.*

IV. ASSORTED GEOMETRIES—PLANE AND FANCY 112

The talking fish and St. Augustine . . . A new alphabet . . . High priests and mumbo jumbo . . . Pure and applied mathematics . . . Euclid and Texas . . . Mathematical tailors . . . Geometry—a game . . . Ghosts, table-tipping, and the land of the dead . . . Fourth-dimension flounders . . . Henry More to the rescue . . . Fourth dimension—a new gusher . . . A cure for arthritis . . . Syntax suffers a setback . . . The physicist's delight . . . Dimensions and manifolds . . . Distance formulae . . . Scaling blank walls . . . Four-dimensional geometry defined . . . Moles and tesseracts . . . A four-dimensional fancy . . . Romance of flatland . . . Three-dimensional cats and two-dimensional kings . . . Gallant Gulliver and the gloves . . . Beguiling voices and strange footprints . . . Non-Euclidean geometry . . . Space credos and millinery . . . Private and public space . . . Rewriting our textbooks . . . The prince and the Boethians . . . The flexible fifth . . . The mathematicians unite—nothing to lose but their chains . . . Lobachevsky breaks a link . . . Riemann breaks another . . . Checks and double checks in mathematics . . . The tractrix and the pseudosphere . . . Great circles and bears . . . The skeptic persists—and is stepped on . . . Geodesics . . . Seventh Day Adventists . . . Curvature . . . Lobachevskian Eiffel Towers and Riemannian Holland Tunnels.

V. PASTIMES OF PAST AND PRESENT TIMES 156

Puzzle acorns and mathematical oaks . . . Charlemagne and crossword puzzles . . . Mark Twain and the "farmer's daughter" . . . The syntax of puzzles . . . Carolyn Flaubert and the cabin boy . . . A wolf, a goat, and a head of cabbage . . . Brides and cuckolds . . . I'll be switched . . . Poisson, the misfit . . . High finance; or, The international beer wolf . . . Lions and poker players . . . The decimal system . . . Casting out nines . . . Buddha, God, and the binary scale . . . The march of culture; or, Russia, the home of the binary system . . . The Chinese rings . . . The tower of Hanoi . . . The ritual of Benares: or, Charley horse in the Orient . . . Nim, Sissa Ben Dahir, and Josephus . . . Bismarck plays the boss . . . The 15 puzzle plague . . . The spider and the fly . . . A nightmare of relatives . . . The magic square . . . Take a number from 1 to 10 . . . Fermat's last theorem . . . Mathematics' lost legacy.

VI. PARADOX LOST AND PARADOX REGAINED 193

Great paradoxes and distant relatives . . . Three species of paradox . . . Paradoxes strange but true . . . Wheels that move faster on top than on bottom . . . The cycloid family . . . The curse of transportation; or, How locomotives can't make up their minds . . . Reformation of geometry . . . Ensuing troubles . . . Point sets—the Arabian Nights of mathematics . . . Hausdorff spins a tall tale . . . Messrs. Banach and Tarski rub the magic lamp . . . Baron Munchhausen is stymied by a pea . . . Mathematical fallacies . . . Trouble from a bubble; or, Dividing by zero . . . The infinite—troublemaker par excellence . . . Geometrical fallacies . . . Logical paradoxes—the folk tales of mathematics . . . Deluding dialectics of the poacher and the prince; of the introspective barber; of the number 111777; of this book and Confucius; of the Hon. Bertrand Russell . . . Scylla and Charybdis; or, What shall poor mathemathics do?

VII. CHANCE AND CHANCEABILITY 223

The clue of the billiard cue . . . A little chalk, a lot of talk . . . Watson gets his leg pulled by probable inference . . . Finds it all absurdly simple . . . Passionate oysters, waltzing ducks, and the syllogism . . . The twilight of probability . . . Interesting behavior of a modest coin . . . Biological necessity and a pair of dice . . . What is probability? . . . A poll of views: a meteorologist, a bootlegger, a bridge player . . . The subjective view—based on insufficient reason, contains an element of truth . . . The jackasses on Mars . . . The statistical view . . . What happens will probably happen . . . Experimental eurythmics; or, Pitching pennies . . . Relative frequencies . . . The adventure of the dancing men . . . Scheherezade and John Wilkes Booth—a challenge to statistics . . . The red and the black . . . Charles Peirce predicts the weather . . . How far is "away"? . . . Herodotus explains . . . The calculus of chance . . . The benefits of gambling . . . De Mere and Pascal . . . Mr. Jevons omits an acknowledgment . . . The study of craps —the very guide of life . . . Dice, pennies, permutations, and combinations . . . Measuring probabilities . . . D'Alembert drops the ball . . . Count Buffon plays with a needle . . . The point . . . A black ball and a white ball . . . The binomial theorem . . . The calculus of probability re-examined . . . Found to rest on hypothesis . . . Laplace needs no hypothesis . . . Twits Napoleon, who does . . . The Marquis de Condorcet has high hopes . . . M. le Marquis omits a factor and loses his head . . . Fourier of the Old Guard . . . Dr. Darwin of the New . . . The syllogism scraps a standby . . . Mr. Socrates may not die . . . Ring out the old logic, ring in the new.

VIII. RUBBER-SHEET GEOMETRY 265

Seven bridges over a stein of beer . . . Euler shivers . . . Is warmed by news from home . . . Invents topology . . . Dissolves the dilemma of Sunday strollers . . . Babies' cribs and Pythagoreans . . . Talismen and queer figures . . . Position is everything in topology . . . Da Vinci and Dali . . . Invariants . . . Transformations . . . The immutable derby . . . Competition for the caliph's cup; or, Sifting out the suitors by science . . . Mr. Jordan's theorem . . . Only seems *idiotic . . . Deformed circles . . . Odd facts concerning Times Square and a balloonist's head . . . Eccentric deportment of several distinguished gentlemen at Princeton . . . Their passion for pretzels . . . Their delving in doughnuts . . . Enforced modesty of readers and authors . . . The ring . . . Lachrymose recital around a Paris* pissoir . . . "*Who staggered how many times around the walls of what?*" . . . *In and out the doughnut . . . Gastric surgery—from doughnut to sausage in a single cut . . . N-dimensional pretzels . . . The Möbius strip . . . Just as black as it is painted . . . Foments industrial discontent . . . Never takes sides . . . Bane of painter and paintpot alike . . . The iron rings . . . Mathematical cotillion; or, How on earth do I get rid of my partner? . . . Topology—the pinnacle of perversity; or, Removing your vest without your coat . . . Down to earth—map coloring . . . Four-color problem . . . Euler's theorem . . . The simplest universal law . . . Brouwer's puzzle . . . The search for invariants.*

IX. CHANGE AND CHANGEABILITY 299

The calculus and cement . . . Meaning of change and rate of change . . . Zeno and the movies . . . "Flying Arrow" local—stops at all points . . . Geometry and genetics . . . The arithmetic men dig pits . . . Lamentable analogue of the boomerang . . . History of the calculus . . . Kepler . . . Fermat . . . Story of the great rectangle . . . Newton and Leibniz . . . Archimedes and the limit . . . Shrinking and swelling; or, "Will the circle go the limit?" . . . Brief dictionary of mathematics and physics . . . Military idyll; or, The speed of the falling bomb . . . The calculus at work . . . The derivative . . . Higher derivatives and radius of curvature . . . Laudable scholarship of automobile engineers . . . The third derivative as a shock absorber . . . The derivative finds its mate . . . Integration . . . Kepler and the bungholes . . . Measuring lengths; or, The yawning regress . . . Methods of approximation . . . Measuring areas under curves . . . Method of rectangular strips . . . The definite $\int$ *. . . Indefinite* $\int$ *. . . One the inverse of the other . . . The outline of history and the descent of man: or,* $y=e^x$ *. . . Sickly curves and orchidaceous* *ones . . . The snowflake . . . Infinite perimeters and postage stamps . . . Anti-snowflake . . . Super-colossal pathological specimen—the curve that fills space . . . The unbelieveable crisscross.*

EPILOGUE, MATHEMATICS AND THE IMAGINATION 357

Introduction

The fashion in books in the last decade or so has turned increasingly to popular science. Even newspapers, Sunday supplements and magazines have given space to relativity, atomic physics, and the newest marvels of astronomy and chemistry. Symptomatic as this is of the increasing desire to know what happens in laboratories and observatories, as well as in the awe-inspiring conclaves of scientists and mathematicians, a large part of modern science remains obscured by an apparently impenetrable veil of mystery. The feeling is widely prevalent that science, like magic and alchemy in the Middle Ages, is practiced and can be understood only by a small esoteric group. The mathematician is still regarded as the hermit who knows little of the ways of life outside his cell, who spends his time compounding incredible and incomprehensible theories in a strange, clipped, unintelligible jargon.

Nevertheless, intelligent people, weary of the nervous pace of their own existence—the sharp impact of the happenings of the day—are hungry to learn of the accomplishments of more leisurely, contemplative lives, timed by a slower, more deliberate clock than their own. Science, particularly mathematics, though it seems less practical and less real than the news contained in the latest radio dispatches, appears to be building the one permanent and stable edifice in an age where all others are either crumbling or being blown to bits. This is not to say that science has not also undergone revolutionary changes. But it has happened quietly and honorably. That which is no longer useful has been rejected only after mature deliberation, and the building has been reared steadily on the creative achievements of the past.

Thus, in a certain sense, the popularization of science is a duty to be performed, a duty to give courage and comfort to the men and women of good will everywhere who are gradually losing their faith in the life of reason. For most of the sciences the veil of mystery is gradually being torn asunder. Mathematics, in large measure, remains unrevealed. What most popular books on mathematics have tried to do is either to discuss it philosophically, or to make clear the stuff once learned and already forgotten. In this respect our purpose in writing has been somewhat different. "Haute vulgarisation" is the term applied by the French to that happy result which neither offends by its condescension nor leaves obscure in a mass of technical verbiage. It has been our aim to extend the process of "haute vulgarisation" to those outposts of mathematics which are mentioned, if at all, only in a whisper; which are referred to, if at all, only by name; to show by its very diversity something of the character of mathematics, of its bold, untrammeled spirit, of how, as both an art and a science, it has continued to lead the creative faculties beyond even imagination and intuition. In the compass of so brief a volume there can only be snapshots, not portraits. Yet, it is hoped that even in this kaleidoscope there may be a stimulus to further interest in and greater recognition of the proudest queen of the intellectual world.

MATHEMATICS AND THE IMAGINATION

I will not go so far as to say that to construct a history of thought without profound study of the mathematical ideas of successive epochs is like omitting Hamlet from the play which is named after him. That would be claiming too much. But it is certainly analogous to cutting out the part of Ophelia. This simile is singularly exact. For Ophelia is quite essential to the play, she is very charming,—and a little mad. Let us grant that the pursuit of mathematics is a divine madness of the human spirit, a refuge from the goading urgency of contingent happenings.

—ALFRED NORTH WHITEHEAD,
Science and the Modern World.

New Names for Old

For out of olde feldes, as men seith,
Cometh al this newe corn fro yeer to yere;
And out of olde bokes, in good feith,
Cometh al this newe science that men lere.

—CHAUCER

EVERY ONCE in a while there is house cleaning in mathematics. Some old names are discarded, some dusted off and refurbished; new theories, new additions to the household are assigned a place and name. So what our title really means is new *words* in mathematics; not new names, but new words, new terms which have in part come to represent new concepts and a reappraisal of old ones in more or less recent mathematics. There are surely plenty of words already in mathematics as well as in other subjects. Indeed, there are so many words that it is even easier than it used to be to speak a great deal and say nothing. It is mostly through words strung together like beads in a necklace that half the population of the world has been induced to believe mad things and to sanctify mad deeds. Frank Vizetelly, the great lexicographer, estimated that there are 800,000 words in use in the English language. But mathematicians, generally quite modest, are not satisfied with these 800,000; let us give them a few more.

We can get along without new names until, as we advance in science, we acquire new ideas and new forms.

A peculiar thing about mathematics is that it does not use so many long and hard names as the other sciences. Besides, it is more conservative than the other sciences in that it clings tenaciously to old words. The terms used by Euclid in his *Elements* are current in geometry today. But an Ionian physicist would find the terminology of modern physics, to put it colloquially, pure Greek. In chemistry, substances no more complicated than sugar, starch, or alcohol have names like these: Methylpropenylenedihydroxycinnamenylacrylic acid, or, 0-anhydrosulfaminobenzoine, or, protocatechuicaldehydemethylene. It would be inconvenient if we had to use such terms in everyday conversation. Who could imagine even the aristocrat of science at the breakfast table asking, "Please pass the 0-anhydrosulfaminobenzoic acid," when all he wanted was sugar for his coffee? Biology also has some tantalizing tongue twisters. The purpose of these long words is not to frighten the exoteric, but to describe with scientific curtness what the literary man would take half a page to express.

In mathematics there are many easy words like "group," "family," "ring," "simple curve," "limit," etc. But these ordinary words are sometimes given a very peculiar and technical meaning. In fact, here is a booby-prize definition of mathematics: *Mathematics is the science which uses easy words for hard ideas.* In this it differs from any other science. There are 500,000 known species of insects and every one has a long Latin name. In mathematics we are more modest. We talk about "fields," "groups," "families," "spaces," although much more meaning is attached to these words than ordinary conversation implies. As its use becomes more and more technical, nobody can guess the mathematical meaning of a word any more than one could guess that a "drug store" is a place where they sell ice-cream sodas and umbrellas. No one could guess the meaning of the word "group" as it is used in mathematics. Yet it is so important that whole courses are given on the theory of "groups," and hundreds of books are written about it.

Because mathematicians get along with common words, many amusing ambiguities arise. For instance, the word "function" probably expresses the most important idea in the whole history of mathematics. Yet, most people hearing it would think of a "function" as meaning an evening social affair, while others, less socially minded, would think of their livers. The word "function" has at least a dozen meanings, but few people suspect the mathematical one. The mathematical meaning (which we shall elaborate upon later) is expressed most simply by a *table*. Such a table gives the relation between two variable quantities when the value of one variable quantity is determined by the value of the other. Thus, one variable quantity may express the years from 1800 to 1938, and the other, the number of men in the United States wearing handle-bar mustaches; or one variable may express in decibels the amount of noise made by a political speaker, and the other, the blood pressure units of his listeners. You could probably never guess the meaning of the word "ring" as it has been used in mathematics. It was introduced into the newer algebra within the last twenty years. The theory of rings is much more recent than the theory of groups. It is now found in most of the new books on algebra, and has nothing to do with either matrimony or bells.

Other ordinary words used in mathematics in a peculiar sense are "domain," "integration," "differentiation." The uninitiated would not be able to guess what they represent; only mathematicians would know about them. The word "transcendental" in mathematics has not the meaning it has in philosophy. A mathematician would say: The number π, equal to 3.14159 . . . , is transcendental, because it is not the root of any algebraic equation with integer coefficients.

Transcendental is a very exalted name for a small number, but it was coined when it was thought that transcendental numbers were as rare as quintuplets. The work of Georg Cantor in the realm of the infinite has since proved that of all the numbers in mathematics, the transcendental ones are the most common, or, to use the word in a slightly different sense, the least transcendental. We shall talk of this later when we speak of another famous transcendental number, e, the base of the natural logarithms. Immanuel Kant's "transcendental epistemology" is what most educated people might think of when the word transcendental is used, but in that sense it has nothing to do with mathematics. Again, take the word "evolution," used in mathematics to denote the process most of us learned in elementary school, and promptly forgot, of extracting square roots, cube roots, etc. Spencer, in his philosophy, defines evolution as "an integration of matter, and a dissipation of motion from an indefinite, incoherent homogeneity to a definite, coherent heterogeneity," etc. But that, fortunately, has nothing to do with mathematical evolution either. Even in Tennessee, one may extract square roots without running afoul of the law.

As we see, mathematics uses simple words for complicated ideas. An example of a simple word used in a complicated way is the word "simple." "Simple curve" and "simple group" represent important ideas in higher mathematics.

FIG. 1

The above is not a simple curve. A simple curve is a closed curve which does not cross itself and may look like Fig. 2. There are many important theorems about such figures that make the word worth while. Later, we are

FIG. 2

going to talk about a queer kind of mathematics called "rubber-sheet geometry," and will have much more to say about simple curves and nonsimple ones. A French mathematician, Jordan, gave the fundamental theorem: every simple curve has one inside and one outside. That is, every simple curve divides the plane into two regions, one inside the curve, and one outside.

There are some groups in mathematics that are "simple" groups. The definition of "simple group" is really so hard that it cannot be given here. If we wanted to get a clear idea of what a simple group was, we should probably have to spend a long time looking into a great many books, and then, without an extensive mathematical background, we should probably miss the point. First of all, we should have to define the concept "group." Then we should have to give a definition of subgroups, and then of self-conjugate subgroups, and then we should be able to tell what a simple group is. A simple group is simply a group without any self-conjugate subgroups—simple, is it not?

Mathematics is often erroneously referred to as the science of common sense. Actually, it may transcend common sense and go beyond either imagination or intuition. It has become a very strange and perhaps frightening subject from the ordinary point of view, but anyone who penetrates into it will find a veritable fairyland, a fairyland which is strange, but makes sense, if not common sense. From the ordinary point of view mathematics deals with strange things. We shall show you that occasionally it does deal with strange things, but mostly it deals with familiar things in a strange way. If you look at yourself in an ordinary mirror, regardless of your physical attributes, you may find yourself amusing, but not strange; a subway ride to Coney Island, and a glance at yourself in one of the distorting mirrors will convince you that from another point of view you may be strange as well as amusing. It is largely a matter of what you are accustomed to. A Russian peasant came to Moscow for the first time and went to see the sights. He went to the zoo and saw the giraffes. You may find a moral in his reaction as plainly as in the fables of La Fontaine. "Look," he said, "at what the Bolsheviks have done to our horses." That is what modern mathematics has done to simple geometry and to simple arithmetic.

There are other words and expressions, not so familiar, which have been invented even more recently. Take, for instance, the word "turbine." Of course, that is already used in engineering, but it is an entirely new word in geometry. The mathematical name applies to a certain diagram. (Geometry, whatever others may think, is the study of different shapes, many of them very beautiful, having harmony, grace and symmetry. Of course, there are also fat books written on abstract geometry, and abstract space in which neither a diagram nor a shape appears. This is a very important branch of mathematics, but it is not the geometry studied by the Egyptians and the Greeks. Most of us, if we can play chess at all, are content to play it on a board with wooden

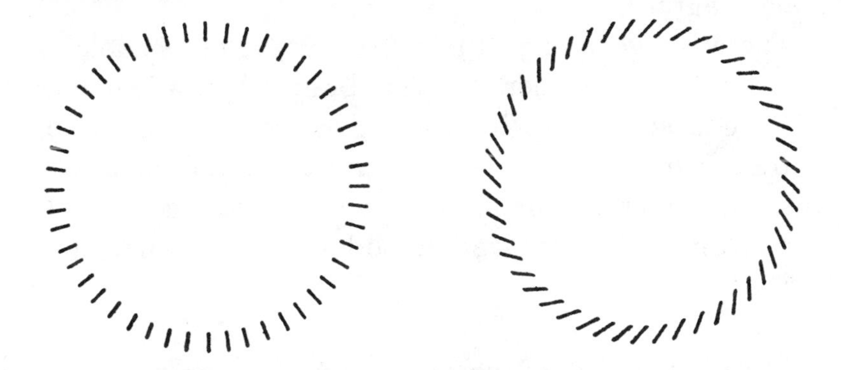

FIG. 3.—Turbines.

chess pieces; but there are some who play the game blindfolded and without touching the board. It might be a fair analogy to say that abstract geometry is like blindfold chess—it is a game played without concrete objects.) Above you see a picture of a turbine, in fact, two of them.

A turbine consists of an infinite number of "elements" filled in continuously. An element is not merely a point;

it is a point with an associated direction—like an iron filing. A turbine is composed of an infinite number of these elements, arranged in a peculiar way: the points must be arranged on a perfect circle, and the inclination of the iron filings must be at the same angle to the circle throughout. There are thus an infinite number of elements of equal inclination to the various tangents of the circle. In the special case where the angle between the direction of the element and the direction of the tangent is zero, what would happen? The turbine would be a circle. In other words, the theory of turbines is a generalization of the theory of the circle. If the angle is ninety degrees, the elements point toward the center of the circle. In that special case we have a normal turbine (see left-hand diagram).

There is a geometry of turbines, instead of a geometry of circles. It is a rather technical branch of mathematics which concerns itself with working out continuous groups of transformations connected with differential equations and differential geometry. The geometry connected with the turbine bears the rather odd name of "turns and slides."

*

The circle is one of the oldest figures in mathematics. The straight line is the simplest line, but the circle is the simplest nonstraight curve. It is often regarded as the limit of a polygon with an infinite number of sides. You can see for yourself that as a series of polygons is inscribed in a circle with each polygon having more sides than its predecessor, each polygon gets to look more and more like a circle.[1]

The Greeks were already familiar with the idea that as a regular polygon increases in the number of its sides, it differs less and less from the circle in which it is inscribed. Indeed, it may well be that in the eyes of an omniscient creature, the circle would look like a polygon with an infinite number of straight sides.[2] However, in the absence of complete omniscience, we shall continue

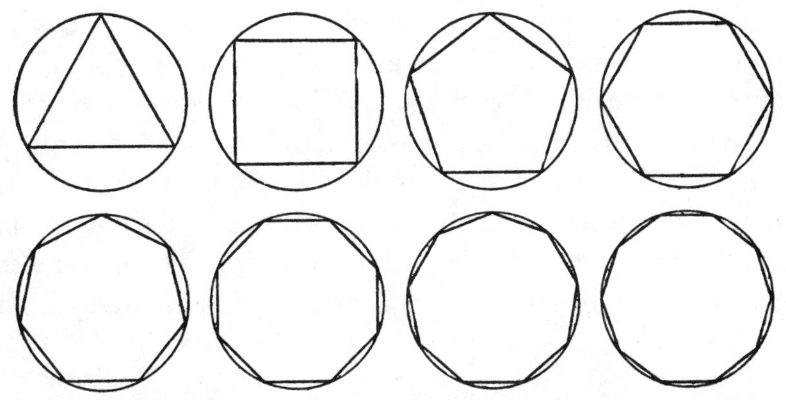

FIG. 4.—The circle as the limit of inscribed polygons.

to regard a circle as being a nonstraight curve. There are some interesting generalizations of the circle when it is viewed in this way. There is, for example, the concept denoted by the word "cycle," which was introduced by a French mathematician, Laguerre. A cycle is a circle with an arrow on it, like this:

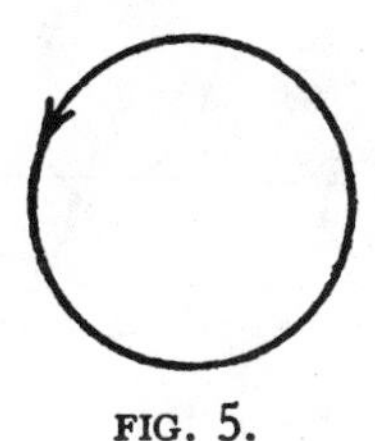

FIG. 5.

If you took the same circle and put an arrow on it in the opposite direction, it would become a different cycle.

The Greeks were specialists in the art of posing problems which neither they nor succeeding generations of mathematicians have ever been able to solve. The three most famous of these problems—the squaring of the circle, the duplication of the cube, and the trisection of an angle—we shall discuss later. Many well-meaning, self-appointed, and self-anointed mathematicians, and a motley assortment of lunatics and cranks, knowing neither history nor mathematics, supply an abundant crop of "solutions" of these insoluble problems each year. However, some of the classical problems of antiquity have been solved. For example, the theory of cycles was used by Laguerre in solving the problem of Apollonius: given three fixed circles, to find a circle that touches them all. It turns out to be a matter of elementary high

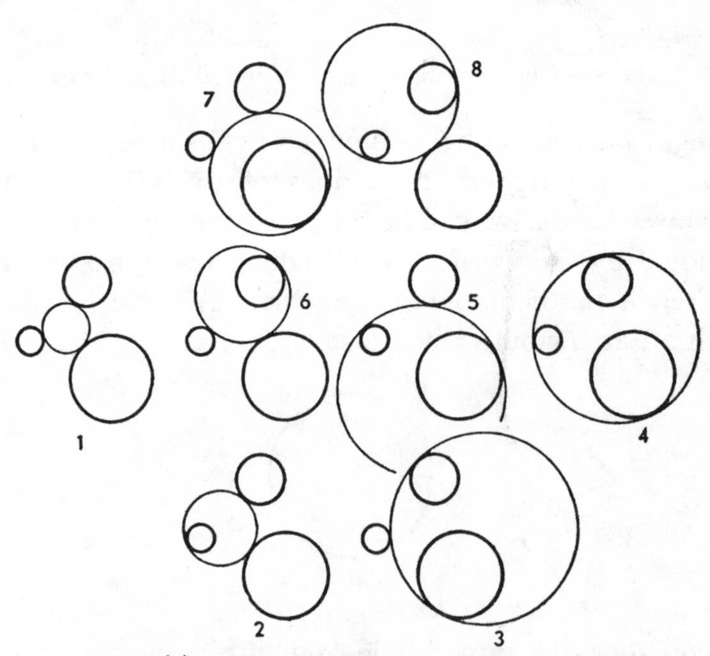

FIG. 6(a).—The eight solutions of the problem of Apollonius. Each lightly drawn circle is in contact with 3 heavily drawn ones.

school geometry, although it involves ingenuity, and any brilliant high school student could work it out. It has eight answers, as shown in Fig. 6(a).

They can all be constructed with ruler and compass, and many methods of solution have been found. Given three *circles*, there will be eight circles touching all of them. Given three *cycles*, however, there will be only one clockwise cycle that touches them all. (Two cycles are said to touch each other only if their arrows agree in direction at the point of contact.) Thus, by using the idea of cycles, we have one definite answer instead of eight. Laguerre made the idea of cycles the basis of an elegant theory.

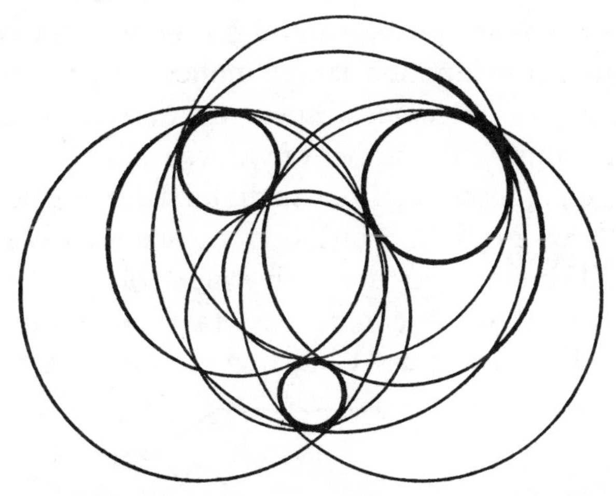

FIG. 6(b).—The eight solutions of Appolonius merged into one diagram.

Another variation of the circle introduced by the eminent American mathematician, C. J. Keyser, is obtained by taking a circle and removing one point.[3] This creates a serious change in conception. Keyser calls it "a pathocircle," (from pathological circle). He has used it in discussing the logic of axioms.

We have made yet another change in the concept of circle, which introduces another word and a new diagram. Take a circle and instead of leaving one point out, simply emphasize one point as the initial point. This is to be called a "clock." It has been used in the theory of polygenic functions. "Polygenic" is a word recently introduced into the theory of complex functions —about 1927. There was an important word, "monogenic," introduced in the nineteenth century by the famous French mathematician, Augustin Cauchy, and used in the classical theory of functions. It is used to denote functions that have a single derivative at a point, as in the differential calculus. But most functions, in the complex domain, have an infinite number of derivatives at a point. If a function is not monogenic, it can never be bigenic, or trigenic. Either the derivative has one value or an infinite number of values—either monogenic or polygenic, nothing intermediate. Monogenic means one rate of growth. Polygenic means many rates of growth. The complete derivative of a polygenic function is represented by a congruence (a double infinity) of clocks, all with different starting points, but with the

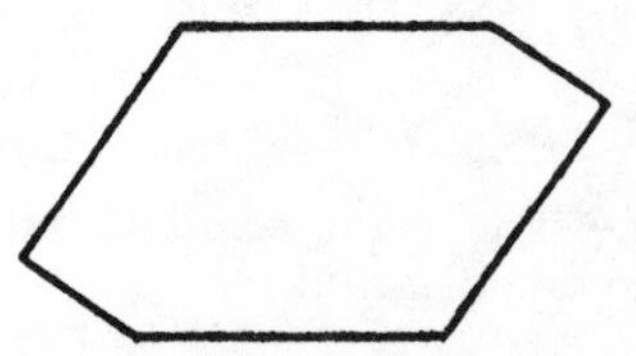

FIG. 7.—The parhexagon.

same uniform rate of rotation. It would be useless to attempt to give a simplified explanation of these concepts. (The neophyte will have to bear with us over a few intervals like this for the sake of the more experienced mathematical reader.)

The going has been rather hard in the last paragraph, and if a few of the polygenic seas have swept you overboard, we shall throw you a hexagonal life preserver. We may consider a very simple word that has been introduced in elementary geometry to indicate a certain kind of hexagon. The word on which to fix your attention is "parhexagon." An ordinary hexagon has six arbitrary sides. A parhexagon is that kind of hexagon in which any side is both equal and parallel to the side opposite to it (as in Fig. 7).

If the opposite sides of a quadrilateral are equal and parallel, it is called a parallelogram. By the same reasoning that we use for the word parhexagon, a parallelogram might have been called a parquadrilateral.

Here is an example of a theorem about the parhexagon: take any irregular hexagon, not necessarily a parhexagon, ABCDEF. Draw the diagonals AC, BD, CE, DF, EA, and FB, forming the six triangles, ABC, BCD, CDE, DEF, EFA, and FAB. Find the six centers of gravity, A′, B′, C′, D′, E′, and F′ of these triangles. (The center of gravity of a triangle is the point at which the triangle would balance if it were cut out of cardboard and supported only at that point; it coincides with the

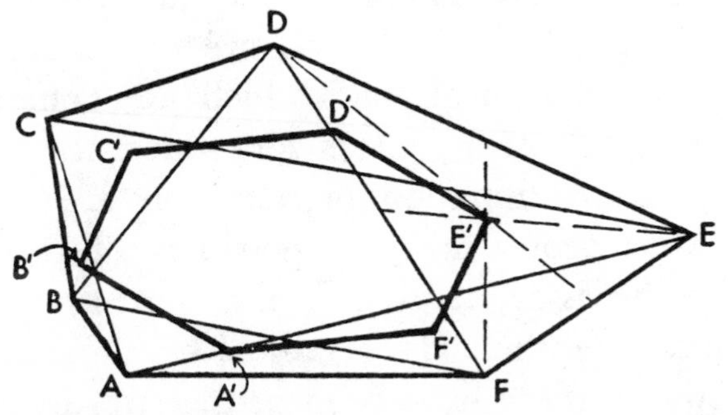

FIG. 8.—*ABCDEF* is an irregular hexagon. *A′B′C′D′E′F′* is a parhexagon.

point of intersection of the medians.) Draw A′B′, B′C′, C′D′, D′E′, E′F′, and F′A′. Then the new inner hexagon A′B′C′D′E′F′ will always be a parhexagon.

The word radical, favorite call to arms among Republicans, Democrats, Communists, Socialists, Nazis, Fascists, Trotskyites, etc., has a less hortatory and bellicose character in mathematics. For one thing, everybody knows its meaning: i.e., square root, cube root, fourth root, fifth root, etc. Combining a word previously defined with this one, we might say that the extraction of a root is the evolution of a radical. The square root of 9 is 3; the square root of 10 is greater than 3, and the most famous and the simplest of all square roots, the first incommensurable number discovered by the Greeks, the square root of 2, is 1.414. . . . There are also composite radicals—expressions like $\sqrt{7} + \sqrt[5]{10}$. The symbol for a radical is not the hammer and sickle, but a sign three or four centuries old, and the idea of the mathematical radical is even older than that. The concept of the "hyperradical," or "ultraradical," which means something higher than a radical, but lower than a transcendental, is of recent origin. It has a symbol which we shall see in a moment. First, we must say a few words about radicals in general. There are certain numbers and functions in mathematics which are not expressible in the language of radicals and which are generally not well understood. Many ideas for which there are no concrete or diagrammatic representations are difficult to explain. Most people find it impossible to think without words; it is necessary to give them a word and a symbol to pin their attention. Hyperradical or ultraradical, for which hitherto there have been neither words, nor symbols, fall into this category.

We first meet these ultraradicals, not in Mexico City, but in trying to solve equations of the fifth degree. The Egyptians solved equations of the first degree perhaps 4000 years ago. That is, they found that the solution of the equation $ax + b = 0$, which is represented in geometry by a straight line, is $x = \frac{-b}{a}$. The quadratic equation $ax^2 + bx + c = 0$ was solved by the Hindus and the Arabs with the formula $x = \frac{-b \pm \sqrt{b^2 - 4ac}}{2a}$. The various conic sections, the circle, the ellipse, the parabola, and the hyperbola, are the geometric pictures of quadratic equations in two variables.

Then in the sixteenth century the Italians solved the equations of third and fourth degree, obtaining long formulas involving cube roots and square roots. So that by the year 1550, a few years before Shakespeare was born, the equation of the first, second, third, and fourth degrees had been solved. Then there was a delay of 250 years, because mathematicians were struggling with the equation of the fifth degree—the general quintic. Finally, at the beginning of the nineteenth century, Ruffini and Abel showed that equations of the fifth degree could not be solved with radicals. The general quintic is thus not like the general quadratic, cubic or biquadratic. Nevertheless, it presents a problem in algebra which theoretically can be solved by algebraic operations. Only, these operations are so hard that they cannot be expressed by the symbols for radicals. These new higher things are

FIG. 9.—A portrait of two ultra-radicals.

named "ultraradicals," and they too have their special symbols (shown in Fig. 9).

With such symbols combined with radicals, we can solve equations of the fifth degree. For example, the solution of $x^5 + x = a$ may be written $x = \sqrt{\widehat{a}}$ or $x = \int\overline{a}$. The usefulness of the special symbol and name is apparent. Without them the solution of the quintic equation could not be compactly expressed.

*

We may now give a few ideas somewhat easier than those with which we have thus far occupied ourselves. These ideas were presented some time ago to a number of children in kindergarten. It was amazing how well they understood everything that was said to them. Indeed, it is a fair inference that kindergarten children can enjoy lectures on graduate mathematics as long as the mathematical concepts are clearly presented.

It was raining and the children were asked how many raindrops would fall on New York. The highest answer was 100. They had never counted higher than 100 and what they meant to imply when they used that number was merely something very, very big—as big as they could imagine. They were asked how many raindrops hit the roof, and how many hit New York, and how many single raindrops hit all of New York in 24 hours. They soon got a notion of the bigness of these numbers even though they did not know the symbols for them. They were certain in a little while that the number of raindrops was a great deal bigger than a hundred. They were asked to think of the number of grains of sand on the beach at Coney Island and decided that the number of grains of sand and the number of raindrops were about the same. But the important thing is that they realized that the number was *finite, not infinite*. In this respect they showed their distinct superiority over many scientists who to this day use the word infinite when they mean some big number, like a billion billion.

Counting, something such scientists evidently do not realize, is a precise operation.* It may be wonderful but there is nothing vague or mysterious about it. If you count something, the answer you get is either perfect or all wrong; there is no half way. It is very much like catching a train. You either catch it or you miss it, and if you miss it by a split second you might as well have come a week late. There is a famous quotation which illustrates this:

> "Oh, the little more, and how much it is!
> And the little less, and what worlds away!"

A big number is big, but it is definite and it is finite. Of course in poetry, the finite ends with about three thousand; any greater number is infinite. In many poems, the poet will talk to you about the infinite number of stars. But, if ever there was a hyperbole, this is it, for nobody, not even the poet, has ever seen more than three thousand stars on a clear night, without the aid of a telescope.

With the Hottentots, infinity begins at three.† Ask a Hottentot how many cows he owns, and if he has more than three he'll say "many." The number of raindrops

* No one would say that 1 + 1 is "about equal to 2." It is just as silly to say that a billion billion is not a finite number, simply because it is big. Any number which can be named, or conceived of in terms of the integers is finite. *Infinite means something quite different*, as we shall see in the chapter on the googol.

† Although, in all fairness, it must be pointed out that some of the tribes of the Belgian Congo can count to a million and beyond.

falling on New York is also "many." It is a large finite number, but nowhere near infinity.

Now here is the name of a very large number: "Googol."* Most people would say, "A googol is so large that you cannot name it or talk about it; it is so large that it is infinite." Therefore, we shall talk about it, explain exactly what it is, and show that it belongs to the very same family as the number 1.

A googol is this number which one of the children in the kindergarten wrote on the blackboard:

100000000000000000000000000000000000000000000000
000000000000000000000000000000000000000000000000
00000

The definition of a googol is: 1 followed by a hundred zeros. It was decided, after careful mathematical researches in the kindergarten, that the number of raindrops falling on New York in 24 hours, or even in a year or in a century, is much less than a googol. Indeed, the googol is a number just larger than the largest numbers that are used in physics or astronomy. All those numbers require less than a hundred zeros. This information is, of course, available to everyone, but seems to be a great secret in many scientific quarters.

A very distinguished scientific publication recently came forth with the revelation that the number of snow crystals necessary to form the ice age was a billion to the billionth power. This is very startling and also very silly. A billion to the billionth power looks like this:

$$1000000000^{1000000000}.$$

A more reasonable estimate and a somewhat smaller number would be 10^{30}. As a matter of fact, it has been estimated that if the entire universe, which you will con-

* Not even approximately a Russian author.

cede is a trifle larger than the earth, were filled with protons and electrons, so that no vacant space remained, the total number of protons and electrons would be 10^{110} (i.e., 1 with 110 zeros after it). Unfortunately, as soon as people talk about large numbers, they run amuck. They seem to be under the impression that since zero equals nothing, they can add as many zeros to a number as they please with practically no serious consequences. We shall have to be a little more careful than that in talking about big numbers.

To return to Coney Island, the number of grains of sand on the beach is about 10^{20}, or more descriptively, 100000000000000000000. That is a large number, but not as large as the number mentioned by the divorcee in a recent divorce suit who had telephoned that she loved the man "a million billion billion times and eight times around the world." It was the largest number that she could conceive of, and shows the kind of thing that may be hatched in a love nest.

Though people do a great deal of talking, the total output since the beginning of gabble to the present day, including all baby talk, love songs, and Congressional debates, totals about 10^{16}. This is ten million billion. Contrary to popular belief, this is a larger number of words than is spoken at the average afternoon bridge.

A great deal of the veneration for the authority of the printed word would vanish if one were to calculate the number of words which have been printed since the Gutenberg Bible appeared. It is a number somewhat larger than 10^{16}. A recent popular historical novel alone accounts for the printing of several hundred billion words.

The largest number seen in finance (though new records are in the making) represents the amount of money in circulation in Germany at the peak of the inflation. It was less than a googol—merely

496,585,346,000,000,000,000.

A distinguished economist vouches for the accuracy of this figure. The number of marks in circulation was very nearly equal to the number of grains of sand on Coney Island beach.

The number of atoms of oxygen in the average thimble is a good deal larger. It would be represented by perhaps 1000000000000000000000000000. The number of electrons, in size exceedingly smaller than the atoms, is much more enormous. The number of electrons which pass through the filament of an ordinary fifty-watt electric lamp in a minute equals the number of drops of water that flow over Niagara Falls in a century.

One may also calculate the number of electrons, not only in the average room, but over the whole earth, and out through the stars, the Milky Way, and all the nebulae. The reason for giving all these examples of very large numbers is to emphasize the fact that no matter how large the collection to be counted, a finite number will do the trick. We will have occasion later on to speak of infinite collections, but those encountered in nature, though sometimes very large, are all definitely finite. A celebrated scientist recently stated in all seriousness that he believed that the number of pores (through which leaves breathe) of all the leaves, of all the trees in all the world, would certainly be infinite. Needless to say, he was not a mathematician. The number of electrons in a single leaf is much bigger than the number of pores of all the leaves of all the trees of all the world. And still the number of all the electrons in the entire universe can be found by means of the physics of Einstein. It is a good deal less than a googol—perhaps one with seventy-nine zeros, 10^{79}, as estimated by Eddington.

Words of wisdom are spoken by children at least as often as by scientists. The name "googol" was invented by a child (Dr. Kasner's nine-year-old nephew) who was asked to think up a name for a very big number, namely, 1 with a hundred zeros after it. He was very certain that this number was not infinite, and therefore equally certain that it had to have a name. At the same time that he suggested "googol" he gave a name for a still larger number: "Googolplex." A googolplex is much larger than a googol, but is still finite, as the inventor of the name was quick to point out. It was first suggested that a googolplex should be 1, followed by writing zeros until you got tired. This is a description of what would happen if one actually tried to write a googolplex, but different people get tired at different times and it would never do to have Carnera a better mathematician than Dr. Einstein, simply because he had more endurance. The googolplex then, is a specific finite number, with so many zeros after the 1 that the number of zeros is a googol. A googolplex is much bigger than a googol, much bigger even than a googol times a googol. A googol times a googol would be 1 with 200 zeros, whereas a googolplex is 1 with a googol of zeros. You will get some idea of the size of this very large but finite number from the fact that there would not be enough room to write it, if you went to the farthest star, touring all the nebulae and putting down zeros every inch of the way.

One might not believe that such a large number would ever really have any application; but one who felt that way would not be a mathematician. A number as large as the googolplex might be of real use in problems of combination. This would be the type of problem in which it might come up scientifically:

Consider this book which is made up of carbon and nitrogen and of other elements. The answer to the question, "How many atoms are there in this book?" would certainly be a finite number, even less than a googol. Now imagine that the book is held suspended by a string, the end of which you are holding. How long will it be necessary to wait before the book will jump up into your hand? Could it conceivably ever happen? One answer might be "No, it will never happen without some external force causing it to do so." But that is not correct. The right answer is that it will almost *certainly* happen *sometime* in less than a googolplex of years—perhaps tomorrow.

The explanation of this answer can be found in physical chemistry, statistical mechanics, the kinetic theory of gases, and the theory of probability. We cannot dispose of all these subjects in a few lines, but we will try. Molecules are always moving. Absolute rest of molecules would mean absolute zero degrees of temperature, and absolute zero degrees of temperature is not only nonexistent, but impossible to obtain. All the molecules of the surrounding air bombard the book. At present the bombardment from above and below is nearly the same and gravity keeps the book down. It is necessary to wait for the favorable moment when there happens to be an enormous number of molecules bombarding the book from below and very few from above. Then gravity will be overcome and the book will rise. It would be somewhat like the effect known in physics as the Brownian movement, which describes the behavior of small particles in a liquid as they dance about under the impact of molecules. It would be analogous to the Brownian movement on a vast scale.

But the probability that this will happen in the near future or, for that matter, on any specific occasion that we might mention, is between $\frac{1}{\text{googol}}$ and $\frac{1}{\text{googolplex}}$. To be reasonably sure that the book will rise, we should have to wait between a googol and a googolplex of years.

When working with electrons or with problems of combination like the one of the book, we need larger numbers than are usually talked about. It is for that reason that names like googol and googolplex, though they may appear to be mere jokes, have a real value. The names help to fix in our minds the fact that we are still dealing with finite numbers. To repeat, a googol is 10^{100}; a googolplex is 10 to the googol power, which may be written $10^{10^{100}} = 10^{\text{googol}}$.

We have seen that the number of years that one would have to wait to see the miracle of the rising book would be less than a googolplex. In that number of years the earth may well have become a frozen planet as dead as the moon, or perhaps splintered to a number of meteors and comets. The real miracle is not that the book will rise, but that with the aid of mathematics, we can project ourselves into the future and predict with accuracy *when* it will probably rise, i.e., some time between today and the year googolplex.

*

We have mentioned quite a few new names in mathematics—new names for old and new ideas. There is one more new name which it is proper to mention in conclusion. Watson Davis, the popular science reporter, has given us the name "mathescope." With the aid of the magnificent new microscopes and telescopes, man, midway between the stars and the atoms, has come a little closer to both. The mathescope is not a physical instrument; it is a purely intellectual instrument, the ever-increasing insight which mathematics gives into the fairyland which lies beyond intuition and beyond imagination. Mathematicians, unlike philosophers, say nothing about ultimate truth, but patiently, like the makers of the great microscopes, and the great telescopes, they grind their lenses. In this book, we shall let you see through the newer and greater lenses which the mathematicians have ground. Be prepared for strange sights through the mathescope!

FOOTNOTES

1. See the Chapter on PIE.—P. 10.
2. See the Chapter on Change and Changeability—Section on Pathological Curves.—P.11.
3. N.B. This is a diagram which the reader will have to imagine, for it is beyond the capacity of any printer to make a circle with one point omitted. A point, having no dimensions, will, like many of the persons on the Lord High Executioner's list, never be missed. So the circle with one point missing is purely conceptual, not an idea which can be pictured.—P.13.

Beyond the Googol

If you do not expect the unexpected, you will not find it;
for it is hard to be sought out, and difficult.

—HERACLITUS

MATHEMATICS MAY well be a science of austere logical propositions in precise canonical form, but in its countless applications it serves as a tool and a language, the language of description, of number and size. It describes with economy and elegance the elliptic orbits of the planets as readily as the shape and dimensions of this page or a corn field. The whirling dance of the electron can be seen by no one; the most powerful telescopes can reveal only a meager bit of the distant stars and nebulae and the cold far corners of space. But with the aid of mathematics and the imagination the very small, the very large—all things may be brought within man's domain.

To count is to talk the language of number. To count to a googol, or to count to ten is part of the same process; the googol is simply harder to pronounce. The essential thing to realize is that the googol and ten are kin, like the giant stars and the electron. Arithmetic—this counting language—makes the whole world kin, both in space and in time.

To grasp the meaning and importance of mathematics, to appreciate its beauty and its value, arithmetic must first be understood, for mostly, since its beginning, mathematics has been arithmetic in simple or elaborate attire. Arithmetic has been the queen and the handmaiden of the sciences from the days of the astrologers of Chaldea and the high priests of Egypt to the present days of relativity, quanta, and the adding machine. Historians may dispute the meaning of ancient papyri, theologians may wrangle over the exegesis of Scripture, philosophers may debate over Pythagorean doctrine, but all will concede that the numbers in the papyri, in the Scriptures and in the writings of Pythagoras are the same as the numbers of today. As arithmetic, mathematics has helped man to cast horoscopes, to make calendars, to predict the risings of the Nile, to measure fields and the height of the Pyramids, to measure the speed of a stone as it fell from a tower in Pisa, the speed of an apple as it fell from a tree in Woolsthorpe, to weigh the stars and the atoms, to mark the passage of time, to find the curvature of space. And although mathematics is also the calculus, the theory of probability, the matrix algebra, the science of the infinite, it is still the art of counting.

*

Everyone who will read this book can count, and yet, what is counting? The dictionary definitions are about as helpful as Johnson's definition of a net: "A series of reticulated interstices." *Learning to compare is learning to count.* Numbers come much later; they are an artificiality, an abstraction. Counting, matching, comparing are almost as indigenous to man as his fingers. Without the faculty of comparing, and without his fingers, it is unlikely that he would have arrived at numbers.

One who knows nothing of the formal processes of counting is still able to compare two classes of objects, to determine which is the greater, which the less. Without knowing anything about numbers, one may ascertain whether two classes have the same number of elements; for example, barring prior mishaps, it is easy to show that we have the same number of fingers on both hands by simply matching finger with finger on each hand.

To describe the process of matching, which underlies counting, mathematicians use a picturesque name. They call it putting classes into a "one-to-one reciprocal correspondence" with each other. Indeed, that is all there is to the art of counting as practiced by primitive peoples, by us, or by Einstein. A few examples may serve to make this clear.

In a monogamous country it is unnecessary to count both the husbands and the wives in order to ascertain the number of married people. If allowances are made for the few gay Lotharios who do not conform to either custom or statute, it is sufficient to count either the husbands or the wives. There are just as many in one class as in the other. The correspondence between the two classes is one-to-one.

There are more useful illustrations. Many people are gathered in a large hall where seats are to be provided. The question is, are there enough chairs to go around? It would be quite a job to count both the people and the chairs, and in this case unnecessary. In kindergarten children play a game called "Going to Jerusalem"; in a room full of children and chairs there is always one less chair than the number of children. At a signal, each child runs for a chair. The child left standing is "out." A chair is removed and the game continues. Here is the solution to our problem. It is only necessary to ask everyone in the hall to be seated. If everyone sits down and no chairs are left vacant, it is evident that there are as many chairs as people. In other words, without actually knowing the number of chairs or people, one does know that the number is the same. The two classes—chairs and people—have been shown to be equal in number by a one-to-one correspondence. To each person corresponds a chair, to each chair, a person.

In counting any class of objects, it is this method alone which is employed. One class contains the things to be counted; the other class is always at hand. It is the class of integers, or "natural numbers," which for convenience we regard as being given in serial order: 1, 2, 3, 4, 5, 6, 7 . . . Matching in one-to-one correspondence the elements of the first class with the integers, we experience a common, but none the less wonderful phenomenon—the last integer necessary to complete the pairings denotes *how many* elements there are.

*

In clarifying the idea of counting, we made the unwarranted assumption that the concept of number was understood by everyone. The number concept may seem intuitively clear, but a precise definition is required. While the definition may seem worse than the disease, it is not as difficult as appears at first glance. Read it carefully and you will find that it is both explicit and economical.

Given a class C containing certain elements, it is possible to find other classes, such that the elements of each may be matched one to one with the elements of C. (Each of these classes is thus called "equivalent to C.") All such classes, including C, whatever the character of their elements, share one property in common: all of them have the same *cardinal number*, which is called the *cardinal number* of the class C.[1]

The cardinal number of the class *C* is thus seen to be the *symbol* representing the set of all classes that can be put into one-to-one correspondence with *C*. For example, the number 5 is simply the name, or symbol, attached to the set of all the classes, each of which can be put into one-to-one correspondence with the fingers of one hand.

Hereafter we may refer without ambiguity to the number of elements in a class as the cardinal number of that class or, briefly, as "its cardinality." The question, "How many letters are there in the word *mathematics?*" is the same as the question, "What is the cardinality of the class whose elements are the letters in the word *mathematics?*" Employing the method of one-to-one correspondence, the following graphic device answers the question, and illustrates the method:

M	A	T	H	E	M	A	T	I	C	S
↕	↕	↕	↕	↕	↕	↕	↕	↕	↕	↕
1	2	3	4	5	6	7	8	9	10	11

It must now be evident that this method is neither strange nor esoteric; it was not invented by mathematicians to make something natural and easy seem unnatural and hard. It is the method employed when we count our change or our chickens; it is the proper method for counting any class, no matter how large, from ten to a googolplex—and beyond.

Soon we shall speak of the "beyond" when we turn to classes which are not finite. Indeed, we shall try *to measure our measuring class*—the integers. One-to-one correspondence should, therefore, be thoroughly understood, for an amazing revelation awaits us: Infinite classes can also be counted, and by the very same means. But before we try to count them, let us practice on some very big numbers—big, but not infinite.

*

"Googol" is already in our vocabulary: It is a big number—one, with a hundred zeros after it. Even bigger is the googolplex: 1 with a googol zeros after it. Most numbers encountered in the description of nature are much smaller, though a few are larger.

Enormous numbers occur frequently in modern science. Sir Arthur Eddington claims that there are, not approximately, but exactly $136 \cdot 2^{256}$ protons,* and an equal number of electrons, in the universe. Though not easy to visualize, this number, as a symbol on paper, takes up little room. Not quite as large as the googol, it is completely dwarfed by the googolplex. None the less, Eddington's number, the googol, and the googolplex are finite.

A veritable giant is Skewes' number, even bigger than a googolplex. It gives information about the distribution of primes[2] and looks like this:

$$10^{10^{10^{34}}}$$

Or, for example, the total possible number of moves in a game of chess is:

$$10^{10^{50}}$$

And speaking of chess, as the eminent English mathematician, G. H. Hardy, pointed out—if we imagine the

* Let no one suppose that Sir Arthur has counted them. But he does have a theory to justify his claim. Anyone with a better theory may challenge Sir Arthur, for who can be referee? Here is his number written out: 15,747,724,136,275,002,577,605,653,961,181,555,468,-044,717,914,527,116,709,366,231,425,076,185,631,031,296—accurate, he says, to the last digit.

entire universe as a chessboard, and the protons in it as chessmen, and if we agree to call any interchange in the position of two protons a "move" in this cosmic game, then the total number of possible moves, of all odd coincidences, would be Skewes' number:

$$10^{10^{10^{34}}}$$

No doubt most people believe that such numbers are part of the marvelous advance of science, and that a few generations ago, to say nothing of centuries back, no one in dream or fancy could have conceived of them.

There is some truth in that idea. For one thing, the ancient cumbersome methods of mathematical notation made the writing of big numbers difficult, if not actually impossible. For another, the average citizen of today encounters such huge sums, representing armament expenditures and stellar distances, that he is quite conversant with, and immune to, big numbers.

But there were clever people in ancient times. Poets in every age may have sung of the stars as infinite in number, when all they saw was, perhaps, three thousand. But to Archimedes, a number as large as a googol, or even larger, was not disconcerting. He says as much in an introductory passage in *The Sand Reckoner*, realizing that a number is not infinite merely because it is enormous.

There are some, King Gelon, who think that the number of the sand is infinite in multitude; and I mean by the sand, not only that which exists about Syracuse and the rest of Sicily, but also that which is found in every region whether inhabited or uninhabited. Again there are some who, without regarding it as infinite, yet think that no number has been named which is great enough to exceed its multitude. And it is clear that they who hold this view, if they imagined a mass made up of sand in other respects as large as the mass of the earth, including in it all the seas and the hollows of the earth filled up to a height equal to that of the highest of the mountains, would be many times further still from recognizing that any number could be expressed which exceeded the multitude of the sand so taken. But I will try to show you by means of geometrical proofs, which you will be able to follow, that, of the numbers named by me and given in the work which I sent to Zeuxippus, some exceed not only the number of the mass of sand equal in magnitude to the earth filled up in the way described, but also that of a mass equal in magnitude to the universe.

The Greeks had very definite ideas about the infinite. Just as we are indebted to them for much of our wit and our learning, so are we indebted to them for much of our sophistication about the infinite. Indeed, had we always retained their clear-sightedness, many of the problems and paradoxes connected with the infinite would never have arisen.

Above everything, we must realize that "very big" and "infinite" are entirely different.* By using the method of one-to-one correspondence, the protons and electrons in the universe can theoretically be counted as easily as the buttons on a vest. Sufficient and more than sufficient for that task, or for the task of counting any finite collection, are the integers. But measuring the

* There is no point where the very big starts to merge into the infinite. You may write a number as big as you please; it will be no nearer the infinite than the number 1 or the number 7. Make sure that you keep this distinction very clear and you will have mastered many of the subtleties of the transfinite.

totality of integers is another problem. To measure such a class demands a lofty viewpoint. Besides being, as the German mathematician Kronecker thought, the work of God, which requires courage to appraise, the class of integers is infinite—which is a great deal more inconvenient. It is worse than heresy to measure our own endless measuring rod!

*

The problems of the infinite have challenged man's mind and have fired his imagination as no other single problem in the history of thought. The infinite appears both strange and familiar, at times beyond our grasp, at times natural and easy to understand. In conquering it, man broke the fetters that bound him to earth. All his faculties were required for this conquest—his reasoning powers, his poetic fancy, his desire to know.

To establish the science of the infinite involves the principle of *mathematical induction.* This principle affirms the power of reasoning by recurrence. It typifies almost all mathematical thinking, all that we do when we construct complex aggregates out of simple elements. It is, as Poincaré remarked, "at once necessary to the mathematician and irreducible to logic." His statement of the principle is: "If a property be true of the number one, and if we establish that it is true of $n + 1$,* provided it be of n, it will be true of all the whole numbers." Mathematical induction is not derived from experience, rather is it an inherent, intuitive, almost instinctive property of the mind. "*What we have once done we can do again.*"

If we can construct numbers to ten, to a million, to a googol, we are led to believe that there is no stopping,

* Where n is any integer.

no end. Convinced of this, we need not go on forever; the mind grasps that which it has never experienced—the infinite itself. Without any sense of discontinuity, without transgressing the canons of logic, the mathematician and philosopher have bridged in one stroke the gulf between the finite and the infinite. The mathematics of the infinite is a sheer affirmation of the inherent power of reasoning by recurrence.

In the sense that "infinite" means "without end, without bound," simply "not finite," probably everyone understands its meaning. No difficulty arises where no precise definition is required. Nevertheless, in spite of the famous epigram that mathematics is the science in which we do not know what we are talking about, at least we shall have to agree to talk about the same thing. Apparently, even those of scientific temper can argue bitterly to the point of mutual vilification on subjects ranging from Marxism and dialectical materialism to group theory and the uncertainty principle, only to find, on the verge of exhaustion and collapse, that they are on the same side of the fence. Such arguments are generally the results of vague terminology; to assume that everyone is familiar with the precise mathematical definition of "infinite" is to build a new Tower of Babel.

Before undertaking a definition, we might do well to glance backwards to see how mathematicians and philosophers of other times dealt with the problem.

The infinite has a double aspect—the infinitely large, and the infinitely small. Repeated arguments and demonstrations, of apparently apodictic force, were advanced, overwhelmed, and once more resuscitated to prove or disprove its existence. Few of the arguments were ever refuted—each was buried under an avalanche of others. The happy result was that the problem never became any clearer.*

*

The warfare began in antiquity with the paradoxes of Zeno; it has never ceased. Fine points were debated with a fervor worthy of the earliest Christian martyrs, but without a tenth part of the acumen of medieval theologians. Today, some mathematicians think the infinite has been reduced to a state of vassalage. Others are still wondering what it is.

Zeno's puzzles may help to bring the problem into sharper focus. Zeno of Elea, it will be recalled, said some disquieting things about motion, with reference to an arrow, Achilles, and a tortoise. This strange company was employed on behalf of the tenet of Eleatic philosophy —that all motion is an illusion. It has been suggested, probably by "baffled critics," that "Zeno had his tongue in cheek when he made his puzzles." Regardless of motive, they are immeasurably subtle, and perhaps still defy solution.†

One paradox—the Dichotomy—states that it is impossible to cover any given distance. The argument: First, half the distance must be traversed, then half of the remaining distance, then again half of what remains,

* No one has written more brilliantly or more wittily on this subject than Bertrand Russell. See particularly his essays in the volume *Mysticism and Logic*.

† To be sure, a variety of explanations have been given for the paradoxes. In the last analysis, the explanations for the riddles rest upon the interpretation of the foundations of mathematics. Mathematicians like Brouwer, who reject the infinite, would probably not accept any of the solutions given.

and so on. It follows that some portion of the distance to be covered always remains, and therefore motion is impossible! A solution of this paradox reads:

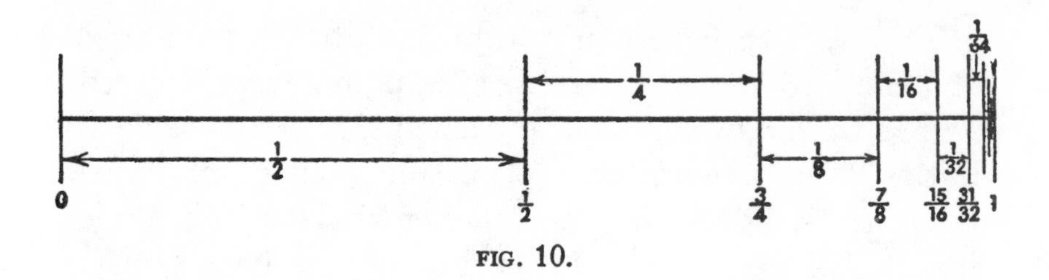

FIG. 10.

The successive distances to be covered form an infinite geometric series:

$$\frac{1}{2} + \frac{1}{4} + \frac{1}{8} + \frac{1}{16} + \frac{1}{32} + \ldots$$ [3]

each term of which is half of the one before. Although this series has an infinite number of terms, its sum is *finite* and equals 1. Herein, it is said, lies the flaw of the Dichotomy. Zeno assumed that any totality composed of an infinite number of parts must, itself, be infinite, whereas we have just seen an infinite number of elements which make up the finite totality—1.

The paradox of the tortoise states that Achilles, running to overtake the tortoise, must first reach the place where it started:—but the tortoise has already departed. This comedy, however, is repeated indefinitely. As Achilles arrives at each new point in the race, the tortoise having been there, has already left. Achilles is as unlikely to catch him as a rider on a carrousel the rider ahead.

Finally: the arrow in flight must be moving every instant of time. But at every instant it must be *somewhere* in space. However, if the arrow must always be in some one place, it cannot at every instant also be in transit, for to be in transit is to be *nowhere.*

Aristotle and lesser saints in almost every age tried to demolish these paradoxes, but not very creditably. Three German professors succeeded where the saints had failed. At the end of the nineteenth century, it seemed that Bolzano, Weierstrass and Cantor had laid the infinite to rest, and Zeno's paradoxes as well.

The modern method of disposing of the paradoxes is not to dismiss them as mere sophisms unworthy of serious attention. The history of mathematics, in fact, recounts a poetic vindication of Zeno's stand. Zeno was, at one time, as Bertrand Russell has said, "A notable victim of posterity's lack of judgement." That wrong has been righted. In disposing of the infinitely small, Weierstrass showed that the moving arrow *is* really always at rest, and that we live in Zeno's changeless world. The work of Georg Cantor, which we shall soon encounter, showed that if we are to believe that Achilles *can* catch the tortoise, we shall have to be prepared to swallow a bigger paradox than any Zeno ever conceived of: THE WHOLE IS NO GREATER THAN MANY OF ITS PARTS!

The infinitely small had been a nuisance for more than two thousand years. At best, the innumerable opinions it evoked deserved the laconic verdict of Scotch juries: "Not proven." Until Weierstrass appeared, the total advance was a confirmation of Zeno's argument against motion. Even the jokes were better. Leibniz, according to Carlyle, made the mistake of trying to explain the infinitesimal to a Queen—Sophie Charlotte of Prussia. She informed him that the behavior of her courtiers made her so familiar with the infinitely small, that she needed no mathematical tutor to explain it. But philosophers and mathematicians, according to Russell, "having less acquaintance with the courts, continued to discuss this topic, though without making any advance."

Berkeley, with the subtlety and humor necessary for an Irish bishop, made some pointed attacks on the infinitesimal, during the adolescent period of the calculus, that had the very best, sharp-witted, scholastic sting. One could perhaps speak, if only with poetic fervor, of the infinitely large, but what, pray, was the infinitely small? The Greeks, with less than their customary sagacity, introduced it in regarding a circle as differing infinitesimally from a polygon with a large number of equal sides. Leibniz used it as the bricks for the infinitesimal calculus. Still, no one knew what it was. The infinitesimal had wondrous properties. It was not zero, yet smaller than any quantity. It could be assigned no quantity or size, yet a sizable number of infinitesimals made a very definite quantity. Unable to discover its nature, happily able to dispense with it, Weierstrass interred it alongside of the phlogiston and other once-cherished errors.

*

The infinitely large offered more stubborn resistance. Whatever it is, it is a doughty weed. The subject of reams of nonsense, sacred and profane, it was first discussed fully, logically, and without benefit of clergy-like prejudices by Bernhard Bolzano. *Die Paradoxien des Unendlichen,* a remarkable little volume, appeared posthumously in 1851. Like the work of another Austrian priest, Gregor Mendel, whose distinguished treatise on the principles of heredity escaped oblivion only by chance, this important book, charmingly written, made no great impression on Bolzano's contemporaries. It is the creation of a clear, forceful, penetrating intelligence. For the first time in twenty centuries the infinite was treated as a problem in science, and not as a problem in theology.

Both Cantor and Dedekind are indebted to Bolzano for the foundations of the mathematical treatment of the infinite. Among the many paradoxes he gathered and explained, one, dating from Galileo, illustrates a typical source of confusion:

Construct a square—*ABCD*. About the point *A* as center, with one side as radius, describe a quarter-circle, intersecting the square at *B* and *D*. Draw *PR* parallel to *AD*, cutting *AB* at *P*, *CD* at *R*, the diagonal *AC* at *N*, and the quarter-circle at *M*.

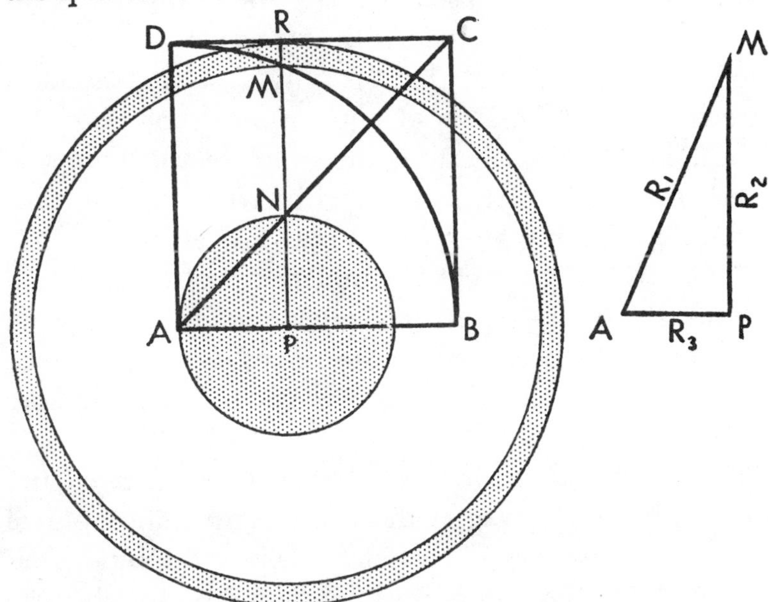

FIG. 11.—Extract triangle *APM* from the figure. It is not hard to see that its three sides equal respectively the radii of the three circles.
Thus

$$R_1^2 - R_2^2 = R_3^2$$

or,

$$\pi R_1^2 - \pi R_2^2 = \pi R_3^2$$

or, the two shaded areas are equal.

By a well-known geometrical theorem, it can be shown that if *PN*, *PM* and *PR* are radii, the following relationship exists:

$$\pi\overline{PN}^2 = \pi\overline{PR}^2 - \pi\overline{PM}^2 \qquad (1)$$

Permit *PR* to approach *AD*. Then the circle with *PN* as radius becomes smaller, and the ring between the circles with *PM* and *PR* as radii becomes correspondingly smaller. Finally, when *PR* becomes identical with *AD*, the radius *PN* vanishes, leaving the point *A*, while the ring between the two circles *PM* and *PR* contracts into one periphery with *AD* as radius. From equation (1) it may be concluded that the *point A* takes up as much area as the *circumference* of the circle with *AD* as radius.

Bolzano realized that there is only an *appearance* of a paradox. The two classes of points, one composed of a single member, the point *A*, the other of the points in the circumference of the circle with *AB* as radius, take up exactly the same amount of area. The area of both is zero! The paradox springs from the erroneous conception that the number of points in a given configuration is an indication of the area which it occupies. Points, finite or infinite in number, have no dimensions and can therefore occupy no area.

Through the centuries such paradoxes had piled up. Born of the union of vague ideas and vague philosophical reflections, they were nurtured on sloppy thinking. Bolzano cleared away most of the muddle, preparing the way for Cantor. It is to Cantor that the mathematics of the infinitely large owes its coming of age.

*

Georg Cantor was born in St. Petersburg in 1845, six years before Bolzano's book appeared. Though born in Russia, he lived the greater part of his life in Germany, where he taught at the University of Halle. While Weierstrass was busy disposing of the infinitesimal, Cantor set himself the apparently more formidable task at the other pole. The infinitely small might be laughed out of existence, but who dared laugh at the infinitely large? Certainly not Cantor! Theological curiosity prompted his task, but the mathematical interest came to subsume every other.

In dealing with the science of the infinite, Cantor realized that the first requisite was to define terms. His definition of "infinite class" which we shall paraphrase, rests upon a paradox. AN INFINITE CLASS HAS THE UNIQUE PROPERTY THAT THE WHOLE IS NO GREATER THAN SOME OF ITS PARTS. That statement is as essential for the mathematics of the infinite as THE WHOLE IS GREATER THAN ANY OF ITS PARTS is for finite arithmetic. When we recall that two classes are equal if their elements can be put into one-to-one correspondence, the latter statement becomes obvious. Zeno would not have challenged it, in spite of his scepticism about the obvious. But what is obvious for the finite is false for the infinite; our extensive experience with finite classes is misleading. Since, for example, the class of men and the class of mathematicians are both finite, anyone realizing that some men are not mathematicians would correctly conclude that the class of men is the larger of the two. He might also conclude that the number of integers, even and odd, is greater than the number of even integers. But we see from the following pairing that he would be mistaken:

$$\begin{array}{ccccccc} 1 & 2 & 3 & 4 & 5 & 6 & 7\ldots \\ \updownarrow & \updownarrow & \updownarrow & \updownarrow & \updownarrow & \updownarrow & \updownarrow \\ 2 & 4 & 6 & 8 & 10 & 12 & 14\ldots \end{array}$$

Under every integer, odd or even, we may write its double—an even integer. That is, we place each of the elements of the class of all the integers, odd and even, into a one-to-one correspondence with the elements of the class composed solely of even integers. This process may be continued to the googolplex and beyond.

Now, the class of integers is infinite. No integer, no matter how great, can describe its cardinality (or numerosity). Yet, since it is possible to establish a one-to-one correspondence between the class of even numbers and the class of integers, we have succeeded in counting the class of even numbers just as we count a finite collection. The two classes being perfectly matched, we must conclude that they have the same cardinality. That their cardinality is the same we *know*, just as we knew that the chairs and the people in the hall were equal in number when every chair was occupied and no one was left standing. Thus, we arrive at the fundamental paradox of all infinite classes:—There exist component parts of an infinite class which are just as great as the class itself. THE WHOLE IS NO GREATER THAN SOME OF ITS PARTS!

The class composed of the even integers *is thinned out* as compared with the class of all integers, but evidently "thinning out" has not the slightest effect on its cardinality. Moreover, there is almost no limit to the number of times this process can be repeated. For instance, there are as many square numbers and cube numbers as there are integers. The appropriate pairings are:

1	2	3	4	5	6 . . .
↕	↕	↕	↕	↕	↕
1	4	9	16	25	36 . . .
1^2	2^2	3^2	4^2	5^2	6^2

1	2	3	4	5	6 . . .
↕	↕	↕	↕	↕	↕
1	8	27	64	125	216 . . .
1^3	2^3	3^3	4^3	5^3	6^3

Indeed, from any denumerable class there can always be removed a denumerably infinite number of denumerably infinite classes without affecting the cardinality of the original class.

*

Infinite classes which can be put into one-to-one correspondence with the integers, and thus "counted," Cantor called *countable*, or *denumerably infinite*. Since all finite sets are countable, and we can assign to each one a number, it is natural to try to extend this notion and assign to the class of all integers a number representing its cardinality. Yet, it is obvious from our description of "infinite class" that no ordinary integer would be adequate to describe the cardinality of the whole class of integers. In effect, it would be asking a snake to swallow itself entirely. Thus, the first of the transfinite numbers was created to describe the cardinality of countable infinite classes. Etymologically old, mathematically new, $\aleph$ (aleph), the first letter of the Hebrew alphabet, was suggested. However, Cantor finally decided to use the compound symbol $\aleph_0$ (Aleph-Null). If asked, "How many integers are there?" it would be correct to reply, "There are $\aleph_0$ integers."

Because he suspected that there were other transfinite numbers, in fact an infinite number of transfinites, and the cardinality of the integers the smallest, Cantor affixed to the first $\aleph$ a small zero as subscript. The cardinality of a denumerably infinite class is therefore referred to as $\aleph_0$ (Aleph-Null). The anticipated transfinite numbers form a hierarchy of alephs: $\aleph_0, \aleph_1, \aleph_2, \aleph_3$ . . .

All this may seem very strange, and it is quite excusable for the reader by now to be thoroughly bewildered. Yet, if you have followed the previous reasoning step by step, and will go to the trouble of rereading, you will see that nothing which has been said is repugnant to straight thinking. Having established what is meant by counting in the finite domain, and what is meant by number, we decided to extend the counting process to infinite classes. As for our right to follow such a procedure, we have the same right, for example, as those who decided that man had crawled on the surface of the earth long enough and that it was about time for him to fly. It is our right to venture forth in the world of ideas as it is our right to extend our horizons in the physical universe. One restraint alone is laid upon us in these adventures of ideas: that we abide by the rules of logic.

Upon extending the counting process it was evident at once that no finite number could adequately describe an infinite class. If any number of ordinary arithmetic describes the cardinality of a class, that class must be finite, even though there were not enough ink or enough space or enough time to write the number out. We shall then require an entirely new kind of number, nowhere to be found in finite arithmetic, to describe the cardinality of an infinite class. Accordingly, the totality of integers was assigned the cardinality "aleph." Suspecting that there were *other* infinite classes with a cardinality *greater* than that of the totality of integers, we supposed a whole hierarchy of alephs, of which the cardinal number of the totality of integers was named Aleph-Null to indicate it was the smallest of the transfinites.

Having had an interlude in the form of a summary, let us turn once more to scrutinize the alephs, to find if, upon closer acquaintance, they may not become easier to understand.

The arithmetic of the alephs bears little resemblance

to that of the finite integers. The immodest behavior of $\aleph_0$ is typical.

A simple problem in addition looks like this:

$$\begin{aligned} \aleph_0 + 1 &= \aleph_0 \\ \aleph_0 + \text{googol} &= \aleph_0 \\ \aleph_0 + \aleph_0 &= \aleph_0 \end{aligned}$$

The multiplication table would be easy to teach, easier to learn:

$$\begin{aligned} 1 \times \aleph_0 &= \aleph_0 \\ 2 \times \aleph_0 &= \aleph_0 \\ 3 \times \aleph_0 &= \aleph_0 \\ n \times \aleph_0 &= \aleph_0 \end{aligned}$$

where *n* represents any finite number.
Also,

$$\begin{aligned} (\aleph_0)^2 &= \aleph_0 \times \aleph_0 \\ &= \aleph_0 \end{aligned}$$

And thus,

$$(\aleph_0)^n = \aleph_0$$

when *n* is a finite integer.

There seems to be no variation of the theme; the monotony appears inescapable. But it is all very deceptive and treacherous. We go along obtaining the same result, no matter what we do to $\aleph_0$, when suddenly we try:

$$(\aleph_0)^{\aleph_0}$$

This operation, at last, creates a new transfinite. But before considering it, there is more to be said about countable classes.

*

Common sense says that there are many more fractions than integers, for between any two integers there is an infinite number of fractions. Alas—common sense is amidst alien corn in the land of the infinite. Cantor discovered a simple but elegant proof that the rational fractions form a denumerably infinite sequence equivalent to the class of integers. Whence, this sequence must have the same cardinality.*

The set of all rational fractions is arranged, not in order of increasing magnitude, but in order of ascending numerators and denominators in an array:

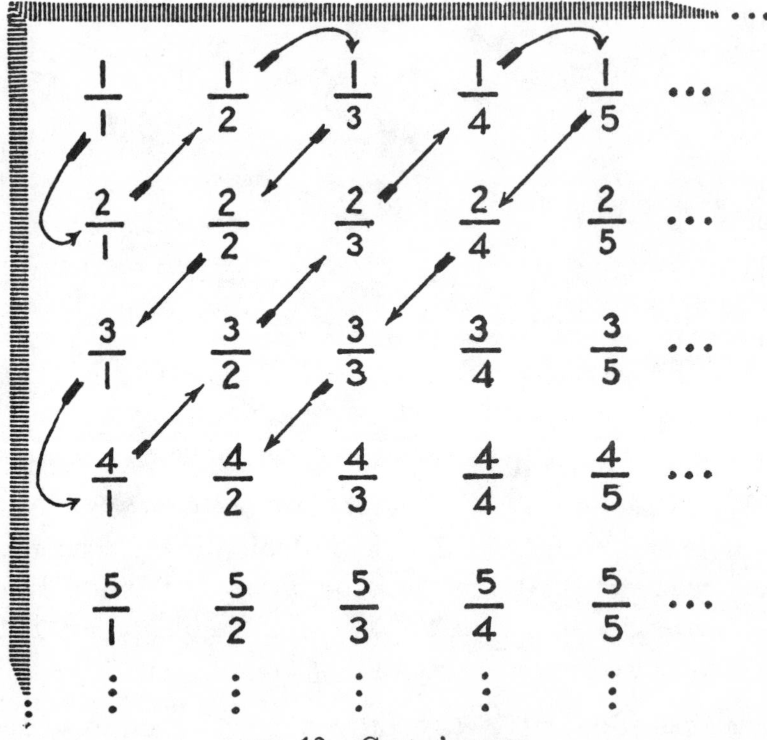

FIG. 12.—Cantor's array.

Since each fraction may be written as a pair of integers, i.e., $\frac{3}{4}$ as (3,4), the familiar one-to-one correspondence

* It has been suggested that at this point the tired reader puts the book down with a sigh—and goes to the movies. We can only offer

with the integers may be effected. This is illustrated in the above array by the arrows.

$$\begin{array}{ccccccccc} 1 & 2 & 3 & 4 & 5 & 6 & 7 & 8 & 9 \\ \updownarrow & \updownarrow & \updownarrow & \updownarrow & \updownarrow & \updownarrow & \updownarrow & \updownarrow & \updownarrow \ldots \\ (1,1) & (2,1) & (1,2) & (1,3) & (2,2) & (3,1) & (4,1) & (3,2) & (2,3) \end{array}$$

Cantor also found, by means of a proof (too technical to concern us here) based on the "height" of algebraic equations, that the class of all algebraic numbers, numbers which are the solutions of algebraic equations with integer coefficients, of the form:

$$a_0x^n + a_1x^{n-1} + \ldots + a_{n-1}x + a_n = 0$$

is denumerably infinite.

But Cantor felt that there were other transfinites, that there were classes which were not countable, which could not be put into one-to-one correspondence with the integers. And one of his greatest triumphs came when he succeeded in showing that there are classes with a cardinality greater than $\aleph_0$.

The class of real numbers composed of the rational and irrational numbers† is such a class. It contains those irrationals which are algebraic as well as those which are not. The latter are called *transcendental numbers.*[4]

in mitigation that this proof, like the one which follows on the non-countability of the real numbers, is tough and no bones about it. You may grit your teeth and try to get what you can out of them, or conveniently omit them. The essential thing to come away with is that Cantor found that the rational fractions are countable but that the set of real numbers is not. Thus, in spite of what common sense tells you, there are no more fractions than there are integers and there are more real numbers between 0 and 1 than there are elements in the whole class of integers.

† Irrational numbers are numbers which *cannot* be expressed as rational fractions. For example, $\sqrt{2}$, $\sqrt{3}$, e, π. The class of real numbers is made up of rationals like 1, 2, 3, $\frac{1}{4}$, $\frac{17}{32}$, and irrationals as above.

Two important transcendental numbers were known to exist in Cantor's time: π, the ratio of the circumference of a circle to its diameter, and e, the base of the natural logarithms. Little more was known about the class of transcendentals: it was an enigma. What Cantor had to prove, in order to show that the class of real numbers was nondenumerable (i.e., too big to be counted by the class of integers), was the unlikely fact that the class of transcendentals was nondenumerable. Since the rational and the algebraic numbers were known to be denumerable, and the sum of any denumerable number of denumerable classes is also a denumerable class, the sole remaining class which could make the totality of real numbers nondenumerable was the class of transcendentals.

He was able to devise such a proof. If it can be shown that the class of real numbers between 0 and 1 is nondenumerable, it will follow *a fortiori* that all the real numbers are nondenumerable. Employing a device often used in advanced mathematics, the *reductio ad absurdum*, Cantor assumed that to be true which he suspected was false, and then showed that this assumption led to a contradiction. He assumed that the real numbers between 0 and 1 were countable and could, therefore, be paired with the integers. Having proved that this assumption led to a contradiction, it followed that its opposite, namely, that the real numbers could *not* be paired with the integers (and were therefore not countable), was true.

To count the real numbers between 0 and 1, it is required that they all be expressed in a uniform way and a method of writing them down in order be devised so that they can be paired one to one with the integers. The first requirement can be fulfilled, for it is possible to express every real number as a nonterminating decimal. Thus, for example: [5]

$$\frac{1}{3} = .3333\ldots \qquad \frac{3}{14} = .21428571428571\ldots$$

$$\frac{1}{9} = .1111111\ldots \qquad \frac{\sqrt{2}}{2} = \frac{1.414\ldots}{2} = .707\ldots$$

Now, the second requirement confronts us. *How shall we make the pairings?* What system may be devised to ensure the appearance of *every* decimal? We did find a method for ensuring the appearance of every rational fraction. Of course, we could not actually write them all, any more than we could actually write all the integers; but the method of increasing numerators and denominators was so explicit that, if we had had an infinite time in which to do it, we could actually have set down all the fractions and have been certain that we had not omitted any. Or, to put it another way: It was always certain and determinate after a fraction had been paired with an integer, what the next fraction would be, and the next, and the next, and so on.

On the other hand, when a real number, expressed as a nonterminating decimal, is paired with an integer, what method is there for determining what the next decimal in order should be? You have only to ask yourself, which shall be the *first* of the nonterminating decimals to pair with the integer 1, and you have an inkling of the difficulty of the problem. Cantor however *assumed* that such a pairing does exist, without attempting to give its explicit form. His scheme was: With the integer 1 pair the decimal $.a_1a_2a_3\ldots$, with the integer 2, $.b_1b_2b_3\ldots$, etc. Each of the letters represents a digit of the nonterminating decimal in which it appears. The determinate array of pairing between the decimals and the integers would then be:

$$
\begin{array}{l}
1 \longleftrightarrow 0.\ a_1\ a_2\ a_3\ a_4\ a_5 \ldots \\
2 \longleftrightarrow 0.\ b_1\ b_2\ b_3\ b_4\ b_5 \ldots \\
3 \longleftrightarrow 0.\ c_1\ c_2\ c_3\ c_4\ c_5 \ldots \\
4 \longleftrightarrow 0.\ d_1\ d_2\ d_3\ d_4\ d_5 \ldots \\
\cdot \quad \cdot \quad \cdot \quad \cdot \quad \cdot \quad \cdot \quad \cdot \\
\cdot \quad \cdot \quad \cdot \quad \cdot \quad \cdot \quad \cdot \quad \cdot \\
\cdot \quad \cdot \quad \cdot \quad \cdot \quad \cdot \quad \cdot \quad \cdot
\end{array}
$$

That was Cantor's array. But at once it was evident that it glaringly exhibited the very contradiction for which he had been seeking. And in this defeat lay his triumph. For no matter *how* the decimals are arranged, by whatever system, by whatever scheme, it is always possible to construct an infinity of others which are not present in the array. The point is worth repeating: having contrived a general form for an array which we believed would include *every* decimal, we find, in spite of all our efforts, that *some* decimals are bound to be omitted. This, Cantor showed by his famous "diagonal proof." The conditions for determining a decimal omitted from the array are simple. It must differ from the first decimal in the array in its first place, from the second decimal in the array in its second place, from the third decimal in its third place, and so on. But then, *it must differ from every decimal in the entire array in* at least one place. If (as illustrated in the figure) we draw a diagonal line through our model array and write a new decimal, each digit of which shall differ from every digit intercepted by the diagonal, this new decimal cannot be found in the array.

$$\begin{array}{lcl}
1 & \longleftrightarrow & 0.\ a_1\ a_2\ a_3\ a_4\ a_5\ \ldots \\
2 & \longleftrightarrow & 0.\ b_1\ b_2\ b_3\ b_4\ b_5\ \ldots \\
3 & \longleftrightarrow & 0.\ c_1\ c_2\ c_3\ c_4\ c_5\ \ldots \\
4 & \longleftrightarrow & 0.\ d_1\ d_2\ d_3\ d_4\ d_5\ \ldots \\
5 & \longleftrightarrow & 0.\ e_1\ e_2\ e_3\ e_4\ e_5\ \ldots \\
\cdot & & \cdot\ \cdot\ \cdot\ \cdot\ \cdot\ \cdot \\
\cdot & & \cdot\ \cdot\ \cdot\ \cdot\ \cdot\ \cdot \\
\cdot & & \cdot\ \cdot\ \cdot\ \cdot\ \cdot\ \cdot
\end{array}$$

The new decimal may be written:—

$$0.\ \alpha_1\ \alpha_2\ \alpha_3\ \alpha_4\ \alpha_5\ldots;$$

where α_1 differs from a_1, α_2 differs from b_2, α_3 from c_3 α_4 from d_4, α_5 from e_5, etc. Accordingly, it will differ from each decimal in at least one place, from the *n*th decimal in at least its *n*th place. This proves conclusively that there is no way of including all the decimals in any possible array, no way of pairing them off with the integers. Therefore, as Cantor set out to prove:

1. The class of transcendental numbers is not only infinite, but also not countable, i.e., nondenumerably infinite.
2. The real numbers between 0 and 1 are infinite and not countable.
3. *A fortiori*, the class of all real numbers is nondenumerable.

*

To the noncountable class of real numbers, Cantor assigned a new transfinite cardinal. It was one of the alephs, but which one remains unsolved to this day. It is suspected that this transfinite, called the "cardinal of the continuum," which is represented by *c* or *C*, is identical with $\aleph_1$. But a proof acceptable to most mathematicians has yet to be devised.

The arithmetic of C is much the same as that of $\aleph_0$. The multiplication table has the same dependable monotone quality. But when C is combined with $\aleph_0$, it swallows it completely. Thus:

$$C + \aleph_0 = C \qquad C - \aleph_0 = C$$
$$C \times \aleph_0 = C \text{ and even } C \times C = C$$

Again, we hope for a variation of the theme when we come to the process of involution. Yet, for the moment, we are disappointed, for $C^{\aleph_0} = C$. But just as $(\aleph_0)^{\aleph_0}$ does not equal $\aleph_0$, so C^C does not equal C.

We are now in a position to solve our earlier problem in involution, for actually Cantor found that $(\aleph_0)^{\aleph_0} = C$. Likewise C^C gives rise to a new transfinite, greater than C. This transfinite represents the cardinality of the class of all one-valued functions. It is also one of the $\aleph$'s, but again, which one is unknown. It is often designated by the letter F.[6] In general, the process of involution, when repeated, continues to generate higher transfinites.

Just as the integers served as a measuring rod for classes with the cardinality $\aleph_0$, the class of real numbers serves as a measuring rod for classes with the cardinality C. Indeed, there are classes of geometric elements which can be measured in no other way except by the class of real numbers.

From the geometric notion of a point, the idea is evolved that on any given line segment there are an infinite number of points. The points on a line segment are also, as mathematicians say, "everywhere dense." This means that between any two points there is an infinitude of others. The concept of two immediately adjoining points is, therefore, meaningless. This property of being "everywhere dense," constitutes one of the essential characteristics of a *continuum.* Cantor, in referring to the "cardinality of the continuum," recognized that it applies alike to the class of real numbers and the class of points on a line segment. Both are everywhere dense, and both have the same cardinality, *C.* In other words, it is possible to pair the points on a line segment with the real numbers.

Classes with the cardinality *C* possess a property similar to classes with the cardinality $\aleph_0$: they may be thinned out without in any way affecting their cardinality. In this connection, we see in very striking fashion another illustration of the principle of transfinite arithmetic, that the whole is no greater than many of its parts. For instance, it can be proved that there are as many points on a line one foot long as there are on a line one yard long. The line segment *AB* in Fig. 13 is three times as long as the line *A′B′*. Nevertheless, it is possible to put the class of all points on the segment *AB* into a one-to-one correspondence with the class of points on the segment *A′B′*.

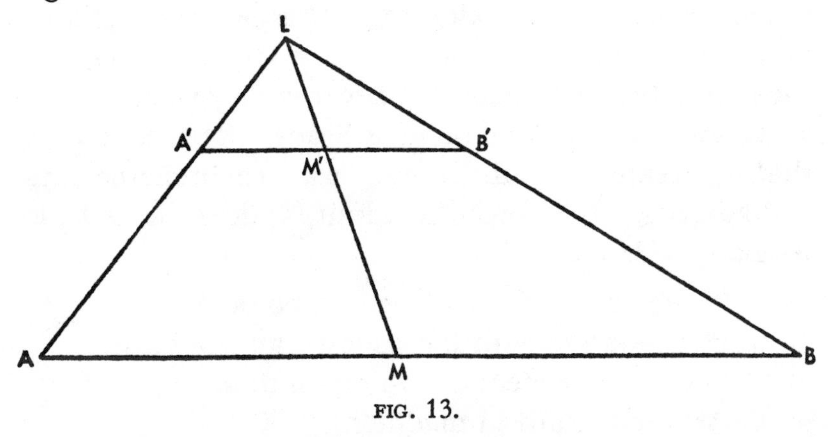

FIG. 13.

Let *L* be the intersection of the lines *AA′* and *BB′*. If then to any point *M* of *AB*, there corresponds a point

M' of $A'B'$, which is on the line LM, we have established the desired correspondence between the class of points on $A'B'$ and those on AB. It is easy to see intuitively and to prove geometrically that this is always possible, and that, therefore, the cardinality of the two classes of points is the same. Thus, since $A'B'$ is smaller than AB, it may be considered a proper part of AB, and we have again established that an infinite class may contain as proper parts, subclasses equivalent to it.

There are more startling examples in geometry which illustrate the power of the continuum. Although the statement that a line one inch in length contains as many points as a line stretching around the equator, or as a line stretching from the earth to the most distant stars, is startling enough, it is fantastic to think that a line segment one-millionth of an inch long has as many points as there are in all three-dimensional space in the entire universe. Nevertheless, this is true. Once the principles of Cantor's theory of transfinites is understood, such statements cease to sound like the extravagances of a mathematical madman. The oddities, as Russell has said, "then become no odder than the people at the antipodes who used to be thought impossible because they would find it so inconvenient to stand on their heads." Even conceding that the treatment of the infinite is a form of mathematical madness, one is forced to admit, as does the Duke in *Measure for Measure:*

> "If she be mad,—as I believe no other,—
> Her madness hath the oddest frame of sense,
> Such a dependency of thing on thing,
> As e'er I heard in madness."

*

Until now we have deliberately avoided a definition

of "infinite class." But at last our equipment makes it possible to do so. We have seen that an infinite class, whether its cardinality is $\aleph_0$, C, or greater, can be thinned out in a countless variety of ways, without affecting its cardinality. In short, the whole is no greater than many of its parts. Now, this property does not belong to finite classes at all; it belongs only to infinite classes. Hence, it is a unique method of determining whether a class is finite or infinite. Thus, our definition reads: *An infinite class is one which can be put into one-to-one reciprocal correspondence with a proper subset of itself.*

Equipped with this definition and the few ideas we have gleaned we may re-examine some of the paradoxes of Zeno. That of Achilles and the tortoise may be expressed as follows: Achilles and the tortoise, running the same course, must each occupy the same number of distinct positions during their race. However, if Achilles is to catch his more leisurely and determined opponent, he will have to occupy *more* positions than the tortoise, in the same elapsed period of time. Since this is manifestly impossible, you may put your money on the tortoise.

But don't be too hasty. There are better ways of saving money than merely counting change. In fact, you had best bet on Achilles after all, for he is likely to win the race. Even though we may not have realized it, we have just finished proving that he could overtake the tortoise by showing that a line a millionth of an inch long has just as many points as a line stretching from the earth to the furthest star. In other words, the points on the tiny line segment can be placed into one-to-one correspondence with the points on the great line, for there is no relation between the number of points on a line and its length. But this reveals the error in thinking that Achilles cannot catch the tortoise. The statement that Achilles must occupy as many distinct positions as the tortoise is correct. So is the statement that he must travel a greater distance than the tortoise in the same time. The only incorrect statement is the inference that since he must occupy the same number of positions as the tortoise he cannot travel further while doing so. Even though the classes of points on each line, which correspond to the several positions of both Achilles and the tortoise are equivalent, the line representing the path of Achilles is much longer than that representing the path of the tortoise. Achilles may travel much further than the tortoise without successively touching more points.

The solution of the paradox involving the arrow in flight requires a word about another type of continuum. It is convenient and certainly familiar to regard time as a continuum. The time continuum has the same properties as the space continuum: the successive instants in any elapsed portion of time, just as the points on a line, may be put into one-to-one correspondence with the class of real numbers; between any two instants of time an infinity of others may be interpolated; time also has the mathematical property mentioned before—it is everywhere dense.

Zeno's argument stated that at every instant of time the arrow was somewhere, in some place or position, and therefore, could not at any instant be in motion. Although the statement that the arrow had at every moment to be in some place is true, the conclusion that, therefore, it could not be moving is absurd. Our natural tendency to accept this absurdity as true springs from our firm conviction that motion is entirely different from rest.

We are not confused about the position of a body when it is at rest—we feel there is no mystery about the state of rest. We should feel the same when we consider a body in motion.

When a body is at rest, it is in one position at one instant of time and at a later instant it is still in the same position. When a body is in motion, there is a one-to-one correspondence between every instant of time and every new position. To make this clear we may construct two tables: One will describe a body at rest, the other, a body in motion. The "rest" table will tell the life history

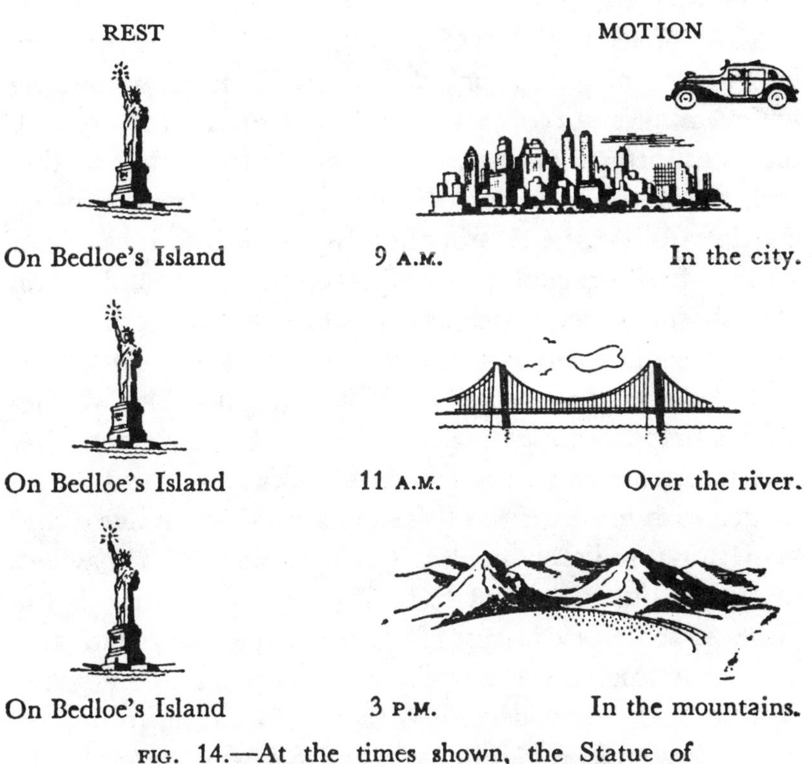

FIG. 14.—At the times shown, the Statue of Liberty is at the point shown, while the taxi's passengers see the different scenes shown at the right.

and the life geography of the Statue of Liberty, while the "motion" table will describe the Odyssey of an automobile.

The tables indicate that to every instant of time there corresponds a position of the Statue of Liberty and of the taxi. There is a one-to-one space-time correspondence for rest as well as for motion.

No paradox is concealed in the puzzle of the arrow when we look at our table. Indeed, it would be strange if there were gaps in the table; if it were impossible, at any instant, to determine exactly what the position of the arrow is.

Most of us would swear by the existence of motion, but we are not accustomed to think of it as something which makes an object occupy different positions at different instants of time. We are apt to think that motion endows an object with the strange property of being continually nowhere. Impeded by the limitations of our senses which prevent us from perceiving that an object in motion simply occupies one position after another and does so rather quickly, we foster an illusion about the nature of motion and weave it into a fairy tale. Mathematics helps us to analyze and clarify what we perceive, to a point where we are forced to acknowledge, if we no longer wish to be guided by fairy tales, that we live either in Mr. Russell's changeless world or in a world where motion is but a form of rest. The story of motion is the same as the story of rest. It is the same story told at a quicker tempo. The story of rest is: "It is here." The story of motion is: "It is here, it is there." Because, in this respect, it resembles Hamlet's father's ghost is no reason to doubt its existence. Most of our beliefs are chained to less substantial phantoms. Motion is perhaps not easy for our senses to grasp, but with the aid of mathematics, its essence may first be properly understood.

*

At the beginning of the twentieth century it was generally conceded that Cantor's work had clarified the concept of the infinite so that it could be talked of and treated like any other respectable mathematical concept. The controversy which arises wherever mathematical philosophers meet, on paper, or in person, shows that this was a mistaken view. In its simplest terms this controversy, so far as it concerns the infinite, centers about the questions: Does the infinite exist? Is there such a thing as an infinite class? Such questions can have little meaning unless the term mathematical "existence" is first explained.

In his famous "Agony in Eight Fits," Lewis Carroll hunted the snark. Nobody was acquainted with the snark or knew much about it except that it existed and that it was best to keep away from a boojum. The infinite may be a boojum, too, but its existence in any form is a matter of considerable doubt. Boojum or garden variety, the infinite certainly does not exist in the same sense that we say, "There are fish in the sea." For that matter, the statement "There is a number called 7" refers to something which has a different existence from the fish in the sea. "Existence" in the mathematical sense is wholly different from the existence of objects in the physical world. A billiard ball may have as one of its properties, in addition to whiteness, roundness, hardness, etc., a relation of circumference to diameter involving the number π. We may agree that the billiard ball and π both exist; we must also agree that the billiard ball and π lead different kinds of lives.

There have been as many views on the problem of existence since Euclid and Aristotle as there have been philosophers. In modern times, the various schools of mathematical philosophy, the Logistic school, Formalists, and Intuitionists, have all disputed the somewhat less than glassy essence of mathematical being. All these disputes are beyond our ken, our scope, or our intention. A stranger company even than the tortoise, Achilles, and the arrow, have defended the existence of infinite classes —defended it in the same sense that they would defend the existence of the number 7. The Formalists, who think mathematics is a meaningless game, but play it with no less gusto, and the Logistic school, which considers that mathematics is a branch of logic—both have taken Cantor's part and have defended the alephs. The defense rests on the notion of self-consistency. "Existence" is a metaphysical expression tied up with notions of being and other bugaboos worse even than boojums. But the expression, "self-consistent proposition" sounds like the language of logic and has its odor of sanctity. A proposition which is not self-contradictory is, according to the Logistic school, a true existence statement. From this standpoint the greater part of Cantor's mathematics of the infinite is unassailable.

New problems and new paradoxes, however, have been discovered arising out of parts of Cantor's structure because of certain difficulties already inherent in classical logic. They center about the use of the word "all." The paradoxes encountered in ordinary parlance, such as "All generalities are false including this one," constitute a real problem in the foundations of logic, just as did the Epimenides paradox whence they sprang. In the Epimenides, a Cretan is made to say that all Cretans are liars, which, if true, makes the speaker a liar for telling the truth. To dispose of this type of paradox the Logistic school invented a "Theory of Types." The theory of types and the axiom of reducibility on which it is based must be accepted as axioms to avoid paradoxes of this kind. In order to accomplish this a reform of classical logic is required which has already been undertaken. Like most reforms it is not wholly satisfactory—even to the reformers—but by means of their theory of types the last vestige of inconsistency has been driven out of the house that Cantor built. The theory of transfinites may still be so much nonsense to many mathematicians, but it is certainly consistent. The serious charge Henri Poincaré expressed in his aphorism, "La logistique n'est plus stérile: elle engendre la contradiction," has been successfully rebutted by the logistic doctrine so far as the infinite is concerned.

To Cantor's alephs then, we may ascribe the same existence as to the number 7. An existence statement free from self-contradiction may be made relative to either. For that matter, there is no valid reason to trust in the finite any more than in the infinite. It is as permissible to discard the infinite as it is to reject the impressions of one's senses. It is neither more, nor less scientific to do so. In the final analysis, this is a matter of faith and taste, but *not* on a par with rejecting the belief in Santa Claus. Infinite classes, judged by finite standards, generate paradoxes much more absurd and a great deal less pleasing than the belief in Santa Claus; but when they are judged by the appropriate standards, they lose their odd appearance, behave as predictably as any finite integer.

At last in its proper setting, the infinite has assumed a respectable place next to the finite, just as real and just as dependable, even though wholly different in character. Whatever the infinite may be, it is no longer a purple cow.

FOOTNOTES

1. We distinguish cardinal from *ordinal numbers*, which denote the relation of an element in a class to the others, with reference to some system of order. Thus, we speak of the *first* Pharaoh of Egypt, or of the *fourth* integer, in their customary order, or of the *third* day of the week, etc. These are examples of ordinals.—P. 30.
2. For the definition of primes, see the Chapter on PIE.—P. 32.
3. This series is said to CONVERGE TO A LIMIT—1. Discussion of this concept must be postponed to the chapters on PIE and the calculus.—P. 38.
4. A transcendental number is one which is not the root of an algebraic equation with integer coefficients. See PIE.—P. 49.
5. Any terminating decimal, such as .4, has a nonterminating form .3999. . .—P. 51.
6. A simple geometric interpretation of the class of all one-valued functions F is the following: With each point of a line segment, associate a color of the spectrum. The class F is then composed of all possible combinations of colors and points that can be conceived.—P. 54.

PIE (π, i, e)
Transcendental and Imaginary

In order to reach the Truth, it is necessary, once in one's life, to put everything in doubt—so far as possible.

—DESCARTES

PERHAPS PURE science begins where common sense ends; perhaps, as Bergson says, "Intelligence is characterized by a natural lack of comprehension of life." [1] But we have no paradoxes to preach, no epigrams to sell. It is only that the study of science, particularly mathematics, often leads to the conclusion that one need only say that a thing is unbelievable, impossible, and science will prove him wrong. Good common sense makes it plain that the earth is flat and stands still, that the Chinese and the Antipodeans walk about suspended by their feet like chandeliers, that parallel lines never meet, that space is infinite, that negative numbers are as real as negative cows, that -1 has no square root, that an infinite series must have an infinite sum, or that it must be possible with ruler and compass alone to construct a square exactly equal in area to a given circle.

Just how far have we been carried by common sense in arriving at these conclusions? Not very far! Yet some of the statements seem quite plausible, even inescapable. It would be wrong to say that science has proved that all are false. We may still cling to the Euclidean hypothesis that parallel lines never meet and remain always equidistant, as long as we remember it is merely a hypothesis, but the statements about the squaring of the circle, the square root of -1, and about infinite series belong in a different category.

The circle *can not* be squared with ruler and compass. -1 *has* a square root. An infinite series *can have* a finite sum. Three symbols, π, i, e, have enabled mathematicians to prove these statements, three symbols which represent the fruits of centuries of mathematical research. How do they stand up to common sense?

*

The most famous problem in the entire history of mathematics is the "squaring of the circle." Two other problems which challenged Greek geometers, the "duplication of the cube" and the "trisection of an angle," may, as a matter of interest, be briefly considered with the first, even though squaring the circle alone involves π.

In the infancy of geometry, it was discovered that it was possible to measure the area of a figure bounded by straight lines. Indeed, geometry was devised for that very purpose—to measure the fields in the valley of the Nile, where each year the floods from the rising river obliterated every mark made by the farmer to indicate which fields were his and which his neighbor's. Measuring areas bounded by curved lines presented greater difficulties, and an effort was made to reduce every problem of this type to one of measuring areas with straight boundaries. Clearly, if a square can be constructed with the area of a given circle, by measuring the area of the square, that of the circle is determined. The expression "squaring the circle" derives its name from this approach.

The number π is the ratio of the circumference of a circle to its diameter. The area of a circle of radius r is given by the formula πr^2. Now the area of a square with side of length A is A^2. Thus, the algebraic statement: $A^2 = \pi r^2$ expresses the equivalence in area between a given square and a circle. Taking square roots of both sides of this equation yields $A = r\sqrt{\pi}$. As r is a known quantity, the problem of squaring the circle is, in effect, the computation [2] of the value of π.

Since mathematicians have succeeded in computing π with extraordinary exactitude, what then is meant by the statement, "It is impossible to square the circle"? Unfortunately, this question is still shrouded in many misapprehensions. But these would vanish if the problem were understood.

*

Squaring the circle is proclaimed *impossible*, but what does "impossible" mean in mathematics? The first steam vessel to cross the Atlantic carried, as part of its cargo, a book that "proved" it was impossible for a steam vessel to cross anything, much less the Atlantic. Most of the savants of two generations ago "proved" that it would be forever impossible to invent a practical heavier-than-air flying machine. The French philosopher, Auguste Comte, demonstrated that it would always be impossible for the human mind to discover the chemical constitution of the stars. Yet, not long after this statement was made the spectroscope was applied to the light of the stars, and we now know more about their chemical constitution, including those of the distant nebulae, than we know about the contents of our medicine chest. As just one illustration, helium was discovered in the sun before it was found in the earth.

Museums and patent offices are filled with cannons, clocks, and cotton gins, already obsolete, each of which confounded predictions that their invention would be impossible. A scientist who says that a machine or a project is impossible only reveals the limitations of his day. Whatever the intentions of the prophet, the prediction has none of the qualities of prophecy. "It is impossible to fly to the moon" is meaningless, whereas "We have not yet devised a means of flying to the moon" is not.

Statements about impossibility in mathematics are of a wholly different character. A problem in mathematics which may not be solved for centuries to come is not always impossible. "Impossible" in mathematics means *theoretically* impossible, and has nothing to do with the present state of our knowledge. "Impossible" in mathematics does *not* characterize the process of making a silk purse out of a sow's ear, or a sow's ear out of a silk purse; it *does* characterize an attempt to prove that 7 times 6 is 43 (in spite of the fact that people not good at arithmetic often achieve the impossible). By the rules of arithmetic 7 times 6 is 42, just as by the rules of chess, a pawn must make at least 5 moves before it can be queened.

Where theoretical proof that a problem cannot be solved is lacking, it is legitimate to attempt a solution, no matter how improbable the hope of success. For centuries the construction of a regular polygon of 17 sides was rightly considered difficult, but falsely considered impossible, for the nineteen-year-old Gauss in 1796 succeeded in finding an elementary construction.[3] On the other hand, many famous problems, such as Fermat's Last Theorem,[4] have defied solution to this day in spite of heroic researches. To determine whether we have the right to say that squaring the circle, trisecting the angle, or duplicating the cube is *impossible*, we must find logical proofs, involving purely mathematical reasoning. Once such proofs have been adduced, to continue the search for a solution is to hunt for a three-legged biped.[5]

*

Having determined what mathematicians mean by impossible, the bare statement, "It is impossible to square the circle" still remains meaningless. To give it meaning we must specify *how* the circle is to be squared. When Archimedes said, "Give me a place to stand and I will move the earth," he was not boasting of his physical powers but was extolling the principle of the lever. When it is said that the circle cannot be squared, all that is meant is that this *cannot be done with ruler and compass alone*, although with the aid of the integraph or higher curves the operation does become possible.

Let us repeat the problem: It is required to construct a square equal in area to a given circle, by means of an exact theoretical plan, using only two instruments: the ruler and compass. By a ruler is meant a straightedge, that is, an instrument for drawing a straight line, not for measuring lengths. By a compass is meant an instrument with which a circle with any center and any radius can be drawn. These instruments are to be used a finite number of times, so that limits or converging processes with an infinite number of steps may not be employed.[6] The construction, by purely logical reasoning, depending only on Euclid's axioms and theorems, is to be absolutely exact.

The concepts of "limit" and "convergence" are more fully explained elsewhere,[7] but a word about them here is in place.

Consider the familiar series $1 + \frac{1}{2} + \frac{1}{4} + \frac{1}{8} + \frac{1}{16} + \frac{1}{32} + \ldots$ The sum of the first 5 terms of this series is 1.9375; the sum of the first 10 terms is 1.9980 . . . ; the sum of the first 15 is 1.999781 . . . What is readily apparent is that this series tends to choke off, i.e., the additional terms which are added become so small that even a vast number will not cause the series to grow beyond a finite bound. In this instance the bound, or limit, is 2. Such a series which chokes off is said to "*converge*"[8] to a "*limit.*"

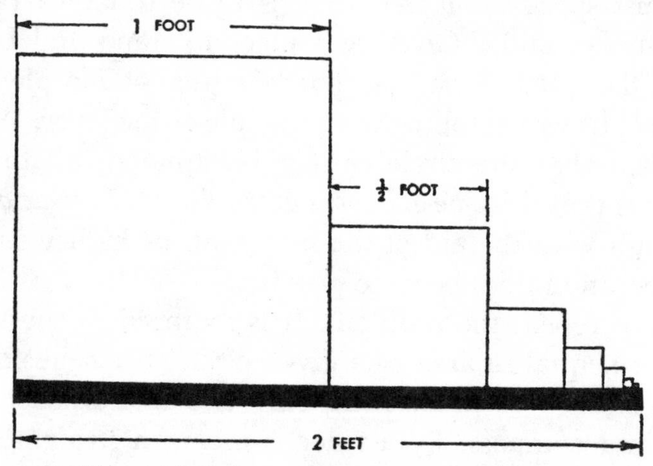

FIG. 15.—An infinite number of terms with a finite sum. If the width of the first block is one foot, the width of the second $\frac{1}{2}$ foot, of the third $\frac{1}{4}$ foot, of the fourth $\frac{1}{8}$ foot, and so one, then an infinite number of blocks rests on the 2-foot bar, that is:

$$1 + \frac{1}{2} + \frac{1}{4} + \frac{1}{8} + \frac{1}{16} + \ldots = 2.$$

The geometric analogues of the concepts of limit and convergence are equally fruitful. A circle may be regarded as the limit of the polygons with increasing number of sides which may be successively inscribed in it, or circumscribed about it, and its area as the common limit of both of these sets of polygons.

This is not a rigorous definition of limit and convergence, but too often mathematical rigor serves only to bring about another kind of rigor—*rigor mortis* of mathematical creativeness.

To return to squaring the circle: the Greeks, and later mathematicians, sought an exact construction with ruler and compass, but always failed. As we shall see later, all ruler and compass constructions are geometric equivalents of first- and second-degree *algebraic* equations and combinations of such equations. But the German mathematician Lindemann, in 1882, published a proof that π is a *transcendental* number and thus any equation which it satisfies cannot be algebraic and surely not algebraic of first or second degree. It follows that the statement, "The squaring of the circle is impossible with ruler and compass alone," *is* meaningful.

So far as the other two problems are concerned, thanks in part to the work of "the marvelous boy . . . who perished in his prime," the sixteen-year-old Galois, it was established about one hundred years ago that the duplication of the cube and the trisection of an angle are also impossible with ruler and compass. We may allude to them briefly.

There is a story among the Greeks that the problem of duplicating the cube originated in a visit to the Delphic oracle. There was an epidemic raging at the time, and the oracle said the epidemic would cease only if a cubical altar to Apollo were doubled in size. The masons and architects made the mistake of *doubling* the side of the cube, but that made the volume *eight* times as great. Of course the oracle was not satisfied, and the Greek mathematicians, on re-examining the problem began to see that the right answer involved, not doubling the side, but multiplying it by the cube root of 2. This could not be done geometrically with ruler and compass. They finally succeeded by using other instruments and higher curves. The oracle was appeased and the epidemic ceased. You may believe the story or not, much as you choose, but you cannot "duplicate the cube."[9]

The trisection of an angle has received a good deal of attention in the newspapers during the past few years because monographs continue to crop up which claim to solve the problem completely. The fallacies contained in these "solutions" are of four kinds: they are sometimes merely approximate and not exact; instruments other than the ruler and compass are occasionally used, either wittingly or unwittingly; at times there is a logical fallacy in the intended proof; and often only special and not general angles are considered. An angle can be bisected but not trisected by elementary geometry, since the first problem involves merely square roots, while the second involves cube roots, which, as we have stated, cannot be constructed with ruler and compass.

*

The difficulty in squaring the circle, as stated at the outset, lies in the nature of the number π. This remarkable number, as Lindemann proved, cannot be the root of an algebraic equation with integer coefficients.[10] It is therefore not expressible by rational operations, or by the extraction of square roots, and as only such operations can be translated into an equivalent ruler and compass construction, it is impossible to square the circle. The parabola is a more complicated curve than a circle, but nevertheless, as Archimedes knew, any area bounded by a parabola and a straight line can be determined by rational operations, and hence the "parabola can be squared."

Lindemann's proof is too technical to concern us here. If, however, we consider the history and development of π, we shall be in a better position to understand its purpose without being compelled to master its difficulties.

If a triangle is *inscribed* in a circle (Fig. 16), the area of the inscribed triangle will be less than the area of the circle:

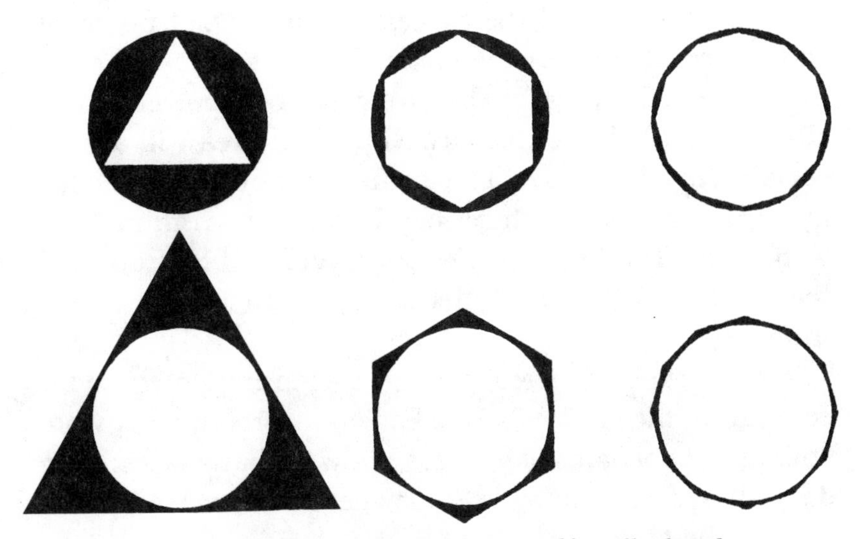

FIG. 16.—The circle as the limit of inscribed and circumscribed polygons.

The difference between the area of the circle and the triangle are the three shaded portions of the circle. Now consider the same circle with a triangle *circumscribed* about it (Fig. 16). The area of the circumscribed triangle will be greater than the area of the circle. The three shaded portions of the triangle again represent the difference in area. It may readily be seen that if the number of sides of the inscribed figure is doubled, the area of the resulting hexagon will be less than the area of the circle, but closer to it than the area of the inscribed triangle. Similarly, if the number of sides of the circumscribed triangle is doubled, the area of the circumscribed hexagon will still be greater than the area of the circle but, again, closer to it than the area of the circumscribed triangle. By well-known, simple, geometric methods, employing only ruler and compass, the number of sides of the inscribed and circumscribed polygons can be doubled as many times as desired. The area of the successively inscribed polygons will approach that of the circle, but will always remain *slightly less;* the area of the circumscribed polygons will also approach that of the circle but their area will always remain *slightly greater.* The common value approached by both is the area of the circle. In other words, the circle is the *limit* of these two series of polygons. If the radius of the circle is equal to 1, its area, which equals πr^2, is simply π.

This method of increasing and decreasing polygons for computing the value of π was known to Archimedes, who, employing polygons of 96 sides, showed that π is less than $3\frac{1}{7}$ and greater than $3\frac{10}{71}$. Somewhere in between lies the area of the circle.

Archimedes' approximation for π is considerably closer than that given in the Bible. In the Book of Kings, and in Chronicles, π is given as 3. Egyptian mathematicians gave a somewhat more accurate value—3.16. The familiar decimal—3.1416, used in our schoolbooks, was already known at the time of Ptolemy in 150 A.D.

Theoretically, Archimedes' method for computing π by increasing the number of sides of the polygons can be extended indefinitely, but the requisite calculations soon become very cumbersome. None the less, during the Middle Ages such calculations were zealously carried out.

Francisco Vieta, the most eminent mathematician of the sixteenth century, though not a professional, made a great advance in the calculation of π in determining its value to ten decimal places. In addition to giving the formula:

$$\pi = \frac{2}{\sqrt{\frac{1}{2}}\cdot\sqrt{\frac{1}{2}+\frac{1}{2}\sqrt{\frac{1}{2}}}\cdot\sqrt{\frac{1}{2}+\frac{1}{2}\sqrt{\frac{1}{2}+\frac{1}{2}\sqrt{2}}}\ \cdots},$$

a nonterminating product, and making many other important mathematical discoveries, Vieta rendered service to King Henry IV of France, in the war against Spain, by deciphering intercepted letters addressed by the Spanish Crown to its governors of the Netherlands. The Spaniards were so impressed that they attributed his discovery of the cipher key to magic. It was neither the first nor the last time that the efforts of mathematicians were branded as necromancy.

In 1596 Ludolph van Ceulen, the German mathematician, long a resident in Holland, calculated 35 decimal places for π. Instead of the epitaph, "died at 40, buried at 60," appropriate where cerebration ceases just when life is supposed to begin, van Ceulen, who worked on π almost to the day of his death at the age of 70, requested that the 35 digits of π which he had computed be inscribed as a fitting epitaph on his tombstone. This was actually done. The value he gave for π is, in part, 3.14159 26535 89793 23846... In memory of his achievement the Germans still call this number the Ludolphian number. We propose to call π the Archimedean number.

*

The number π reached maturity with the invention of the calculus by Newton and Leibniz. The Greek method was abandoned and the purely algebraic device of convergent infinite series, products, and continued fractions came into vogue. John Wallis (1616–1703), the

Englishman, contributed one of the most famous products:

$$\frac{\pi}{2} = \frac{2}{1} \times \frac{2}{3} \times \frac{4}{3} \times \frac{4}{5} \times \frac{6}{5} \times \frac{6}{7} \times \frac{8}{7} \times \frac{8}{9} \times \ldots$$

Leibniz' infinite series, unlike Wallis' product for π, is a sum:

$$\frac{\pi}{4} = 1 - \frac{1}{3} + \frac{1}{5} - \frac{1}{7} + \frac{1}{9} - \frac{1}{11} + \frac{1}{13} - \frac{1}{15} + \ldots$$

The successive products and sums of the terms of these series yield values of π as accurate as desired. These processes, typical of the powerful methods of approximation used not only in mathematics but in the other sciences, although much less cumbersome than the method employed by the Greeks, still entail a great deal of calculation. The products of Wallis' series are:

$$\frac{2}{1} = 2, \frac{2}{1} \times \frac{2}{3} = \frac{4}{3}, \frac{2}{1} \times \frac{2}{3} \times \frac{4}{3} = \frac{16}{9}, \frac{2}{1} \times \frac{2}{3} \times \frac{4}{3} \times \frac{4}{5} = \frac{64}{45}, \text{ etc.}$$

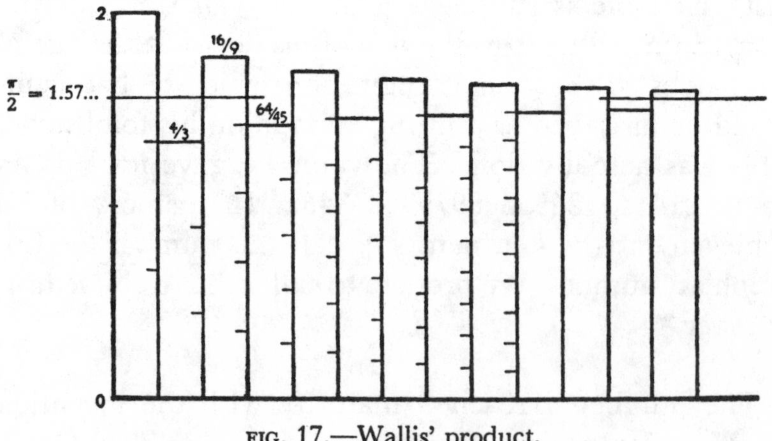

FIG. 17.—Wallis' product.

$$\frac{\pi}{2} = 1.57\ldots$$

$$\frac{\pi}{2} = \frac{2}{1} \times \frac{2}{3} \times \frac{4}{3} \times \frac{4}{5} \times \frac{6}{5} \times \ldots$$

Taking the successive sums of Leibniz' series, we obtain:

$$1,\ 1 - \frac{1}{3} = \frac{2}{3},\ 1 - \frac{1}{3} + \frac{1}{5} = \frac{13}{15},\ 1 - \frac{1}{3} + \frac{1}{5} - \frac{1}{7} = \frac{76}{105}, \text{ etc.}$$

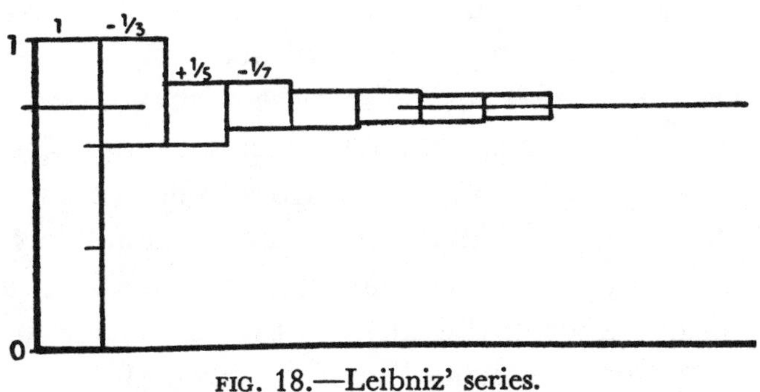

FIG. 18.—Leibniz' series.

$$\frac{\pi}{4} = 0.785 \ldots$$

$$\frac{\pi}{4} = 1 - \frac{1}{3} + \frac{1}{5} - \frac{1}{7} + \frac{1}{9} - \frac{1}{11} + \ldots$$

After taking the first 50 terms of these series, the next 50 will not yield an appreciably more accurate value of π, for the series converge rather slowly. The rapidly convergent series

$$\frac{\pi}{4} = 4\left(\frac{1}{5} - \frac{1}{3 \cdot 5^3} + \frac{1}{5 \cdot 5^5} - \frac{1}{7 \cdot 5^7} + \ldots\right) - \left(\frac{1}{239} - \frac{1}{3 \cdot 239^3} + \frac{1}{5 \cdot 239^5} - \frac{1}{7 \cdot 239^7} + \ldots\right)$$

is much more useful, and is frequently employed in modern mathematics. Its relation to π was established by Machin (1680–1752). Using even more rapidly converging series, Abraham Sharp, in 1699, calculated π to 71 decimal places. Dase, a lightning calculator employed by Gauss, worked out 200 places in 1824. In 1854, Richter computed 500 places, and finally, in 1873, Shanks, an English mathematician, achieved a curious kind of immortality by determining π to 707 decimal places. Even today it would require 10 years of calculation to determine π to 1000 places. Yet that does not seem like a waste of time as compared with the billions of hours spent by millions of people on crossword puzzles and contract bridge, to say nothing of political debates.

Of course Shank's result has no conceivable use in applied science. No more than 10 decimal places for π are ever needed in the most precise work. The famous American astonomer and mathematician, Simon Newcomb, once remarked, "Ten decimal places are sufficient to give the circumference of the earth to the fraction of an inch, and thirty decimals would give the circumference of the whole visible universe to a quantity imperceptible with the most powerful telescope."

Why, then, has so much time and effort been devoted to the calculation of π? The reason is twofold. First, by studying infinite series mathematicians hoped they might find some clue to its transcendental nature. Second, the fact that π, a purely geometric ratio, could be evolved out of so many arithmetic relationships—out of infinite series, with apparently little or no relation to geometry—was a never-ending source of wonder and a never-ending stimulus to mathematical activity.

Who would imagine—that is, who but a mathematician—that the number expressing a fundamental relation between a circle and its diameter could grow out of the curious fraction communicated by Lord Brouncker (1620–1684) to John Wallis?

$$\pi = \cfrac{4}{1 + \cfrac{1^2}{2 + \cfrac{3^2}{2 + \cfrac{5^2}{2 + 7^2 \ldots}}}}$$

But just such relations between infinite series and π illustrate the profound connection between most mathematical forms, geometric or algebraic. It is mere coincidence, a mere accident that π is defined as the ratio of the circumference of a circle to its diameter. No matter how mathematics is approached, π forms an integral part.[11] In his *Budget of Paradoxes,* Augustus De Morgan illustrated how little the usual definition of π suggests its origin. He was explaining to an actuary what the chances were that, at the end of a given time a certain proportion of a group of people would be alive, and quoted the formula employed by actuaries which involves π. On explaining the geometric meaning of π, the actuary, who had been listening with interest, interrupted and exclaimed, "My dear friend, that must be a delusion. What can a circle have to do with the number of people alive at the end of a given time?"

To recapitulate briefly, the problem of squaring the circle turns out to be an impossible construction with ruler and compass alone. The only constructions possible with these instruments correspond to first- and second-degree algebraic equations. Lindemann proved that π is not only not the root of a first- or second-degree algebraic equation, but is not the root of any algebraic equation (with integer coefficients), no matter how great the degree; therefore π is transcendental. Here, then, is the end of every hope of proving this classical problem in the intended way. Here is mathematical impossibility.

*

When the Greek philosophers found that the square root of 2 is not a rational number,[12] they celebrated the discovery by sacrificing 100 oxen. The much more profound discovery that π is a transcendental number deserves a greater sacrifice. Again mathematics triumphed over common sense. π, a finite number—the ratio of the circumference of a circle to its diameter—is accurately expressible only as the sum or product of an infinite series of wholly different and apparently unrelated numbers. The area of the simplest of all geometric figures, the circle, cannot be determined by finite (Euclidean) means.

e

In the seventeenth century, perhaps the greatest of all for the development of mathematics, there appeared a work which in the history of British science can be placed second only to Sir Isaac Newton's monumental *Principia.* In 1614, John Napier of Merchiston issued his *Mirifici Logarithmorum Canonis Descriptio,* ("A Description of the Admirable Table of Logarithms"), the first treatise on logarithms.[13] To Napier, who also invented the decimal point, we are indebted for an invention which is as important to mathematics as Arabic numerals, the concept of zero, and the principle of positional notation.[14] Without these, mathematics would probably not have advanced much beyond the stage to which it had been brought 2000 years ago. Without logarithms the computations accomplished daily with ease by every mathematical tyro would tax the energies of the greatest mathematicians.

Since e and logarithms have the same genealogical tree and were brought up together, we may for the moment turn our attention to logarithms to ascertain something of the nature of the number e.

Stupendous calculations being required to construct trigonometric tables for navigation and astronomy, Napier was prompted to invent some device to facilitate these computations. Although contemporaries like Vieta and Ceulen vied with each other in performing almost unbelievably difficult feats of arithmetic, it was at best a labor of love, an exalted drudgery and self-immolation, with love's labor often lost as the result of one small slip.

Napier succeeded in achieving his purpose, in abbreviating the operations of multiplication and division, operations "so fundamental in their nature that to shorten them seems impossible." Nevertheless, by means of logarithms, every problem in multiplication and division, no matter how elaborate, reduces to a relatively easy one in addition and subtraction. Multiplying and dividing googols and googolplexes becomes as easy as adding a simple column of figures.

Like many another of the profound and fecund inventions of mathematics, the underlying idea was so simple that one wonders why it had not been thought of earlier. Cajori recounts that Henry Briggs (1556–1631), professor of geometry at Oxford, "was so struck with admiration of Napier's book, that he left his studies in London to do homage to the Scottish philosopher. Briggs was delayed in his journey, and Napier complained to a common friend, 'Ah, John, Mr. Briggs will not come.' At that very moment knocks were heard at the gate, and Briggs was brought into the lord's chamber. Almost one quarter of an hour was spent, each beholding the other without speaking a word. At last Briggs began: 'My lord, I have undertaken this long journey purposely to see your person, and to know by what engine of wit or ingenuity you came first to think of this most excellent help in astronomy, viz. the logarithms; but, my lord, being by you found out, I wonder nobody found it out before, when now known it is so easy.' "

Napier's conception of logarithms was based on an ingenious and well-known idea: a comparison between 2 moving points, one of which generates an arithmetical, the other a geometric progression.

The two progressions:

Arithmetical— 0 1 2 3 4 5 6 7 8...

Geometric— 1 2 4 8 16 32 64 128 256...

bear to each other this interesting relationship: If the terms of the arithmetical progression are regarded as exponents (powers) of 2, the corresponding terms of the geometric progression represent the quantity resulting from the indicated operation. Thus,[15] $2^0 = 1$, $2^1 = 2$, $2^2 = 4$, $2^3 = 8$, $2^4 = 16$, $2^5 = 32$, etc. Furthermore, to determine the value of the product $2^2 \times 2^3$, it is only necessary to add the exponents, obtaining $2^{2+3} = 2^5$, which is the desired product. Calling 2 the *base*, *each term in the arithmetical progression is the* LOGARITHM *of the corresponding term of the geometric progression.*

Napier explained this notion geometrically as follows: A point S moves along a straight line, AB, with a velocity at each point S_1 proportional to the remaining distance S_1B. Another point R moves along an unlimited line, CD, with a uniform velocity equal to the initial velocity of S. If both points start from A and C at the same time, then the *logarithm* of the number measured by the distance S_1B is measured by the distance CR_1.

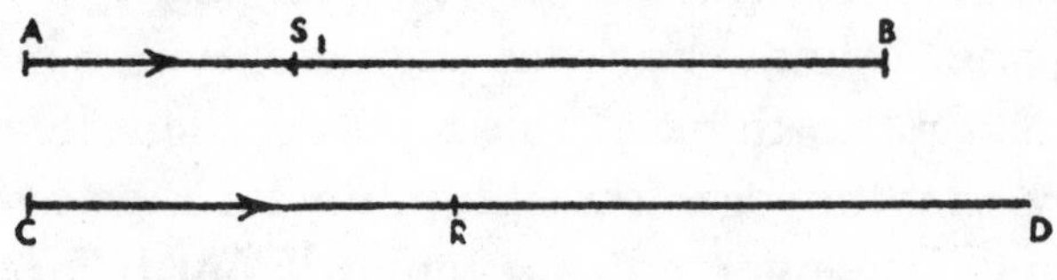

FIG. 19.—Napier's dynamic interpretation of logarithms.

By this method, as S_1B decreases, its logarithm CR_1 increases. But it soon became apparent that it was advantageous to define the logarithm of 1 as zero, and to have the logarithm grow with the number. Napier changed his system accordingly.

One of the fruits of the higher education is the illuminating view that a logarithm is merely a number that is found in a table. We shall have to widen the curriculum. If a, b, and c are three numbers related by the equation $a^b = c$, then b, the exponent of a, is the logarithm of *c to the base a*. In other words, the logarithm of a number to the base a is the power to which a must be raised to obtain that number. In the example, $2^3 = 8$, the logarithm of 8 to the base 2 is 3. Or $10^2 = 100$, and the logarithm of 100 to the base 10 equals 2. The concise way of expressing this is: $3 = \log_2 8$, $2 = \log_{10} 100$. The simple table below gives all the essential properties of logarithms:

(1) $\log_a (b \times c) = \log_a b + \log_a c.$

(2) $\log_a \left(\frac{b}{c}\right) = \log_a b - \log_a c.$

(3) $\log_a b^c = c \times \log_a b.$

(4) $\log_a \sqrt[c]{b} = \left(\frac{1}{c}\right) \log_a b.$

Equations (1) and (2) indicate how to multiply or divide two numbers; nothing more is required than to add or subtract their respective logarithms. The result obtained is the logarithm of the product, or quotient. Equations (3) and (4) show that with the aid of logarithms the operations of raising to powers and extracting roots may be replaced by the much simpler ones of multiplication and division.

Extensive tables of logarithms were soon constructed to the base 10 and to the Napierian or natural base e. So widely were these tables distributed that mathematicians all over Europe were able to avail themselves of the use of logarithms within a very short time of their invention. Kepler was one who not only saw the tables of Napier but himself advanced their development; he was thus one of the first of the legion of scientists whose contributions to knowledge were greatly facilitated by logarithms.

The two systems of logs to the two bases, 10 and e (the Briggs and the natural base respectively), are the principal ones still in use, with e predominating.[16] Like π, the number e is transcendental and like π it is what P. W. Bridgman names a "program of procedure," rather than a number, since it can never be completely expressed (1) in a finite number of digits, (2) as the root of an algebraic equation with integer coefficients, (3) as a nonterminating but repeating decimal.[17] It can only be expressed with accuracy as the limit of a convergent infinite series or of a continued fraction. The simplest and most familiar infinite series giving the value of e is:

$$e = 1 + \frac{1}{1!} + \frac{1}{2!} + \frac{1}{3!} + \frac{1}{4!} + \frac{1}{5!} + \frac{1}{6!} + \frac{1}{7!} \cdots$$ [18]

Accordingly, its value may be approximated as closely as we please by taking additional terms of the series. To the tenth decimal place $e = 2.7182818285$. A glance at the table below will indicate how an infinite convergent series behaves as more and more of its terms are summed.

$$(1)\ 1 + \frac{1}{1!} \qquad = 2.$$

$$(2)\ 1 + \frac{1}{1!} + \frac{1}{2!} \qquad = 2.5$$

$$(3)\ 1 + \frac{1}{1!} + \frac{1}{2!} + \frac{1}{3!} = 2.6666666\ldots$$

$$(4)\ 1 + \frac{1}{1!} + \frac{1}{2!} + \frac{1}{3!} + \frac{1}{4!} = 2.7083334\ldots$$

$$(5)\ 1 + \frac{1}{1!} + \frac{1}{2!} + \frac{1}{3!} + \frac{1}{4!} + \frac{1}{5!} = 2.7166666\ldots$$

$$(6)\ 1 + \frac{1}{1!} + \ldots + \frac{1}{6!} = 2.7180555\ldots$$

$$(7)\ 1 + \frac{1}{1!} + \ldots + \frac{1}{7!} = 2.7182539\ldots$$

$$(8)\ 1 + \frac{1}{1!} + \ldots + \frac{1}{8!} = 2.7182787\ldots$$

$$(9)\ 1 + \frac{1}{1!} + \ldots + \frac{1}{9!} = 2.7182818\ldots$$

Upon taking a few more terms, e looks like this:

$$2.71828182845904523536028 74\ldots$$

Euler, who undoubtedly had the Midas touch in mathematics, not only invented the symbol e and calculated its value to 23 places, but gave several very interesting expressions for it, of which these two are the most important:

$$(1)\ e = 2 + \cfrac{1}{1 + \cfrac{1}{2 + \cfrac{2}{3 + \cfrac{3}{4 + \cfrac{4}{5 + 5\ldots}}}}}$$

$$(2)\ \sqrt{e} = 1 + \cfrac{1}{1 + \cfrac{1}{1 + \cfrac{1}{1 + \cfrac{1}{5 + \cfrac{1}{1 + \cfrac{1}{1 + \cfrac{1}{1 + \cfrac{1}{9 \ldots}}}}}}}}$$

The need for navigational tables was not alone responsible for the development of logarithms. Big business, particularly banking, played its part as well. A remarkable series, the limiting value of which is e, arises in the preparation of tables of compound interest. This series is obtainable from the expansion of $\left(1 + \frac{1}{n}\right)^n$ as n becomes infinite. The origin of this important expression is interesting.

Suppose your bank pays 3 per cent interest yearly on deposits. If this interest is added at the end of each year, for a period of three years, the total amount to your credit, assuming an original capital of \$1.00, is given by the formula: $(1 + .03)^3$. If the interest is *compounded* semiannually, after the three-year period the total of principal plus interest would be $\left(1 + \frac{.03}{2}\right)^{2 \times 3}$.

Imagine however that you are fortunate enough to find a philanthropic bank which decides to pay 100 per cent interest a year. Then the amount to your credit at the end of the year will be $(1 + 1)^1 =$ \$2.00. If the interest is compounded semiannually, the amount will be

$(1 + \frac{1}{2})^{1\times2} = 2\frac{1}{4}$ or \$2.25. If it is compounded quarterly, it will be $(1 + \frac{1}{4})^{1\times4} = \$2.43$. It seems clear that the more often the interest is compounded, the more money you will have in the bank. By a further stretch of the imagination, you may conceive of the possibility that the philanthropic bank decides to compound the interest *continuously*, that is to say at *every* instant throughout the year. How much money will you then have at the end of the year? No doubt a fortune. At least, that is what you would suspect, even allowing for what you know about banks. Indeed you might become, not a millionaire, not a billionaire, but more nearly what could be described as an "infinitaire." Alas, banish all delusions of grandeur, for the process of compounding interest continuously, at every instant, generates an infinite series which *converges* to the limit e. The sum on deposit after this hectic year, with its apparent promise of untold riches, would be not quite \$2.72. For, if one takes the trouble to expand $\left(1 + \frac{1}{n}\right)^n$, as n becomes very large,[19] the successive values thus obtained approximate to the value of e, and where n becomes infinite, $\left(1 + \frac{1}{n}\right)^n$ actually yields the infinite series for e:

$$e = 1 + \frac{1}{1!} + \frac{1}{2!} + \frac{1}{3!} + \frac{1}{4!} + \frac{1}{5!} + \cdots$$

Besides serving as the base for the natural logarithms, e is a number useful everywhere in mathematics and applied science. No other mathematical constant, not even π, is more closely connected with human affairs. In economics, in statistics, in the theory of probability, and in the exponential function, e has helped to do one thing and to do that better than any number yet discovered. It has played an integral part in helping mathematicians describe and predict what is for man the most important of all natural phenomena—that of growth.

The exponential function, $y = e^x$, is the instrument used, in one form or another, to describe the behavior of growing things. For this it is uniquely suited: *it is the only function of x with a rate of change with respect to x equal to the function itself.*[20] A function, it will be remembered, is a table giving the relation between two variable quantities, where a change in one implies some change in the other. The cost of a quantity of meat is a function of its weight; the speed of a train, a function of the quantity of coal consumed; the amount of perspiration given off, a function of the temperature. In each of these illustrations, a change in the second variable: weight, quantity of coal consumed, and temperature, is correlated with a change in the first variable: cost, speed, and volume of perspiration. The symbolism of mathematics permits functional relationships to be simply and concisely expressed. Thus, $y = x$, $y = x^2$, $y = \sin x$, $y = \operatorname{csch} x$, $y = e^x$ are examples of functions.

A function is not only adequate to describe the behavior of a projectile in flight, a volume of gas under changes of pressure, an electric current flowing through a wire, but also of other processes which entail change, such as growth of population, growth of a tree, growth of an amoeba, or as we have just seen, growth of capital and interest. What is peculiar to every organic process is that the *rate* of growth is proportional to the *state* of growth. The bigger something is, the faster it grows. Under ideal conditions, the larger the population of a country becomes, the faster it increases. The rate of speed of many chemical reactions is proportional to the quantity of the reacting substances which are present. Or, the amount of heat given off by a hot body to the surrounding medium is proportional to the temperature. The rate at which the total quantity of a radioactive substance diminishes at any instant, owing to emanations, is proportional to the total quantity present at that instant. All these phenomena, which either are, or resemble, organic processes, may be accurately described by a form of the exponential function (the simplest being $y = e^x$), for this has the property that its rate of change is proportional to the rate of change of its variable.

*

A universe in which e and π were lacking, would not, as some anthropomorphic soul has said, be inconceivable. One could hardly imagine that the sun would fail to rise, or the tides cease to flow for lack of π and e. But without these mathematical artifacts, what we know about the sun and the tides, indeed our ability to describe all natural phenomena, physical, biological, chemical or statistical, would be reduced to primitive dimensions.

i

Alice was criticizing Humpty Dumpty for the liberties he took with words: "When I use a word," Humpty replied, in a scornful tone, "it means just what I choose it to mean—neither more nor less." "The question is," said Alice, "whether you *can* make a word mean so many different things." "The question is," said Humpty, "which is to be master, that's all."

Those who are troubled (and there are many) by the word "imaginary" as it is used in mathematics, should hearken unto the words of H. Dumpty. At most, of course, it is a small matter. In mathematics familiar words are repeatedly given technical meanings. But as Whitehead has so aptly said, this is confusing only to minor intellects. When a word is precisely defined, and signifies only one thing, there is no more reason to criticize its use than to criticize the use of a proper name. Our Christian names may not suit us, may not suit our friends, but they occasion little misunderstanding. Confusion arises only when the same word packs several meanings and is what Humpty D. calls a "portmanteau."

Semantics, a rather fashionable science nowadays, is devoted to the study of the proper use of words. Yet there is much more need for semantics in other branches of knowledge than in mathematics. Indeed, the larger part of the world's troubles today arise from the fact that some of its more voluble magnificoes are definitely antisemantic.

An imaginary number is a precise mathematical idea. It forced itself into algebra much in the same way as did the negative numbers. We shall see more clearly how imaginary numbers came into use if we consider the development of their progenitors—the negatives.

Negative numbers appeared as roots of equations as soon as there were equations, or rather, as soon as mathematicians busied themselves with algebra. Every equation of the form $ax + b = 0$, where a and b are greater than zero, has a negative root.

The Greeks, for whom geometry was a joy and algebra a necessary evil, rejected negative numbers. Unable to fit them into their geometry, unable to represent them by pictures, the Greeks considered negative numbers no numbers at all. But algebra needed them if it were to grow up. Wiser than the Greeks, wiser than Omar Khayyám,[21] the Chinese and the Hindus recognized negative numbers even before the Christian era. Not as learned in geometry, they had no qualms about numbers of which they could draw no pictures. There is a repetition of that indifference to the desire for concrete representation of abstract ideas in the contemporary theories of mathematical physics, (relativity, the mechanics of quanta, etc.) which, although understandable as symbols on paper, defy diagrams, pictures, or adequate metaphors to explain them in terms of common experience.

Cardan, eminent mathematician of the sixteenth century, gambler, and occasional scoundrel, to whom algebra is vastly indebted, first recognized the true importance of negative roots. But his scientific conscience twitted him to the point of calling them "fictitious." Raphael Bombelli of Bologna carried on from where Cardan left off. Cardan had talked about the square roots of negative numbers, but he failed to understand the concept of imaginaries. In a work published in 1572, Bombelli pointed out that imaginary quantities were essential to the solution of many algebraic equations. He saw that equations of the form $x^2 + a = 0$, where a is any number greater than 0, could *not* be solved except with the aid of imaginaries. In trying to solve a simple equation $x^2 + 1 = 0$, there are two alternatives. Either the equation is meaningless, which is absurd, or x is the square root of -1, which is equally absurd. But mathematics thrives on absurdities, and Bombelli helped it along by accepting the second alternative.

*

Three hundred and fifty years have gone by since Bombelli made his choice. Philosophers, scientists, and those with that minor-key quality of mind known as plain common sense have criticized, in ever-increasing diminuendo, the concept of the imaginary. All of these worthies are dead, most of them forgotten, while imaginary numbers flourish wickedly and wantonly over the whole field of mathematics.

Occasionally, even the masters snickered. Leibniz thought: "Imaginary numbers are a fine and wonderful refuge of the Holy Spirit, a sort of amphibian between being and not being." Even the mighty Euler said that numbers like the square root of minus one "are neither nothing, nor less than nothing, which necessarily constitutes them imaginary, or impossible." He was quite right, but what he omitted to say was that imaginaries were useful and essential to the development of mathematics. And so they were allotted a place in the number domain with all the rights, privileges, and immunities thereunto appertaining. In time, the fears and queasiness about their essence all but vanished, so that the judgment of Gauss is the judgment of today:

Our general arithmetic, so far surpassing in extent the geometry of the ancients, is entirely the creation of modern times. Starting originally from the notion of absolute integers, it has gradually enlarged its domain. To integers have been added fractions, to rational quantities, the irrational, to positive, the negative, and to the real, the imaginary. This advance, however, had always been made at first with timorous and hesitating steps. The early algebraists called the negative roots of equations false roots, and this is indeed the case when the problem to which they relate has been stated in such a form that the character of the quantity sought allows of no opposite. But just as in general arithmetic no one would hesitate to admit fractions, although there are so many countable things where a fraction has no meaning, so we would not deny to negative numbers the rights accorded to positives, simply because innumerable things admit of no opposite. The reality of negative numbers is sufficiently justified since in innumerable other cases they find an adequate interpretation.

This has long been admitted, but the imaginary quantities, formerly, and occasionally now, improperly called impossible, as opposed to real quantities—are still rather tolerated than fully naturalized; they appear more like an empty play upon symbols, to which a thinkable substratum is unhesitatingly denied, even by those who would not depreciate the rich contribution which this play upon symbols has made to the treasure of the relations of real quantities.[22]

*

Imaginary numbers, like four-dimensional geometry, developed from the logical extension of certain processes. The process of extracting roots is called evolution. It is an apt name, for imaginary numbers were literally evolved out of the extension of the process of extracting roots. If $\sqrt{4}$, $\sqrt{7}$, $\sqrt{11}$ had meaning, why not $\sqrt{-4}$, $\sqrt{-7}$, $\sqrt{-11}$? If $x^2 - 1 = 0$ had a solution, why not $x^2 + 1 = 0$? The recognition of imaginaries was much like the United States recognizing Soviet Russia—the existence was undeniable, all that was required was formal sanction and approval.

$\sqrt{-1}$ is the best-known imaginary. Euler represented it by the symbol "i" which is still in use.[23] It is idle to be concerned with the question, "What number when multiplied by itself equals -1?" Like all other numbers, i is a symbol which represents an abstract but very precise idea. It obeys all the rules of arithmetic with the added convention that $i \times i = -1$. Its obedience to these rules and its manifold uses and applications justify its existence regardless of the fact that it may be an anomaly.

The formal laws of operation for i are easy:

Since the rule of signs provides:

$$\left.\begin{array}{l}(+1) \times (+1) = +1 \\ (+1) \times (-1) = -1\end{array}\right\} \left\{\begin{array}{l}(-1) \times (+1) = -1 \\ (-1) \times (-1) = +1\end{array}\right.$$

Accordingly:

$$i \times (+1) = \sqrt{-1}$$
$$i \times (-1) = -\sqrt{-1}$$
$$-i \times (-1) = +i$$
$$= \sqrt{-1}$$
$$i \times i = i^2$$
$$= -1$$
$$i \times i \times i = i^3$$
$$= (\sqrt{-1})(\sqrt{-1})^2$$
$$= (\sqrt{-1}) \cdot (-1)$$
$$= -\sqrt{-1}$$
$$i \times i \times i \times i = i^4$$
$$= (\sqrt{-1})^2 (\sqrt{-1})^2$$
$$= (-1) \times (-1)$$
$$= +1$$
$$i \times i \times i \times i \times i = i^5$$
$$= (\sqrt{-1})^2 (\sqrt{-1})^2 (\sqrt{-1})$$
$$= (-1) \times (-1) \times \sqrt{-1}$$
$$= (+1) \times \sqrt{-1}$$
$$= \sqrt{-1}, \text{ etc.*}$$

* From which we may construct a convenient table:

i^1	$= \sqrt{-1} = i$	i^2	$= \sqrt{-1} \cdot \sqrt{-1} = -1$
i^3	$= -1 \cdot \sqrt{-1} = -i$	i^4	$= (\sqrt{-1})^2 \cdot (\sqrt{-1})^2 = +1$
i^5	$= +1 \cdot \sqrt{-1} = i$	i^6	$= +1 \cdot (\sqrt{-1})^2 = -1$
i^7	$= -1 \cdot \sqrt{-1} = -i$	i^8	$= -1 \cdot (\sqrt{-1})^2 = +1$
↓		↓	

The table shows that *odd* powers of i are equal to $-i$, or $+i$, and *even* powers of i are equal to -1 or $+1$.

Extension of the use of imaginaries has led to complex numbers of the form $a + ib$, where a and b are *real* numbers (as distinguished from imaginaries). Thus $3 + 4i$, $1 - 7i$, $2 + 3i$ are examples of complex numbers.

The enormously fruitful field of function theory is a direct consequence of the development of complex numbers. While this is a subject too technical and specialized, we shall have occasion to mention complex numbers again when we explain the geometric representation of imaginaries. To that end, we must turn for a moment to that mathematical idea which, as Boltzmann once said seems almost cleverer than the man who invented it—the science of Analytical Geometry.

*

Program music is distinguished from absolute music, which owes its coherence to structure, in that the purpose of the former is to tell a story. In a certain sense, analytical geometry can be distinguished from the geometry of the Greeks as program music from absolute music. Geometry, practical in its origin, was cultivated and developed for its own sake both as a logical discipline and as a study of form. Geometry was a manifestation of a striving for the ideal. Shapes and forms that were beautiful, harmonious, and symmetric were appreciated and eagerly studied. But the Greeks cultivated the practical only as long as it had a beautiful side; beyond that, their mathematics was hampered by their aesthetics.

There was left to Descartes the task of writing the program music of mathematics, of devising a geometry which tells a story. When it is said that every algebraic equation has a picture, we are describing the relation between analytical geometry and algebra. And just as program music is as important and significant in itself as the stories it illustrates, so analytical geometry has its own dignity and importance—is an autonomous mathematical discipline.

*

The Jesuit Fathers were often very wise: at their school at La Flêche, young René Descartes was permitted, because of his delicate health, to remain in bed each day until noon. What McGuffey would have prophesied about the future of such a child is not difficult to imagine. But Descartes did not turn out a complete profligate. Indeed, his delightful habit of staying in bed until noon bore at least one remarkable fruit. Analytical geometry came to him one morning as he lay pleasantly in bed.

It is powerful, this idea of a co-ordinate geometry, yet easy to understand. Consider two lines (axes) in a plane: xx', yy', intersecting at right angles at a point R:

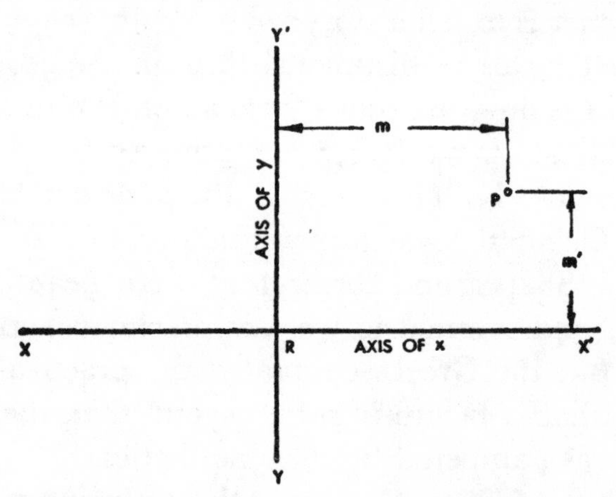

FIG. 20.—The point P has the co-ordinates (m, m').

Any point in the entire plane may then be uniquely determined by its perpendicular distance from the lines xx' and yy'. The point P, for example, by the distances m and m'. Thus, a *pair* of numbers representing scalar distances along xx' and yy' will determine every point in the plane, and conversely, every point in the plane determines a pair of numbers. These numbers are called the *co-ordinates* of the point.

All distances on xx' measured to the right of R are called positive, to the left of R, negative. Similarly, all distances measured on yy' above R are positive, all distances below, negative. The point of intersection, the *origin*, is designated by the co-ordinates (0, 0). The con-

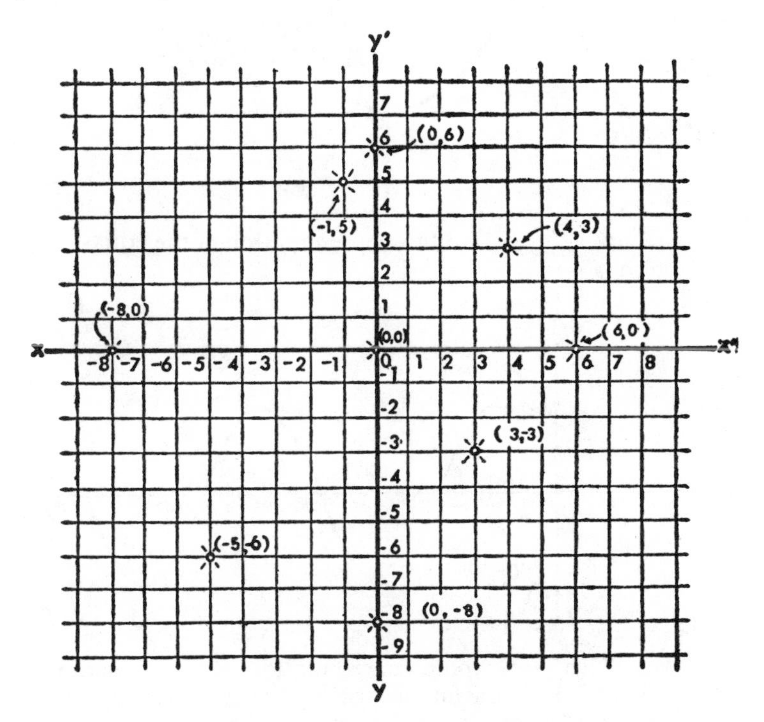

FIG. 21.—The co-ordinate axes in the real plane.

vention for writing co-ordinates is to put down the distance *from* the yy' axis (i.e. the distance *along* the xx' axis) first, the distance *from* the xx' axis, *along* the yy' axis, second; thus: (0, 0), (4, 3), (– 1, 5), (6, 0), (0, 6),

(– 5, – 6), (3, – 3), (– 8, 0), (0, – 8) are the coordinates of the points in Fig. 21.

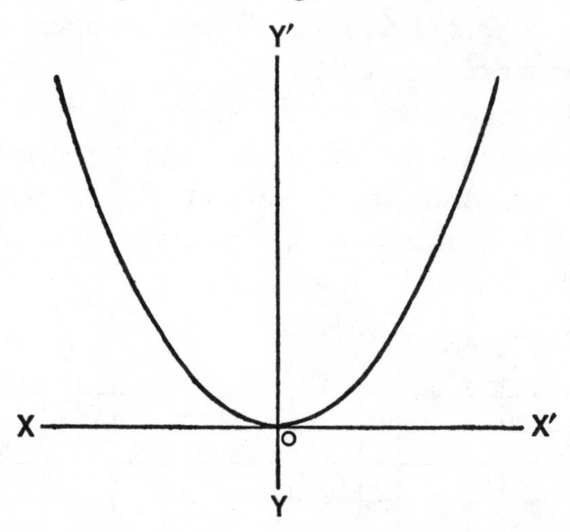

FIG. 22(a).—Graphic representation of the equation $y = x^2$.

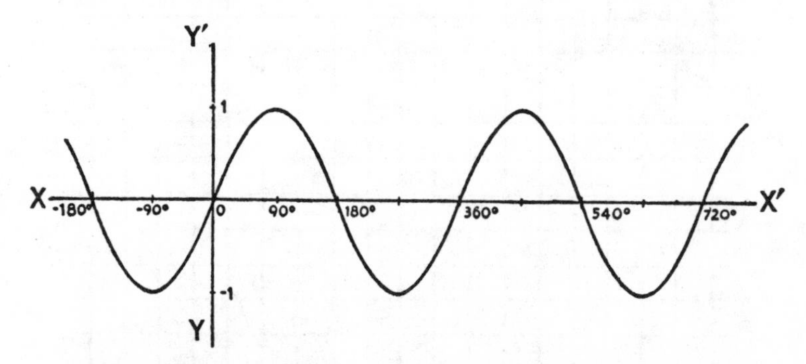

FIG. 22(b).—Graphic representation of the equation $y = \sin x$. This is the famous wave curve used to represent many regular and periodic phenomena, i.e., electrical current, the motion of a pendulum, radio transmission, sound and light waves, etc. (For the meaning of $\sin x$, see note 2 in the chapter on the calculus.)

Coupling this notion with that of a function, it is not difficult to see how an equation may be pictured in the plane of analytic geometry. When x and y are functionally related, to each value of x there corresponds a value of y, which two values determine a point in the plane. The totality of such number pairs, that is, all the values of y corresponding to all the values of x, when joined by a smooth curve as in Figs. 22(a,b,c), make up the geometrical portrait of an equation.

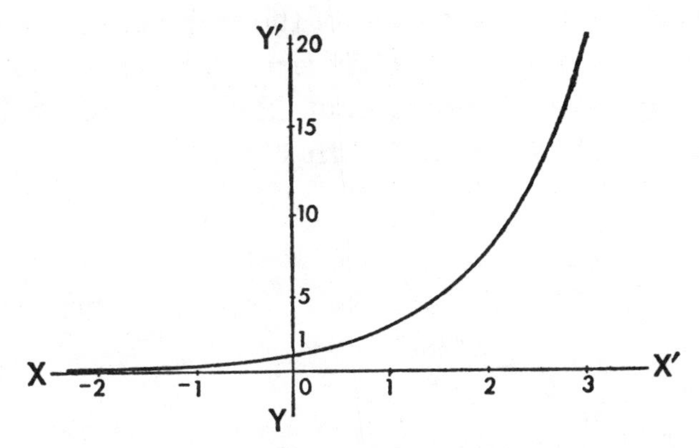

FIG. 22(c).—Graphic representation of the equation $y = e^x$. This curve illustrates the property common to all phenomena of growth: rate of growth is proportional to state of growth.

Employing co-ordinate geometry, how shall we represent an imaginary number like $\sqrt{-1}$? A theorem in

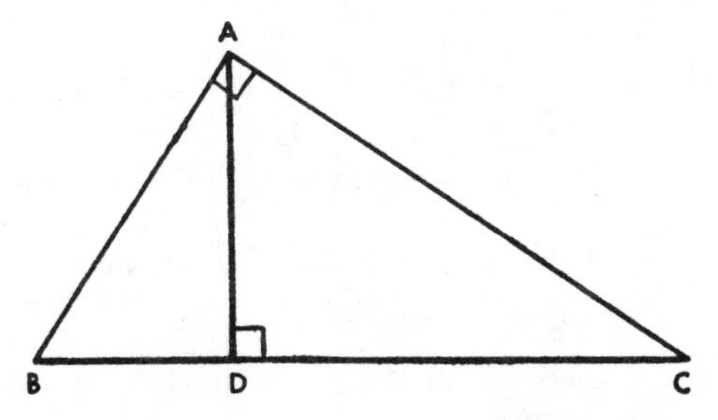

FIG. 23.—Length $AD = \sqrt{BD \times DC}$ = geometric mean of BD and DC.

elementary geometry, relating to the geometric mean, furnishes the clue (see Fig. 23).

In the right triangle ABC, the perpendicular AD divides BC into two portions: BD, DC. The length of the perpendicular AD equals $\sqrt{BD \times DC}$, and is called the *geometric mean* of BD and DC. (Fig. 23.)

A Norwegian surveyor, Wessel, and a Parisian bookkeeper, Argand, at the close of the eighteenth and beginning of the nineteenth centuries, independently found that imaginary numbers could be represented by the application of this theorem. In Fig. 24:

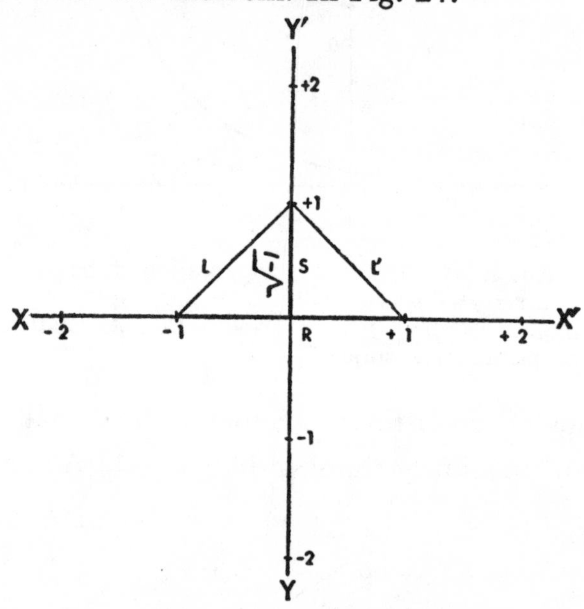

FIG. 24.—Geometric interpretation of i.

the distance S, from the origin R to $+1$, is the geometric mean of the triangle, bounded by the sides L and L', and the base formed by that portion of the xx' axis from -1 to 1.

Then $S = \sqrt{-1 \cdot +1} = \sqrt{-1} = i$

Here, then, is a geometric representation of an imaginary number.

Extending this idea, Gauss built up the entire complex plane. In the complex plane every point represented by a complex number of the form $x + iy$ corresponds to the point in the plane fixed by the co-ordinates x and y. In other words, a complex number may be regarded as a pair of real numbers with the addition of the number i. The use of i appears only on performing the operations of multiplication and division. Conceive of a line joining the point $(a + ib)$ to the origin R. Then the operation of multiplying by -1 is equivalent to rotating that line about the origin through 180° and shifting the point from $(+a +ib)$ to $(-a -ib)$. The effect of multiplying a number by i is such that when performed twice, i^2 is obtained, which is equivalent to multiplication by -1.

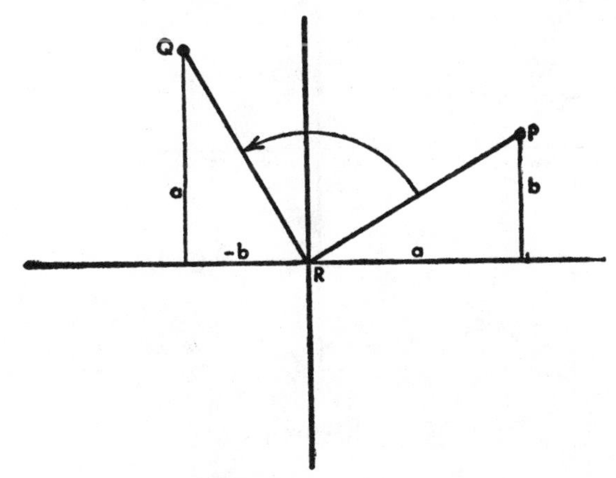

FIG. 25.—Multiplication by i is a rotation through 90°.

Let $P = (a + ib)$.

$$\begin{aligned} \text{Then, } P \times i &= (a + ib) \times i \\ &= (a \times i) + (b \times i \times i) \\ &= ia + b \cdot -1 \\ &= -b + ia \\ &= Q. \end{aligned}$$

Therefore, multiplication by i is a rotation through only 90°.

Complex numbers may be added, subtracted, multiplied, and divided, just as though they were real numbers. The formal rules of these operations (the most interesting being the substitution of -1 for i^2) are illustrated in the examples below.

(1) $x + iy = x' + iy'$ if, and only if $x = x'$ and $y = y'$
(2) $(x + iy) + (x' + iy') = (x + x') + i(y + y')$
(3) $(x + iy) - (x' + iy') = (x - x') + i(y - y')$
(4) $(x + iy)(x' + iy') = (xx' - yy') + i(xy' + yx')$
(5) $(x + iy)/(x' + iy') = \left[\frac{xx' + yy'}{(x')^2 + (y')^2}\right] + i\left[\frac{yx' - xy'}{(x')^2 + (y')^2}\right]$

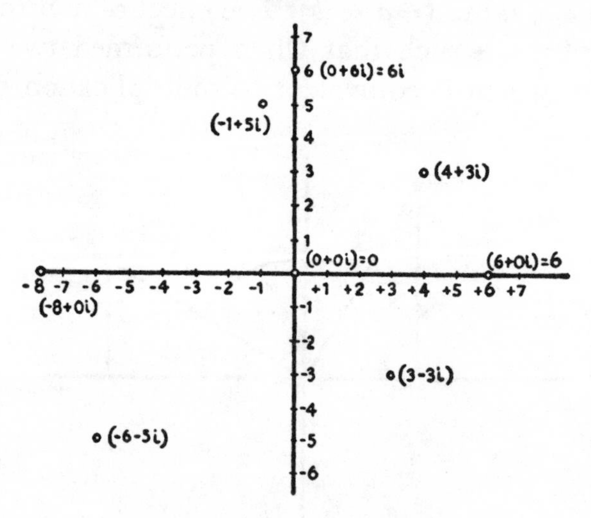

FIG. 26.—The complex plane.

Figure 26 shows the same points in the plane given in Fig. 21, except that for the co-ordinates of x and y of each point we have substituted the corresponding complex number $x + iy$.

By virtue of the peculiar properties of i, complex numbers may be used to represent both magnitude and direction. With their aid some of the most essential notions in physics such as velocity, force, acceleration, etc., are conveniently represented.

Enough has now been said to indicate the general nature of i, its purpose and importance in mathematics, its challenge to and final victory over the cherished tenets of common sense. Undaunted by its paradoxical appearance, mathematicians used it as they used π and e. The result has been to make possible almost the entire edifice of modern physical science.*

*

One thing remains. There is a famous formula—perhaps the most compact and famous of all formulas—developed by Euler from a discovery of the French mathematician, De Moivre: $e^{i\pi} + 1 = 0$. Elegant, concise and full of meaning, we can only reproduce it and not stop to inquire into its implications. It appeals equally to the mystic, the scientist, the philosopher, the mathematician. For each it has its own meaning. Though known for over a century, De Moivre's formula came to Benjamin Peirce, one of Harvard's leading mathematicians in the nineteenth century, as something of a revelation. Having discovered it one day, he turned to his students and made a remark which supplies in dramatic quality and appreciation what it may lack in learning and sophistica-

* Let us have this much balm for the reader who has bravely gone through the pages on analytical geometry and complex numbers. The average college course on analytic geometry (not including complex numbers) takes six months. It is, therefore, a little too much to expect that it can be learned in about five pages. On the other hand, if the basic idea has been put over, that every number, every equation of algebra, can be graphically represented, the harrowing details may be left to more intrepid adventurers.

tion: "Gentlemen," he said, "that is surely true, it is absolutely paradoxical; we cannot understand it, and we don't know what it means, but we have proved it, and therefore, we know it must be the truth."

When there is so much humility and so much vision everywhere, society will be governed by science and not by its clever people.

APPENDIX

BIRTH OF A CURVE

(1) Let us consider the equation $y = x^2$. Take a few sample values of x and find the corresponding values of y, arranging the results in a table:

x	y
0	0
1	1
2	4
3	9
4	16

That is $2^2 = 4$, $3^2 = 9$, etc. Plotting these points on the co-ordinate plane, we obtain Fig. A.

(2) Now, what about the negative values of x? We see, for example, $(-2)^2 = -2 \times -2 = 4$. This is evidently true for all values of x; thus there corresponds to every point plotted in Fig. A another point which is its mirror image, the axis OY being the mirror. Adding these gives the second figure (Fig. B).

(3) The arrangement of the points suggests that we draw a smooth curve through them. (Fig. C.)

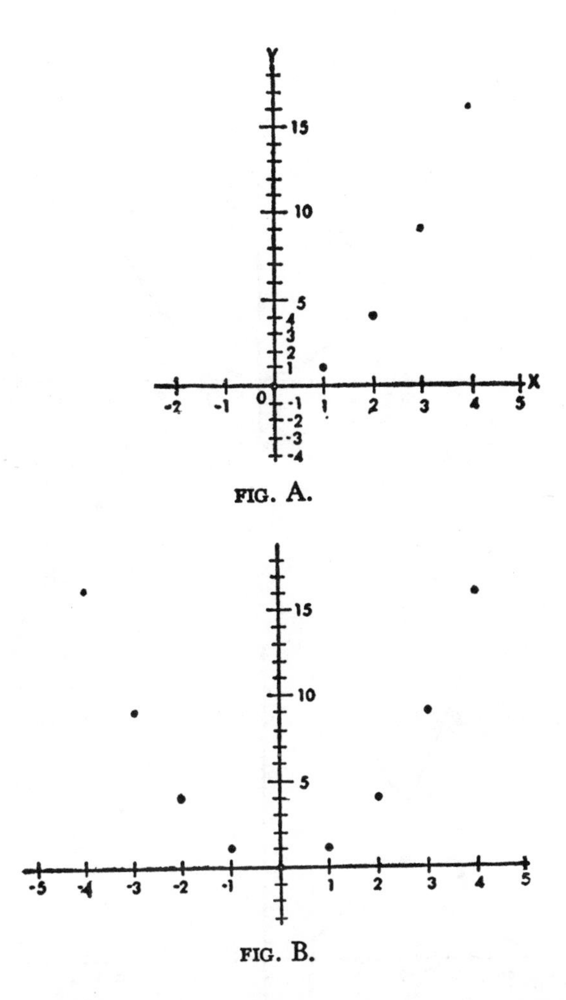

FIG. A.

FIG. B.

But does this curve embrace other points which arise in our functional table. Let us test this, tabulating some fractional values of x.

x	y
$\frac{1}{2}$	$\frac{1}{4}$
$1\frac{1}{2}$	$2\frac{1}{4}$
2.3	5.29
2.7	7.29

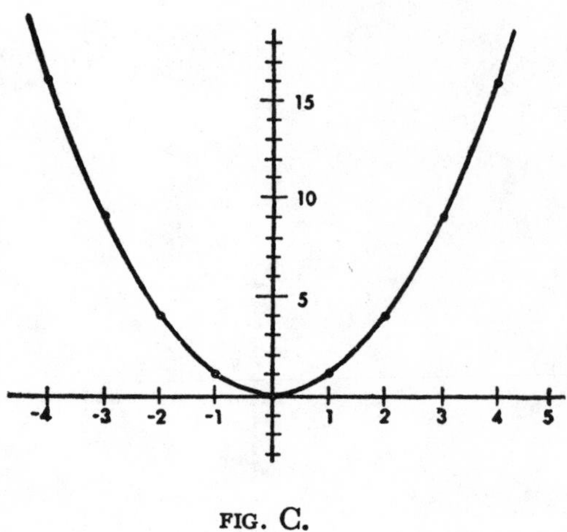

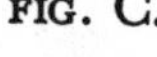
FIG. C.

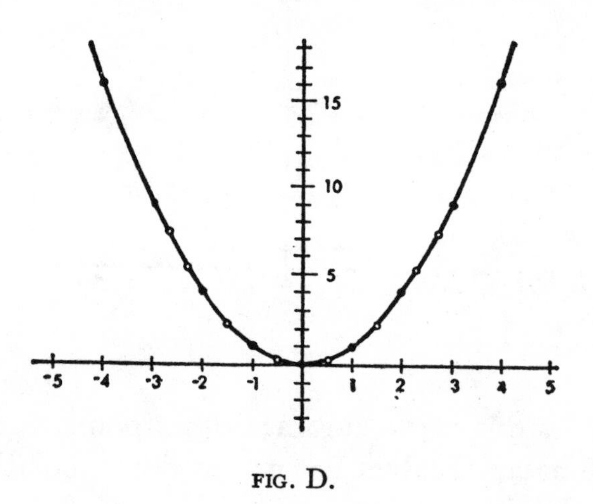

FIG. D.

If we plot these new points, it may be seen that they all lie on the curve (Fig. D). Indeed, if we continue further, we would find that *every* point which might arise in the table will lie on the curve; the totality of such points will form the curve known as the parabola.

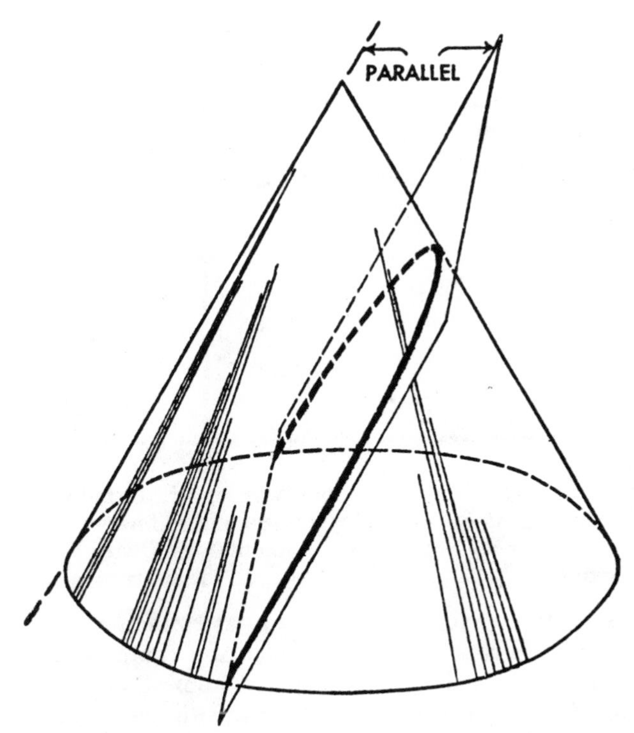

The parabola is formed by the section of a cone cut by a plane parallel to the opposite edge.

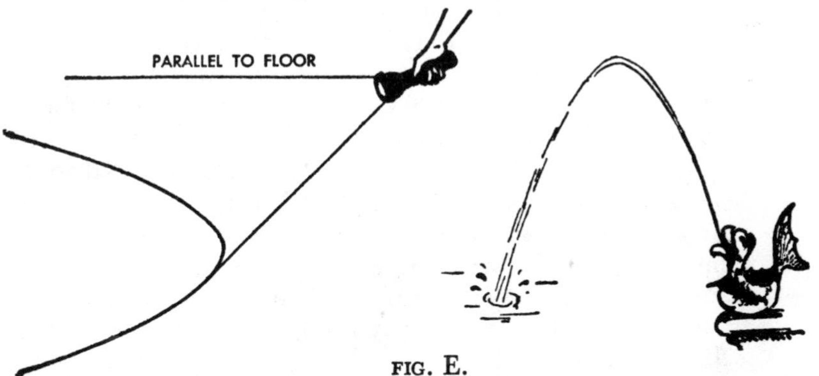

FIG. E.

You can make a parabola for yourself with the help of a flashlight, holding it so that the upper boundary of the beam will be parallel to the floor.

A jet of water forms a parabola. So does the path of a projectile. But the curve formed by a loop of string held at the ends, hanging freely, is *not* a parabola, but a *catenary*.

FOOTNOTES

1. Henri Bergson, *Creative Evolution.*—P. 65.
2. It is a simple matter geometrically to determine the square root of a given length.—P. 67.

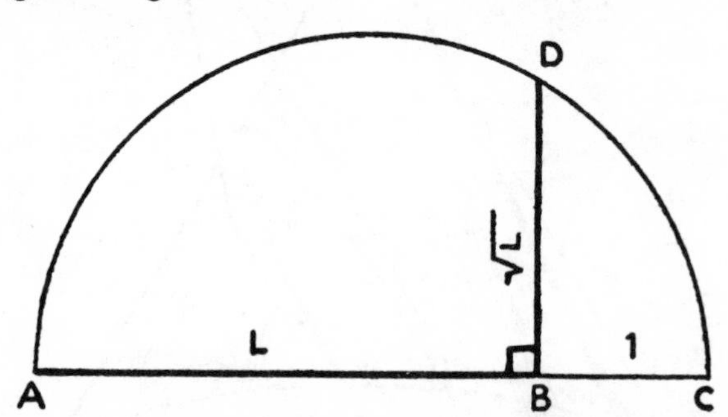

FIG. 27.—Let AB be the given length. Extend it to C so that $BC = 1$. Draw a semicircle having AC as diameter. Erect a perpendicular at B meeting semicircle at D. BD is the required square root of L.

3. Gauss made an exhaustive study to determine what other polygons could be constructed with ruler and compass. The Greeks had been able to construct regular polygons of 3 and 5 sides, but not those with 7, 11, or 13 sides. Gauss, with marvelous precocity, gave the formula which showed what polygons were constructible in the classical way. It had been thought that only regular polygons, the number of whose sides could be expressed by the forms: 2^n, $2^n \times 3$, $2^n \times 5$, $2^n \times 15$ (where n is an integer), could be so constructed. Gauss' formula proves that polygons with a *prime number* of sides may be constructed as follows: Let P be the number of sides and n any integer up to 4, then $P = 2^{2^n} + 1$. If $n = 0$, 1, 2, 3, 4; $P = 3$, 5, 17, 257, 65537. Where n is greater than 4, there are no known primes of the form $2^{2^n} + 1$.

   (A *prime number* is one which is not evenly divisible by any number other than 1 or itself. Thus, 2, 3, 5, 7, 11, 13, 17 are examples of primes. A famous proof of Euclid, which appears in his *Elements*, shows that the number of primes is infinite. See p. 192.) (Footnote 21.)

   It is an amazing fact that of all the possible polygons, the number of whose sides is prime, only the five given above are known to be constructible with ruler and compass.—P. 68.
4. See Chap. 5.—P. 68.
5. As long ago as 1775, the Paris Academy was so overwhelmed

with pretended solutions from circle squarers, angle trisectors, and cube duplicators, that a resolution was passed that no more would be accepted. But at that time the impossibility of these solutions was only suspected and not yet mathematically demonstrated; thus the arbitrary action of the academy can only be explained on the grounds of self-preservation.—P. 69.

6. Limits and converging processes with an infinite number of steps, as we shall soon see, were used in computing π.—P. 69.
7. See the chapter on the calculus.—P. 69.
8. Most infinite series are *divergent*, that is, the sum of the series *exceeds* any assignable integer. A typical divergent series is $1 + \frac{1}{2} + \frac{1}{3} + \frac{1}{4} + \frac{1}{5} + \ldots$ This series seems to differ very little from the convergent series given in the text, and only the most subtle mathematical operations reveal whether a series is *convergent* or *divergent*.—P. 70.
9. A square can be duplicated by drawing a square on the diagonal of the given square, but a cube cannot be duplicated because the cube root of 2 is involved, and this, like π, is not the root of an algebraic equation of first or second degree, and therefore cannot be constructed with ruler and compass. In four-dimensional space, the figure which corresponds to the cube, called a "tessaract" (see the chapter on Assorted Geometries) *can* be duplicated by ruler and compass, because the fourth root of 2, which is what is required, can be written as the square root of the square root of 2.—P. 72.
10. What is meant by "the root of an algebraic equation with integer coefficients"? A word may suffice to jog the memory of those who have had a course in elementary algebra. The root of an equation is the value that must be substituted for the unknown quantity in the equation in order to satisfy it. Thus, in the equation $x - 9 = 0$, 9 is the root, since if you substitute 9 for x, the equation is satisfied. Similarly -4 and 4 are the roots of the equation $x^2 - 16 = 0$, because when either value is substituted for x, the equation balances. "Algebraic" equations are the kind of equations we have just been talking about. There are also trigonometric equations, differential equations and others, and the term "algebraic" is intended to distinguish equations of the form

$$a_0x^n + a_1x^{n-1} + a_2x^{n-2} + \ldots + a_{n-1}x + a_n = 0.$$

The *coefficients* of an equation are the numbers which appear before the unknown quantity or quantities. In the equation

$$3x^4 + 17x^3 + \sqrt{2}x^2 - ix + \pi = 0$$

3, 17, $\sqrt{2}$, i, and π are the coefficients. This is an example of an algebraic equation with *queer* coefficients. In defining an algebraic equation (see page 49), we demand that n be a positive integer and that the a's be integers.—P. 72.

11. See Buffon's Needle Problem in the chapter on Chance and Chanceability.—P. 79.
12. The $\sqrt{2}$ when written as a decimal is just as complicated as π, for it never repeats, never ends, and there is no known law giving the succession of its digits; yet this complicated decimal is easily obtained with exactitude by a ruler and compass construction. It is the diagonal of a square whose side is equal to 1. —P. 79.
13. Jobst Bürgi of Prague had prepared tables of logarithms before Napier's *Descriptio* appeared. Bürgi however failed to publish his tables until 1620 because, as he explained, he was busy on some other problem.—P. 80.
14. According to the principle of positional notation, the value of a digit depends on its position in relation to the other digits in the number in which they appear.—P. 80.
15. The rules for operating with exponents in multiplication and division are:

A) *Multiplication*

$a^m \times a^n = a^{m+n}$; thus,
$a^3 \times a^2 = a^{3+2} = a^5$; or,
$a^3 \times a^2 = (a \cdot a \cdot a) \times (a \cdot a) = a^5$

B) *Division*

$$\frac{a^m}{a^n} = a^{m-n}$$

$$\frac{a^3}{a^2} = a^{3-2} = a$$

But, if m is equal to n,

$$\frac{a^m}{a^n} = a^0 = ?$$

$$\frac{a^3}{a^3} = a^{3-3} = a^0 = ?$$

$$\frac{a^3}{a^3} = \frac{\not{a} \times \not{a} \times \not{a}}{\not{a} \times \not{a} \times \not{a}} = 1$$

Therefore we agree upon
$a^0 = 1$.—P. 82.

16. Because e possesses certain unique properties valuable in many branches of mathematics, particularly the calculus, because of the relation between logarithmic and exponential functions, e is the "natural" base for the logarithmic system.—P. 84.
17. The first proof that e is transcendental (i.e., not the root of an algebraic equation with integer coefficients), was given by Hermite, the distinguished French mathematician, in 1873, nine years before Lindemann's proof of the transcendental character of π appeared. Since that time several others succeeded in simplifying Hermite's proof. The general method is to "assume e to be the root of an algebraic equation, $f(e) = 0$, and show that a multiplier M can be chosen such that when each side of the equation is multiplied by M, (the value of) $Mf(e)$ is reduced to the sum of an integer *not zero* and a number between 1 and 0, showing that the assumption that e can be the root of an algebraic equation is untenable." See U. G. Mitchell and M. Strain, *in Osiris, Studies in History of Science*, Vol. I.—P. 84.
18. The symbol ! as used in mathematics does not indicate surprise or excitement, although in this case it might not be amiss, since the simplicity and beauty of this series is amazing. ! means "take the factorial of the number after which ! appears." The factorial of a number is the product of its components; thus, $1! = 1$, $2! = 1 \times 2$, $3! = 1 \times 2 \times 3$, $4! = 1 \times 2 \times 3 \times 4$, $5! = 1 \times 2 \times 3 \times 4 \times 5$.—P. 84.
19. Actually n need only be equal to 1000 (i.e., the interest computed thrice daily) to give \$2.72.—P. 87.
20. The derivative of $y = e^x$ is equal to the function itself. For a further discussion of the derivative and of problems involving rate of change, see the chapter on the calculus.—P. 88.
21. Omar Khayyâm, besides being the author of the well-worn "Rubâiyât," was also a mathematician of distinction, but one whose prophetic vision failed for negative numbers.—P. 90.
22. Translated in Dantzig, *Number, the Language of Science* (New York; Macmillan), 1933, p. 190.—P. 93.
23. It was once suggested that appropriate symbols for the two constants, e and i should be [symbol] for e, and [symbol] for i in order to avoid confusion. But printers balked at making new type and the old symbols remained. More often than is realized, such considerations determined the character of mathematical notation.—P. 93.

Assorted Geometries—Plane and Fancy

They say that habit is second nature. Who knows but nature is only first habit?

—PASCAL

AMONG OUR most cherished convictions, none is more precious than our beliefs about space and time, yet none is more difficult to explain. The talking fish of Grimm's fairy tale would have had great difficulty in explaining how it felt to be always wet, never having tasted the pleasure of being dry. We have similar difficulties in talking about space, knowing neither what it is, nor what it would be like not to be in it. Space and time are "too much with us late and soon" for us to detach ourselves and describe them objectively.

"For what is time?" asked Saint Augustine. "Who can easily and briefly explain it? Who even in thought can comprehend it, even to the pronouncing of a word concerning it? But what in speaking do we refer to more familiarly and knowingly than time? And certainly we understand when we speak of it; we understand also when we hear it spoken of by another. What, then, is time? If no one ask of me, I know; if I wish to explain to him who asks, I know not."[1]

And this could as well be said of space. Though space cannot be defined, there is little difficulty in measuring distances and areas, in moving about, in charting vast courses, or in seeing through millions of light years. Everywhere there is overwhelming evidence that space is our natural medium and confronts us with no insuperable problems.

But this professes to be no philosophical treatise and no German *Handbook on an Introduction to the Theory of Space* in 14 volumes. Our intention is to explain in the simplest, most general manner, not the physical space of sense perception, but the space of the mathematician. To that end, all preconceived notions must be cast aside and the alphabet learned anew.

In this chapter we propose to discuss two kinds of geometry—four-dimensional and non-Euclidean. Neither of these subjects is beyond the comprehension of the non-mathematician prepared to do a little straight thinking. To be sure, they have both been described, like the theory of relativity (to which they are in some ways related) in high and mighty mumbo jumbo. High priests in every profession devise elaborate rituals and obscure language as much to conceal their own ineptness as to awe the uninitiate. But the corruptness of the clergy should not deter us. The basic ideas underlying four-dimensional and non-Euclidean geometry are simple, and this we aim to prove.

*

Euclid, in writing the *Elements*, recognized no great obstacles. Starting with certain fundamental ideas (presumably understood by everyone) expressed as postulates and axioms, he built upon these as foundations. This ideal method for developing a logical system has never been improved upon, although occasionally it has been neglected or forgotten with sad results.

Although Euclid's *Elements* constitute an imposing intellectual achievement, they fail to make an important distinction between two types of mathematics—*pure* and *applied*—a distinction which has only come to light in modern theoretical developments in mathematics, logic and physics.

A geometry which treats of the space of experience, is *applied* mathematics. If it says nothing about that space—if, in other words, it is a system composed of abstract notions, elements, and classes, with rules of combination obeying the laws of formal logic, it is *pure* mathematics. Its propositions are of the form: If A is true, then B is true, regardless of what A and B may possibly be.[2] Should a system of pure mathematics be applicable to the physical world, its fruitfulness may be regarded either as mere chance, or as further evidence of the profound connection between the forms of nature and those of mathematics. Yet, in either case, this essential fact must be borne in mind—the fruitfulness of a logical system neither diminishes nor augments its validity.

As applied mathematics, Euclid's geometry is a good approximation within a restricted field. Good enough to help draw a map of Rhode Island, it is not good enough for a map of Texas or the United States, or for the measurement of either atomic or stellar distances. As a system of pure mathematics, its propositions are true in a most general way. That is to say, they have validity only as propositions of logic, only if they have been correctly deduced from the axioms. Other geometries with different postulates are therefore possible—indeed, as many others as the mathematician chooses to devise. All that is necessary is to assemble certain fundamental ideas (classes, elements, rules of combination), declare these to be undefinable, make certain that they are not self-contradictory, and the groundwork has been laid for a new edifice, a new geometry. Whether this new geometry will be fruitful, whether it will prove as useful in surveying or navigation as Euclidean geometry, whether its fundamental ideas measure up to any standard of truth other than self-consistency, doesn't concern the mathematician a jot. The mathematician is the tailor to the gentry of science. He makes the suits, anyone who fits into them can wear them. To put it another way, the mathematician makes the rules of the game; anyone who wishes may play, so long as he observes them. There is no sense in complaining afterwards that the game was without profit.

*

If we wish to pay a mathematical system the highest possible compliment, to indicate that it partakes of the same generality and has the same validity as logic, we may call it a game. A four-dimensional geometry is a game: so is the geometry of Euclid. To object to four-dimensional geometry on the grounds that there are only three dimensions is absurd. Chess can be played as well by those who believe in comrades or dictators as by those who cling to the vanishing glory of kings and queens. What sense is there in objecting to chess on the grounds that kings and queens belong to a past age, and that, in any case, they never did behave like chess pieces —no, not even bishops. What merit is there to the contention that chess is an illogical game because it is impossible to conceive that a private citizen may be crowned queen merely by moving forward five steps.

Perhaps these are ridiculous examples, but they are no more so than the complaints of the faint of heart who say that three dimensions make space and space makes three dimensions, "that is all ye know on earth and all ye need to know." If we can rake the doubters fore, we can rake them aft—indeed, from stem to stern. For there is no proof, in the scientific sense, that space is three-dimensional, or for that matter, that it is four-, five-, six-, or anything but n-dimensional. Space cannot be proved three-dimensional by geometry considered as *pure* mathematics, because pure mathematics is concerned only with its own logical consistency and not with space or anything else. Nor is this the province of applied mathematics, which does not generally inquire into the nature of space, but assumes its existence. All that we have learned from applied mathematics is that it is convenient, but not obligatory, to consider the space of our sense perceptions as three-dimensional.

To the objection that a fourth dimension is beyond imagination we may reply that what is common sense today was abstruse reasoning—even wild speculation—yesterday. For primitive man to imagine the wheel, or a pane of glass, must have required even higher powers than for us to conceive of a fourth dimension.

Someone may still object: "You tell me that four-dimensional geometry is a game. I will believe you. But it seems to be a game that doesn't concern itself with anything real, with anything I have ever experienced." We may answer in the Socratic way with another question. "If a four-dimensional geometry treats of nothing real, what does the plane geometry of Euclid consider? Anything more real? Certainly not! It doesn't describe the space accessible to our senses which we explain in terms of sight and touch. It talks about points that have no dimensions, lines that have no breadth, and planes that have no thickness—all abstractions and idealizations resembling nothing we have ever experienced or encountered."

*

The notion of a fourth dimension, although precise, is very abstract, and for the greatest majority beyond imagination and in the purest realm of conception. The development of this idea is as much due to our rather childish desire for consistency as to anything more profound. In this same striving after consistency and generality, mathematicians developed negative numbers, imaginaries, and the transcendentals. Since no one had ever seen minus three cows, or the square root of minus one trees, it was not without a struggle that these now rather commonplace ideas were introduced into mathematics. The same struggle was repeated to introduce a fourth dimension, and there are still skeptics in the camp of the opposition.

Every possible allegory and fiction was proposed to coax and cajole the doubters, to make the idea of a fourth dimension more palatable. There were the romances which described how impossible a three-dimensional world would seem to creatures in a two-dimensional one, there were stories of ghosts, table-tipping, and the land of the dead. It required illustrations from the land of the living, which were still less comprehensible than the fourth dimension, to win even a partial victory. From this, it should not be inferred that a greater absurdity was enlisted in support of a lesser one.

Beginning as usual with Aristotle, it was proved again and again that a fourth dimension was unthinkable and impossible. Ptolemy pointed out that three mutually perpendicular lines could be drawn in space, but a fourth, perpendicular to these, would be without measure or depth. Other mathematicians, unwilling to risk a heresy greater even than going contrary to the Bible—that is, contradicting Euclid—advised that to go beyond three dimensions was to go "against nature." And the English mathematician, John Wallis, of whom one might properly have expected better things, referred to that "fansie," a fourth dimension, as a "Monster in Nature, less possible than a Chimera or a Centaure."

Unwittingly, a philosopher, Henry More, came to the rescue, although mathematicians today would hardly acknowledge his support. His suggestion was not an unmixed blessing. Ghostly spirits, said More, surely have four dimensions. But Kant delivered an earthly blow by laying down his intuitive notions of space which were hardly compatible with either a four-dimensional or a non-Euclidean geometry.

In the nineteenth century several leading mathematicians espoused the apparently hopeless cause, and behold—a new mathematical gusher. The great paper of Riemann *On the Hypotheses Which Underlie the Foundations of Geometry*, together with the works of Cayley, Veronese, Möbius, Plücker, Sylvester, Bolyai, Grassmann, Lobachevsky, created a revolution in geometry. The geometry of four and even higher dimensions became an indispensable part of mathematics, related to many other branches.

When finally, there came, as for some mysterious reason they always come, direct uses and applications of four-dimensional geometry to mathematical physics, to the physical world, when the unwanted child was suddenly recognized and rechristened "Time, the fourth dimension!" the rejoicing made the cup flow over. Curious and marvelous things were said. The fourth dimension would solve all the awful mysteries of the universe, and ultimately might prove a cure for arthritis. So far in the general jubilation did the mathematicians forget themselves that some of them began to refer to it as "*the* fourth dimension," as though, instead of being merely an idea shaken loose from the ends of their pencils, only the fourth in a class of infinite possibilities, it was a physical reality, like a new element. Thus the lamentable confusion spread from mathematics to grammar, from the principles of the 2 + 2 to the science of the proper uses of the definite and indefinite article.

*

Physicists may consider time to be a fourth dimension, but not the mathematician. The physicist, like other scientists, may find that his latest machine has just the right place for some new mathematical gadget; that does not concern the mathematician. The physicist can borrow new parts for his changing machine every day for all the mathematician cares. If they fit, the physicist says they are useful, they are true, because there is a place for them in the model of his world in the making. When they no longer fit, he may discard them or "destroy the whole machine and build a new one as we are ready to buy a new car when the old one doesn't run well." [3]

The practice of calling time a *dimension* points to the necessity of explaining what is meant by that troublesome word. In this way, too, we shall arrive at a clearer image of four-dimensional geometry.

Instead of referring to "a space," or to "spaces," we shall use the more fashionable and more general term—*manifold*.[4] A manifold bears a rough resemblance to a class. A plane is a class composed of all those points uniquely determined by two co-ordinates. It is therefore a two-dimensional manifold.

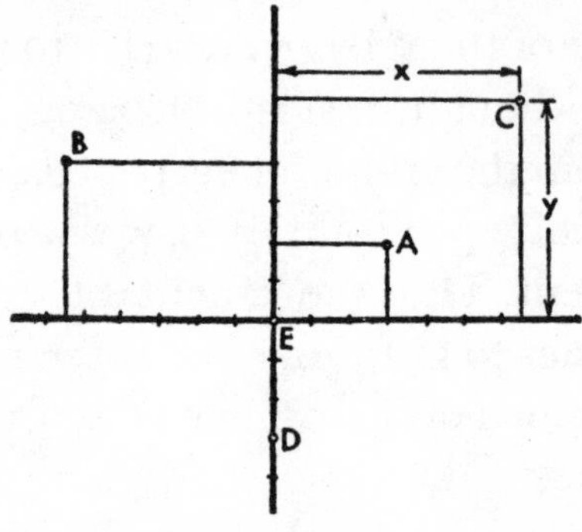

FIG. 28(a).—A two-dimensional manifold. Each point requires a pair of numbers to individualize it.

$$A = (3, 2)$$
$$B = (-5\tfrac{1}{2}, 4)$$
$$C = (x, y)$$
$$D = (0, -3)$$
$$E = (0, 0)$$

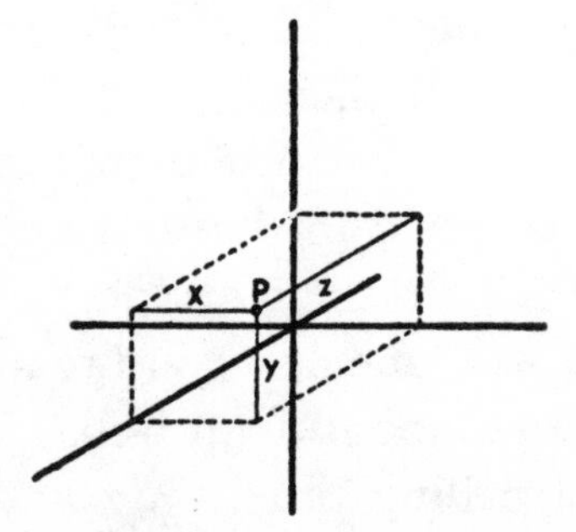

FIG. 28(b).—The same idea can be extended to a three-dimensional manifold (space). Each point requires 3 numbers to individualize it.

Thus, $P = (x, y, z)$

The space studied in three-dimensional analytical geometry may be regarded as a three-dimensional manifold, because exactly three co-ordinates are required to fix every point in it. Generally, if n numbers are necessary to specify, to individualize, each of the members of a manifold, whether it be a space, or any other class, it is called an n-dimensional manifold.

Thus, for the word *dimension*, with its many mysterious connotations and linguistic encrustations, there has been substituted a simple idea—that of a *co-ordinate*. And in place of the physical word *space*, the mathematician introduces the more general and more accurate concept of *class*, or *manifold*.

*

It is now possible, as a consequence of these refinements, to introduce an idea already familiar from our discussion of analytical geometry, which shall serve to uniquely characterize space manifolds. We shall use some geometrical reasoning.

The Pythagorean theorem states that, in a right-angle

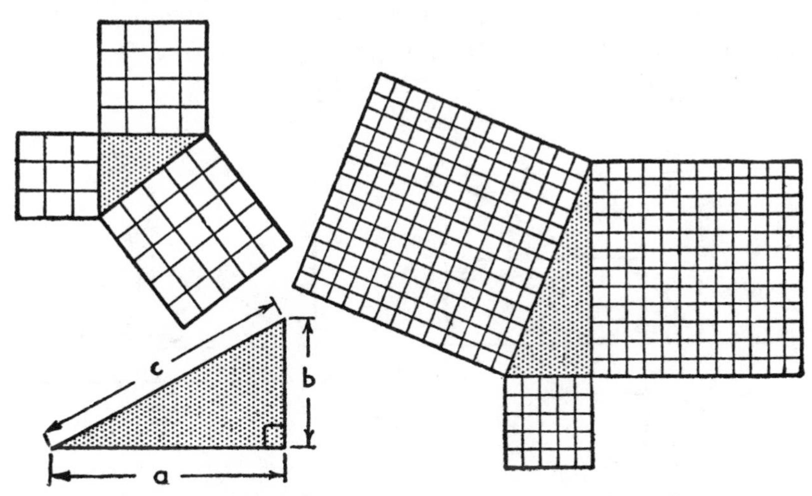

FIG. 29.—The Pythagorean theorem. For any right triangle:

$$c^2 = a^2 + b^2$$

Thus, $5^2 = 3^2 + 4^2$

$$13^2 = 12^2 + 5^2$$

triangle, the length of the hypotenuse equals the square root of the sum of the squares of the other two sides.

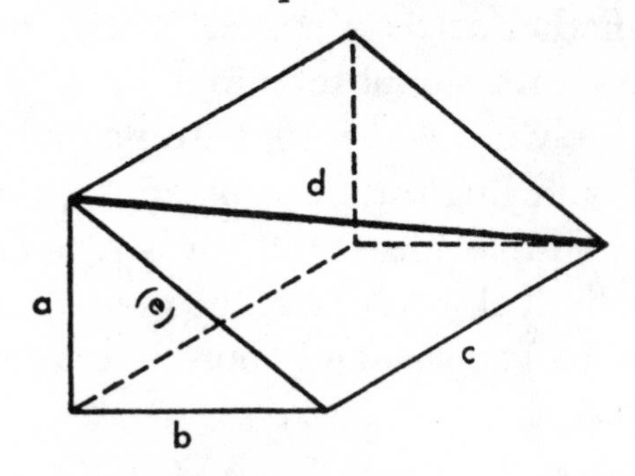

FIG. 30.—The Pythagorean theorem in three dimensions.

$$d^2 = a^2 + b^2 + c^2$$
$$\text{For } d^2 = c^2 + (e)^2$$
$$\text{and } (e)^2 = a^2 + b^2$$

When this is carried over into analytical geometry of two dimensions, the result is the well-known distance formula, according to which the distance between any two points in the plane, having the co-ordinates (x,y) and (x',y') respectively, is $\sqrt{(x - x')^2 + (y - y')^2}$.

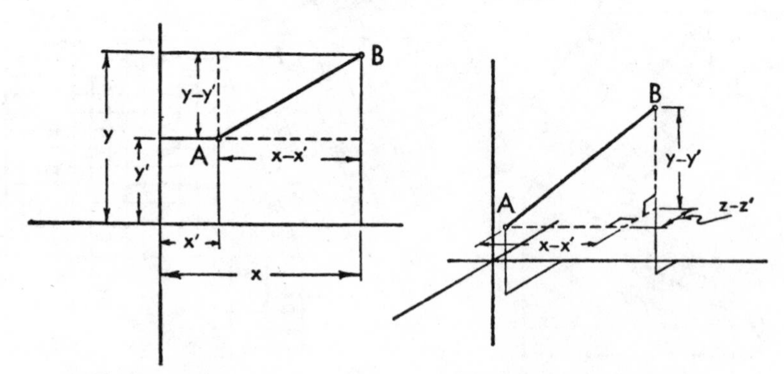

(1) Two dimensions (2) Three dimensions

FIG. 31.

(1) Distance $AB = \sqrt{(x - x')^2 + (y - y')^2}$
(2) Distance $AB = \sqrt{(x - x')^2 + (y - y')^2 + (z - z')^2}$

Similarly, in three-dimensional analytical geometry the distance between any two points having the co-ordinates (x, y, z), (x', y', z') respectively, is

$$\sqrt{(x-x')^2+(y-y')^2+(z-z')^2}.$$

Now, in either two or three dimensions the concept of distance, as both the mathematician and the layman understand it, is the same. The layman is satisfied with an intuitive grasp; the mathematician demands an exact formulation. However, in the higher dimensions, while the layman is halted by a blank wall—the natural limitations of his senses—the mathematician scales the wall using his extended formula as a ladder. Distance in four dimensions means nothing to the layman. Indeed, why should it? For even a four-dimensional space is wholly beyond ordinary imagination. But the mathematician, who rests the concept upon an entirely different base, is not called upon to struggle with the bounds of imagination, but only with the limitations of his logical faculties.

Accordingly, there is no reason for not extending the above formula to 4, 5, 6, . . . or n dimensions. Thus, in a four-dimensional Euclidean manifold, the distance of an element, i.e., point, having the co-ordinates (x, y, z, u) from an element with co-ordinates (x', y', z', u') is

$$\sqrt{(x-x')^2+(y-y')^2+(z-z')^2+(u-u')^2}.$$

This method enables us to define in terms of analytical geometry a 2, 3, 4, . . . or n-dimensional Euclidean manifold. An analogous definition can be given for the manifolds of other geometries, in which case some other distance formula would apply. We have chosen analytical geometry and taken the Pythagorean distance formula to distinguish the Euclidean manifolds.

A condensed definition of a three- and four-dimensional Euclidean manifold in terms of analytical geometry reads: [5]

1. A three-dimensional Euclidean manifold is the class of all number triples: (x, y, z), (x', y', z'), (x'', y'', z''), etc., to any two of which there may uniquely be assigned a *measure* (called the distance between them) defined by the formula $\sqrt{(x - x')^2 + (y - y')^2 + (z - z')^2}$. Certain subclasses of this class are called points, lines, and planes, etc. The theorems derived from these definitions constitute a mathematical system called "Analytical Geometry of Three Dimensions."

2. A four-dimensional Euclidean manifold is the class of all number quadruples: (x, y, z, u), (x', y', z', u'), (x'', y'', z'', u''), etc., to any two of which there may uniquely be assigned a *measure*, (called the distance between them) defined by the formula

$$\sqrt{(x - x')^2 + (y - y')^2 + (z - z')^2 + (u - u')^2}.$$

Certain subclasses of this class are called points, lines, planes, and *hyperplanes*. Analytical four-dimensional Euclidean geometry is the system formed by theorems derived from these definitions.

Note that nothing has been said in either of these definitions about space; neither the space of our sense perceptions, nor the space of the physicist, nor that of the philosopher. All that we have done is to define two systems of mathematics which are logical and self-consistent, which may be played like checkers, or charades, according to stated rules. Anyone who finds a resemblance between his game of checkers or charades and the physical reality of his experience is privileged to point morals and to make capital of his suggestion.

*

But having established that we are in the realm of pure conception, beyond the most elastic bounds of imagination, who is satisfied? Even the mathematician would like to nibble the forbidden fruit, to glimpse what it would be like if he could slip for a moment into a fourth dimension. It's hard to grub along like moles down here below, to hear someone tell of a fourth dimension, to make careful note of it, and then to plow along, giving it no further thought. To make matters worse, books on popular science have made everything so ridiculously simple—relativity, quanta, and what not—that we are shamed by our inability to picture a fourth dimension as something more concrete than time.

Graphic representations of four-dimensional figures have been attempted: it cannot be said these efforts have been crowned with any great success. Fig. 31 illustrates the four-dimensional analogue of the three-dimensional cube, a *hypercube* or *tesseract:* Our difficulties in drawing this figure are in no way diminished by the fact that a three-dimensional figure can only be drawn *in perspective* on a two-dimensional surface—such as this page—, while the four-dimensional object on a two dimensional page is only a perspective of a "perspective."

Yet since a^2 equals the area of a square, a^3 the volume

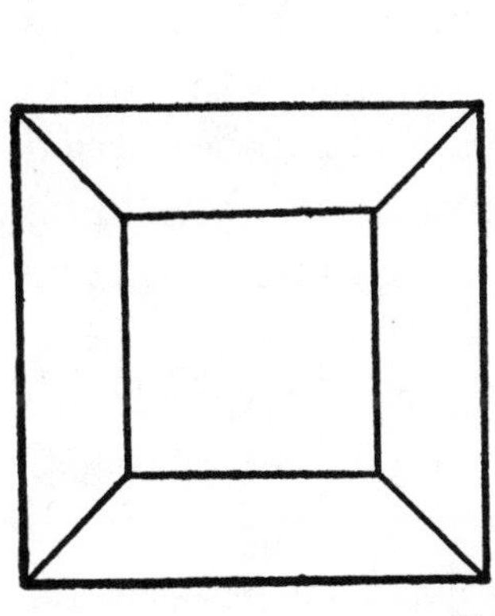
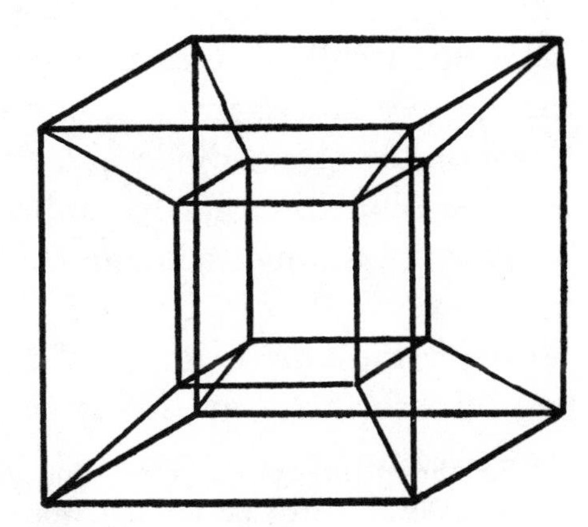

FIG. 31(a).—Cube and tesseract.

of a cube, we feel certain that a^4 describes something, whatever that something may be. Only by analogy can we reason that that "something" is the *hypervolume* (or content) of a tesseract. Reasoning further, we infer that the tesseract is bounded by 8 cubes (or cells), has 16 vertices, 24 faces and 32 edges. But visualization of the tesseract is another story.

Fortunately, without having to rely on distorted diagrams, we may use other means, using familiar objects to help our limping imagination to depict a fourth dimension.

The two triangles *A* and *B* in Fig. 32 are exactly alike.

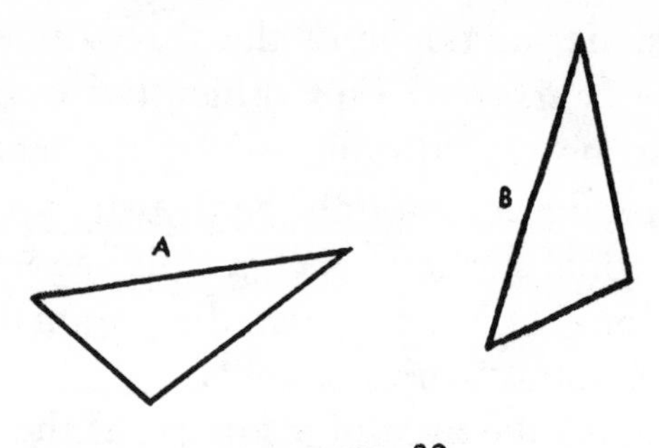

FIG. 32.

Geometrically, it is said they are congruent,* meaning that by a suitable motion, one may be perfectly superposed on the other. Evidently, that motion can be carried out in a plane, i.e., in two dimensions, simply by sliding triangle *A* on top of triangle *B*.† But what about the two triangles *C* and *D* in Fig. 33?

One is the mirror image of the other. There seems to be no reason why by sliding or turning in the plane, *C*

* See the chapter on paradoxes for an exact definition.

† Actually, "sliding on top of" would be impossible in a physical two-dimensional world.

cannot be superimposed on *D*. Strangely enough, this cannot be done. *C* or *D* must be lifted out of the plane, from two dimensions into a third, to effect superposition. Lift *C* up, turn it over, put it back in the plane, and then it can be slid over *D*.

Now, if a third dimension is essential for the solution of certain two-dimensional problems, a fourth dimension would make possible the solution of otherwise unsolvable problems of three dimensions. To be sure, we are in the

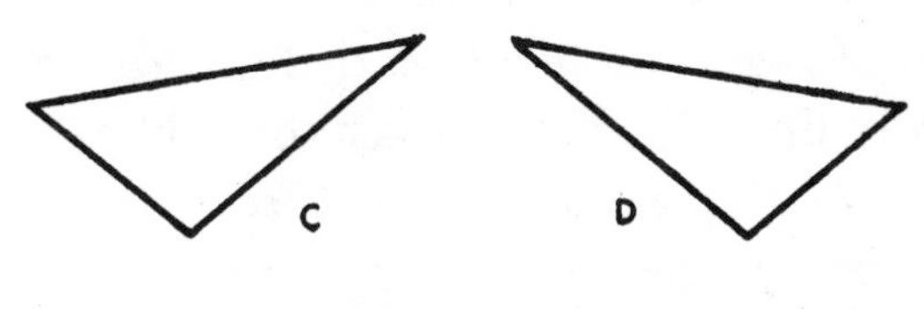

FIG. 33.

realm of fancy, and it need hardly be pointed out that a fourth dimension is not at hand to make Houdinis of us all. Yet, in theoretical inquiries, a fourth dimension

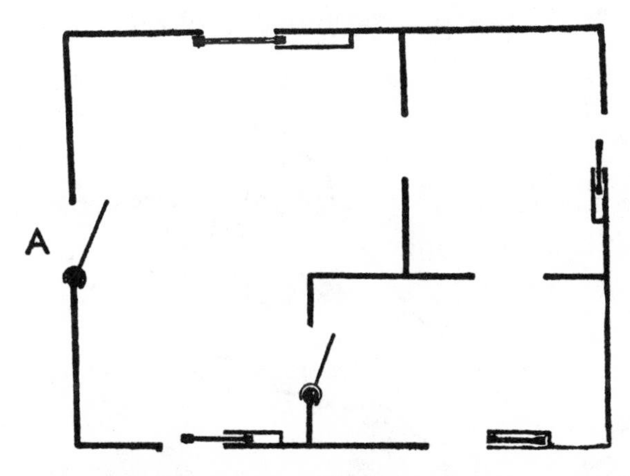

FIG. 34.—This is no blueprint but an actual house in Flatland.

is of signal importance, and part of the warp and woof of modern theoretical physics and mathematics. Examples chosen from these subjects are quite difficult and would be out of place, but some simpler ones in the lower dimensions may prove amusing.

If we lived in a two-dimensional world, so graphically described by Abbott in his famous romance, *Flatland*, our house would be a plane figure, as in Fig. 34. Entering through the door at *A*, we would be safe from our friends and enemies once the door was closed, even though there were no roof over our head, and the walls and windows were merely lines. To climb over these lines would mean getting out of the plane into a third dimension, and of course, no one in the two-dimensional world would have any better idea of how to do that than we know how to escape from a locked safe-deposit vault by means of a fourth dimension. A three-dimensional cat might peek at a two-dimensional king, but he would never be the wiser.

When winter comes to Flatland, its inhabitants wear gloves. Three-dimensional hands look like this:

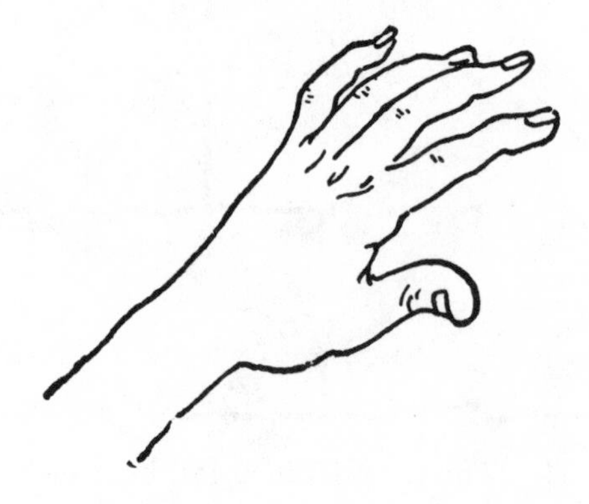

FIG. 35.

and gloves like this:

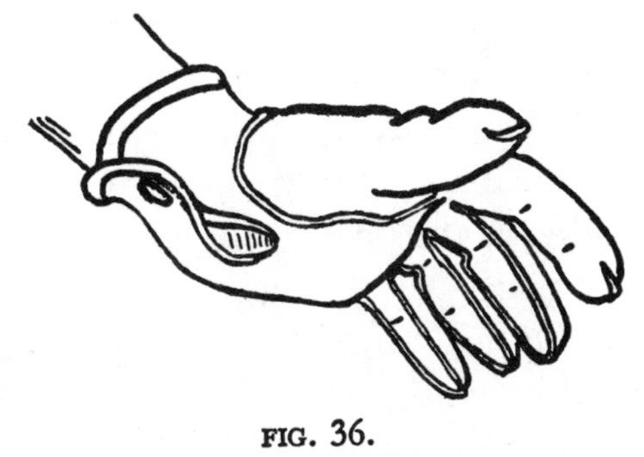

FIG. 36.

In Flatland hands look like this:

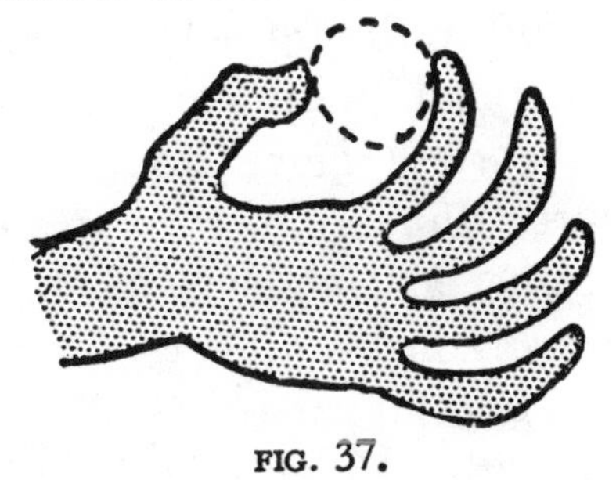

FIG. 37.

and gloves, like this:

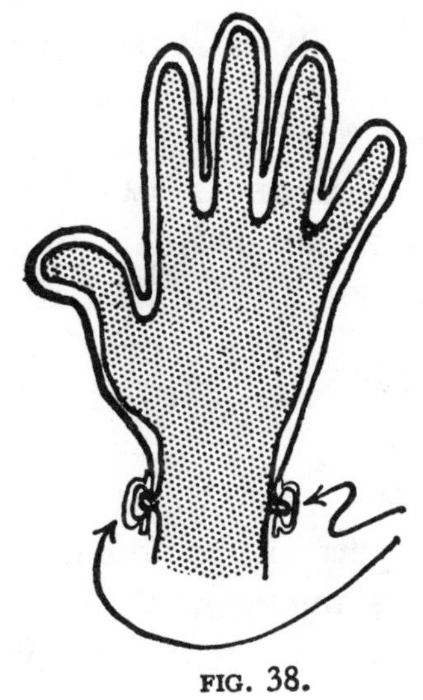

FIG. 38.

Modern science has as yet devised no relief for the man who finds himself with two right gloves instead of a right and a left. In Flatland, the same problem would exist. But there, Gulliver, looking down at its inhabitants from the eminence of a third dimension, would see at once that, just as in the case of the two triangles on page 127, all that is necessary to turn a right glove into a left one is *to lift it up and turn it over*. Of course, no one in Flatland would or could lift a finger to do that, since it involves an extra dimension.

If then, we could be transported into a fourth dimension, there is no end to the miracles we could perform—starting with the rehabilitation of all ill-assorted pairs of gloves. Lift the right glove from three-dimensional space into a fourth dimension, turn it around, bring it back and it becomes a left glove. No prison cell could hold the four-dimensional Gulliver—far more of a menace than a mere invisible man. Gulliver could take a knot and untie it without touching the ends or breaking it, merely by transporting it into a fourth dimension and slipping the solid cord through the extra loophole.

Or he might take two links of a chain apart without breaking them. All this and much more would seem absurdly simple to him, and he would regard our helplessness with the same amusement and pity as we look upon the miserable creatures of Flatland.

*

Our romance must end. If it has aided some readers in making a fourth dimension more real and has satisfied a common anthropomorphic thirst, it has served its purpose. For our own part, we confess that the fables have never made the facts any clearer.

An idea originally associated with ghosts and spirits

needs, if it is to serve science, to be as far removed as possible from fuzzy thinking. It must be clearly and courageously faced if its true essence is to be discovered. But it is even more stupid to reject and deride than to glorify and enshrine it. No concept that has come out of our heads or pens marked a greater forward step in our thinking, no idea of religion, philosophy, or science broke more sharply with tradition and commonly accepted knowledge, than the idea of a fourth dimension.

Eddington has put it very well:[6]

> However successful the theory of a four-dimensional world may be, it is difficult to ignore a voice inside us which whispers: "At the back of your mind, you know that a fourth dimension is all nonsense." I fancy that voice must often have had a busy time in the past history of physics. What nonsense to say that this solid table on which I am writing is a collection of electrons moving with prodigious speed in empty spaces, which relatively to electronic dimensions are as wide as the spaces between the planets in the solar system! What nonsense to say that the thin air is trying to crush my body with a load of 14 lbs. to the square inch! What nonsense that the star cluster which I see through the telescope, obviously there *now*, is a glimpse into a past age 50,000 years ago! Let us not be beguiled by this voice. It is discredited. . . .
>
> We have found a strange footprint on the shores of the unknown. We have devised profound theories, one after another to account for its origin. At last, we have succeeded in reconstructing the creature that made the footprint. And lo! It is our own.

*

We have emphasized the fact that pure geometry is divorced from the physical space which we perceive about us, and we are now prepared to tackle an idea which is slightly tougher. There is no harm, however, in first distinguishing somewhat differently than before between space as it is ordinarily conceived and the space manifolds of mathematics. Perhaps this distinction will help to make our new concept—the non-Euclidean geometries—seem less strange.

We are quite used to thinking of space as infinite, not in the technical mathematical sense of infinite classes, but simply meaning that space is boundless—without end. To be sure, experience teaches us nothing of the kind. The boundaries of a private citizen rarely go much further than the end of his right arm. The boundaries of a nation, as bootleggers once learned, do not go beyond the twelve-mile limit.

Most of what we believe about the infinitude of space comes to us by hearsay, and another part comes from what we think we see. Thus, the stars look as if they were millions of miles away, although on a dark night a candle half a mile off would give the same impression. Moreover, if we imagined ourselves the size of atoms, a pea at a distance of one inch would appear mightier and far more distant than the sun.

The distinction between the space of the individual and "public space" soon becomes apparent. Our personal knowledge of space does not show it to be either infinite, homogeneous, or isotropic. We do not know it to be infinite because we crawl, hop, and fly around in only tiny portions. We do not know it to be homogeneous because a skyscraper in the distance seems much smaller than the end of our nose; and the feather on the hat of the lady in front of us shuts off our vision of the cinema screen. And we know it is not isotropic, that is, it "does not possess the same properties in every direction," [7] because there are blind spots in our vision and our sense of sight is never equally good in all directions.

The notion of physical or "public" space which we abstract from our individual experience is intended to free us from our personal limitations. We say physical space is infinite, homogeneous, isotropic, and Euclidean. These compliments are readily paid to an ideal entity about which very little is actually known. If we were to ask the physicist or astronomer, "What do you think about space?" he might reply: "In order to carry out experimental measurements and describe them with the greatest convenience, the physical scientist decides upon certain conventions with respect to his measuring apparatus and operations performed with it. These are, strictly speaking, conventions with regard to *physical* objects and *physical* operations. However, for practical purposes, it is convenient to assume for them a generality beyond any particular set of objects or operations. They then become, as we say, properties of space. That is what is meant by physical space, which we may define, in brief, as the abstract construct possessing those properties of rigid bodies that are independent of their material content. Physical space is that on which almost the whole of physics is based, and it is, of course, the space of everyday affairs." [8]

On the other hand, the spaces, or more generally the manifolds, which the mathematician considers are constructed without any reference to physical operations, such as measurement. They possess only those properties expressed in the postulates and axioms of the particular geometry in question, as well as those properties deducible from them.

It may well be that the postulates are themselves suggested, in part or in whole, by the physical space of our experience, but they are to be regarded as full-grown and independent. If experiments were to show that some, or all, of our ideas about physical space are wrong (as the theory of relativity has, in fact, done) we would have to rewrite our texts on physics, but not our geometries.

*

But this approach to the concept of space, as well as to geometry, is comparatively recent. There has been no more sweeping movement in the entire history of science than the development of non-Euclidean geometry, a movement which shook to the foundations the age-old belief that Euclid had dispensed eternal truths. Competent and accurate as a measuring tool since Egyptian times, intuitively appealing and full of common sense, sanctified and cherished as one of the richest of intellectual legacies from Greece, the geometry of Euclid stood for more than twenty centuries in lone, resplendent, and irreproachable majesty. It was truly hedged by divinity, and if God, as Plato said, ever geometrized, he surely looked to Euclid for the rules. The mathematicians who occasionally had doubts soon expiated their heresy by votive offerings in the form of further proofs in corroboration of Euclid. Even Gauss, the "Prince of Mathematicians," dared not offer his criticisms for fear of the vulgar abuse of the "Boethians."

Whence came the doubts? Whence the inspiration of those who dared profane the temple? Were not the postulates of Euclid self-evident, plain as the light of day? And the theorems as unassailable as that two plus two equals four? The center of the ever-increasing storm, which finally broke in the nineteenth century was the famous fifth postulate about parallel lines.

This postulate may be restated as follows: "Through

any point in the plane, there is one, and only one, line parallel to a given line."

There is some evidence to show that Euclid, himself, did not regard this postulate as "quite so self-evident" as his others.[9] Philosophers and mathematicians, intent on vindicating him, attempted to show that it was really a *theorem* and thus deducible from his premises. All of these attempts failed for the very good reason which Euclid, much wiser than those who followed him, had already recognized, namely, that the fifth postulate was merely an assumption and hence could not be mathematically proved.

*

More than two thousand years after Euclid, a German, a Russian, and a Hungarian came to shatter two indisputable "facts." The first, that space obeyed Euclid; the second, that Euclid obeyed space. Gauss we credit on faith. Not knowing the extent of his investigations, in deference to his greatness as well as to his integrity, we are hospitable to his statement that he had independently arrived at conclusions resembling those of the Hungarian, Bolyai, some years before Bolyai's father informed Gauss of his son's work.

Lobachevsky, the Russian, and Bolyai, both in the 1830's, presented to the very apathetic scientific world their remarkable theories. They argued that the trouble-making postulate could not be proved, could not be deduced from the other axioms, because it was only a *postulate*. Any other hypothesis about parallels could be substituted in its place, and a different geometry—just as consistent and just as "true"—would follow. All the other postulates of Euclid were to be retained, only, in place of the fifth, a substitution was to be made:

"Through any point in the plane, there are *two lines* parallel to any given line."

Overnight, mathematics had thrown off its chains, and a new line of richly fruitful theoretic and practical inquiry was born.

*

In the figure are two parallel lines:

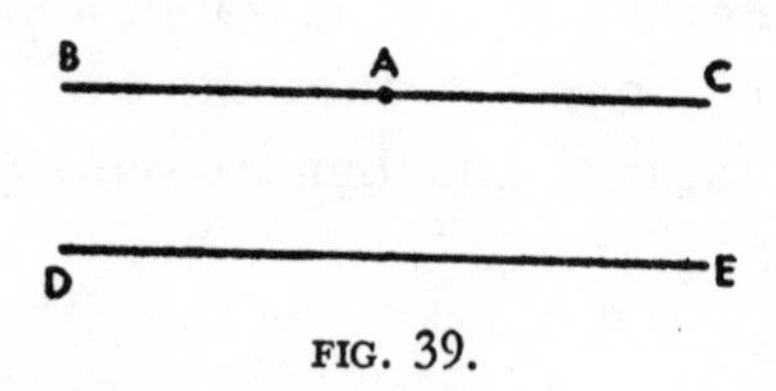

FIG. 39.

How is it possible, you may ask, that another line different from *BC*, yet parallel to *DE* may be drawn through *A?* The answer is that the reader is talking about the physical plane and lines drawn with a pencil. He is haunted by the ghosts of common sense instead of reasoning in terms of pure geometry. *You* might go further and say that in your system, in Euclidean geometry, any line different from *BC* will meet *DE* if sufficiently extended. *We* would reply that that rule holds in your game, not in ours—Lobachevskian geometry. Neither of us, if we are mathematicians, is talking about physical space, but even if we were, there is better ground to believe that we are speaking the truth than you.

Lobachevsky's geometry may be introduced in this way: In Fig. 40 line *AB* is perpendicular to *CD*. If we permit it to rotate about *A* counterclockwise, it will intersect *CD* at various points to the right of *B* until it reaches a limiting position *EF*, when it becomes parallel to *CD*.

Continuing the rotation, it will start to intersect CD to the left of B. Euclid assumed that there is only *one* position for the line, namely EF, when it would be parallel to CD. Lobachevsky assumed that there were two such positions, represented by $A'B'$ and $C'D'$, and further, that all lines falling within the angle Θ, while not parallel to CD, would *never* meet it, no matter how far extended.

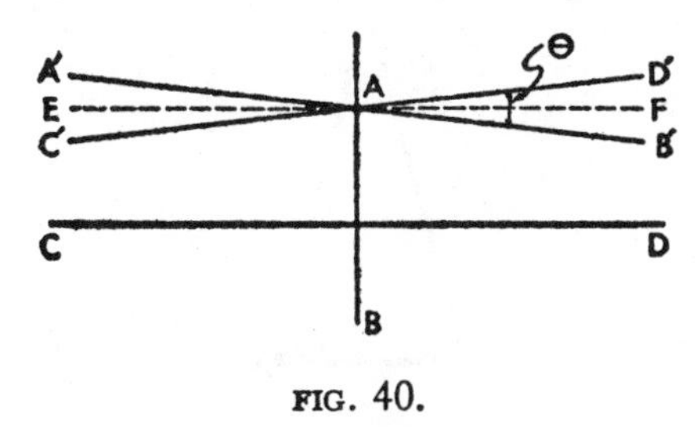

FIG. 40.

Now this is an assumption, and there is no sense in arguing from the diagram that it is evident that if $A'B'$ or $C'D'$ were extended sufficiently far, they would eventually intersect CD. If, as Professor Cohen has pointed out, we rely wholly on our intuition of space, which is finite, there will always be an angle Θ which grows smaller as our space is extended, but which never vanishes, and all lines falling within Θ will fail to intersect the given line.[10]

*

What happens to the geometry of Euclid when its parallel postulate is replaced by that of Lobachevsky? Many of its important theorems, those which in no way depend upon the fifth postulate, are carried over. Thus, in both geometries:

1. If two straight lines intersect, the vertical angles are equal:

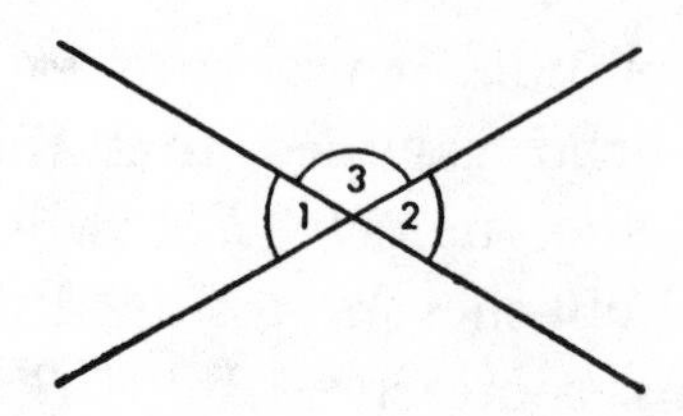

FIG. 41.—Angle 1 = Angle 2 (because each one = 180° − Angle 3).

2. In an isosceles triangle, the base angles are equal:

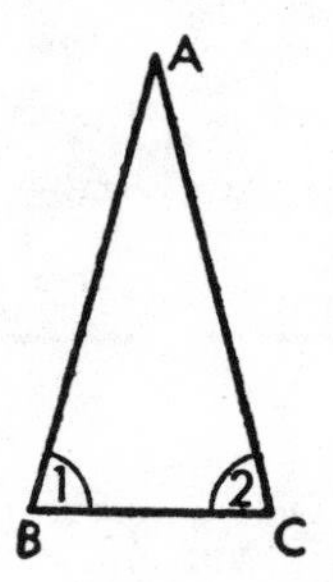

FIG. 42.—If $AB = AC$, then Angle 1 = Angle 2.

3. Through a point, only one perpendicular can be drawn to a straight line:

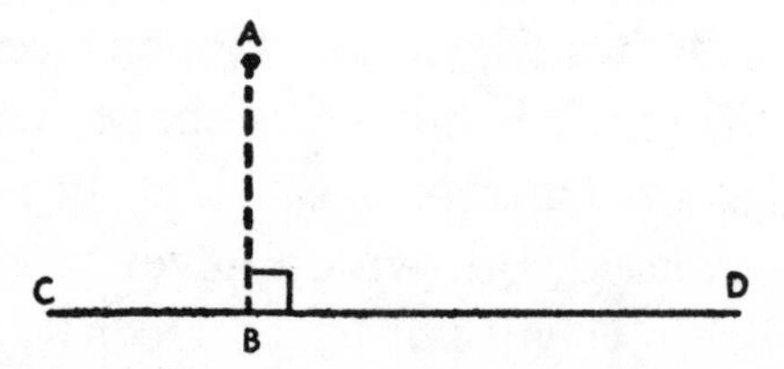

FIG. 43.—Through the point A one and only one perpendicular can be drawn to CD.

On the other hand, some very important theorems of Euclidean geometry are changed when another postulate is substituted for the fifth, with startling results. Thus, in Euclidean geometry, the sum of the angles of every triangle *equals* 180 degrees, whereas in Lobachevsky's geometry, the sum of the angles of every triangle is *less than* 180 degrees. Parallel lines in Euclidean geometry never intersect and remain, no matter how far extended, a constant distance apart. Parallel lines in Lobachevsky's geometry *never meet, but approach each other asymptotically*—that is, the distance between them becomes less as they are further extended.

To cite one more interesting theorem, two triangles in Euclidean geometry may have the same angles but different areas; i.e., one may be a magnification of the other. But in Lobachevsky's geometry, *as a triangle increases in area, the sum of its angles decreases;* thus, only triangles equal in area can have the same angles. (See Fig. 47(b).)

*

The brilliant Riemann, in his famous inaugural lecture *On the Hypotheses Which Underlie the Foundations of Geometry*, proposed still another substitute for Euclid's fifth postulate differing from that of Lobachevsky and Bolyai. This assumption holds: "Through a point in the plane, *no* line can be drawn parallel to a given line." In other words, every pair of lines in the plane must intersect. It should be noted that this contradicts Euclid's tacit supposition that a straight line may be infinitely extended. In this connection, Riemann pointed out the important distinction between infinite and unbounded: Thus, space may be finite though *unbounded*. Moving in any given direction, like the hands of a clock, we can keep going forever, forever retracing our steps. As might be expected, Riemann's hypothesis also affects those theorems of Euclid dependent on the fifth postulate. Both Euclid's and Lobachevsky's geometries state that only one perpendicular can be drawn to a straight line from a given point. But in Riemann's any number of perpendiculars can be drawn from an appropriate point to a given straight line. Again, the sum of the angles of any triangle is *greater* than 180 degrees in Riemann's geometry, and the angles increase as the triangle grows larger. (See Fig. 47(a), page 144.)

*

We thus have three postulate systems: Euclid's, Lobachevsky's, and Riemann's. From these, three geometries have been developed: the first, Euclidean, the other two, non-Euclidean. The non-Euclidean geometries are, of course, vastly indebted to the postulates and the methods of Euclid. So far as the postulates are concerned, they differ only with respect to the parallel postulate. The theorems differ greatly in many respects.

A little earlier we laid down the criterion for every mathematical system—its postulates must be consistent, must lead to no contradictions. But how are we to discover whether the non-Euclidean geometries of Lobachevsky and Riemann are consistent? For that matter, it may well be asked, how are we to be certain that the postulates of Euclid will engender no contradictions? Evidently, we may pile up theorem after theorem without encountering any, but that is no proof that at some future time one may not arise. Is it that we are no better off than if we were

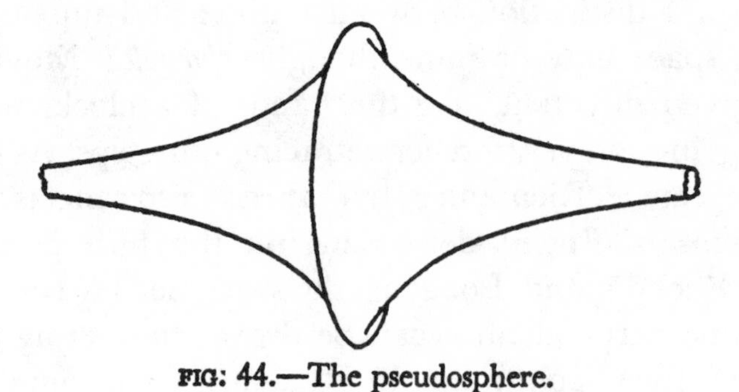

FIG: 44.—The pseudosphere.

verifying an hypothesis in physics or any other experimental science?

Fortunately mathematicians have devised a trick which satisfies their conscience on this score. It consists in showing, for example, in non-Euclidean geometry, that a set of entities which exist in Euclidean geometry would sat-

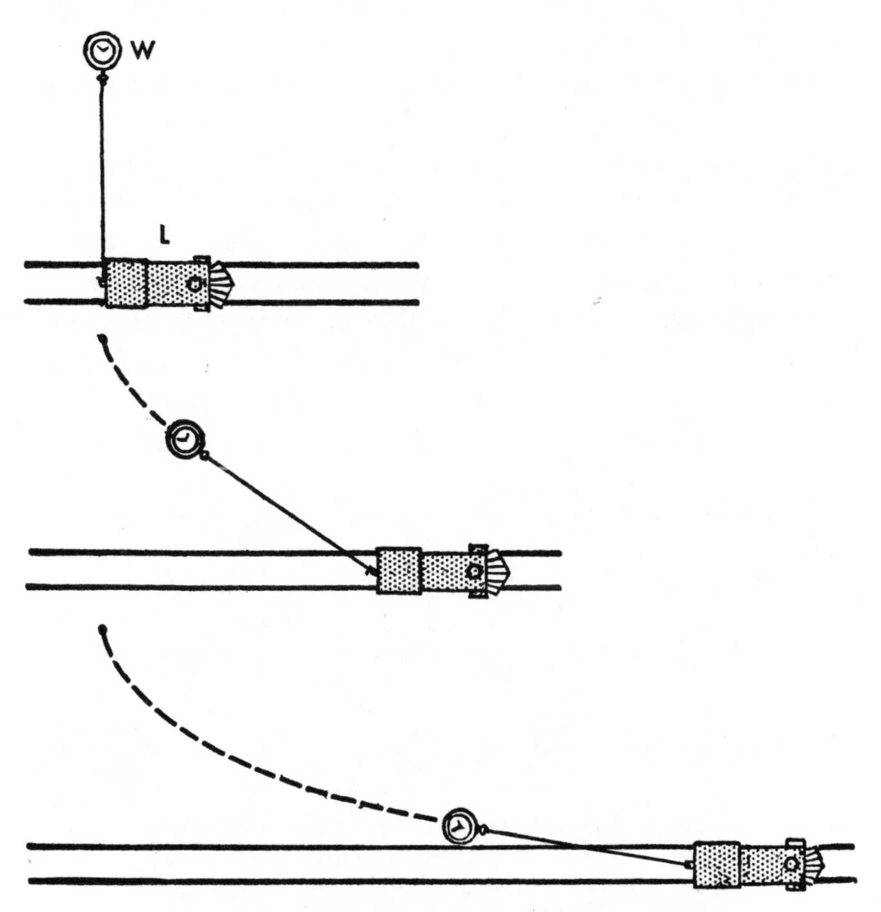

FIG. 45(a).—One way of generating the tractrix. The toy locomotive *L* is tied to the watch *W*, the string being perpendicular to the track. When the locomotive starts pulling, the path of the watch is a tractrix.

isfy the non-Euclidean theorems. It is assumed that these entities, themselves, are "free from contradictions, and that they in effect, fully embody the axioms," [11] and the latter are therefore shown to involve no inconsistencies. Let us take separate examples from Lobachevsky's and Riemann's geometries to illustrate what is meant.

Figure 44 illustrates a surface generated by revolving the curve known as the *tractrix* about a horizontal line.

The tractrix itself may be obtained as follows: On a pair of mutually perpendicular axes, as in Cartesian geometry, imagine a chain lying along YY'. To one end of this chain there is attached a watch; the other end coincides with the point of origin 0. Keep the chain taut, and pull the free end slowly along the X axis, to the right of 0. Then repeat this procedure to the left. The path of the watch in either case generates a tractrix. If this curve is now revolved about the line XX', a "double trumpet surface," as E. T. Bell calls it, is formed.

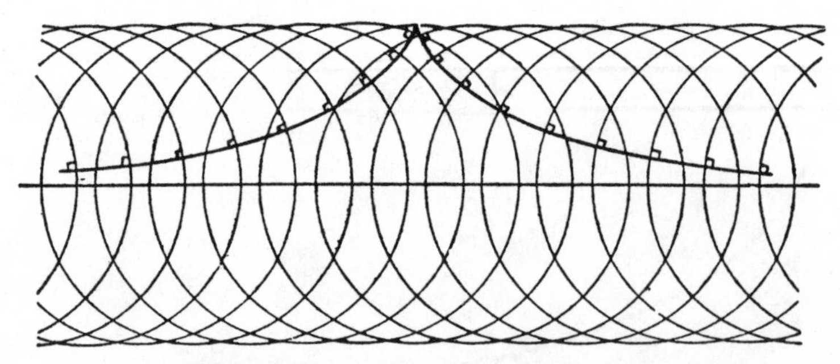

FIG. 45(b).—The tractrix is also that curve which is perpendicular to a family of equal circles with their centers on a straight line.

This surface Beltrami named a *pseudosphere*. We find that the geometry applicable to a pseudosphere is that of Lobachevsky. For example, on the pseudosphere, through a given point two lines may be drawn parallel to a third line, which will approach them asymptotically without ever intersecting.[12] Thus, Lobachevsky's geometry is satisfied by an entity from Euclid's geometry, thus complying with the criterion of consistency.

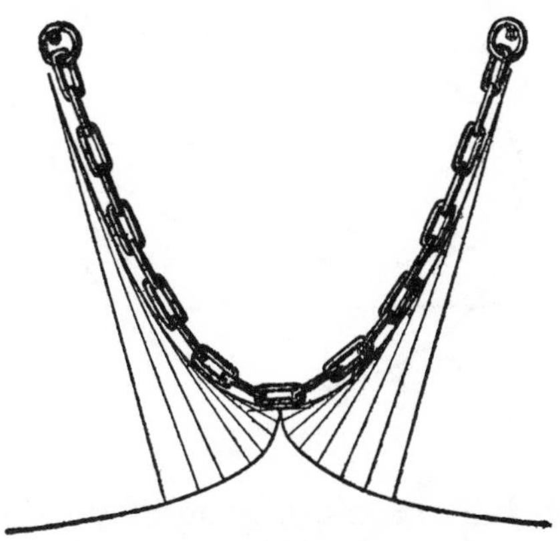

FIG. 45(c).—The curve formed by a freely hanging chain is called the catenary. If the tangents to a catenary (lines just touching it) are drawn, the curve which is perpendicular to them and which meets the catenary at its lowest point is again the tractrix.

The geometry of Riemann is applicable to a very familiar object—the sphere. It may be seen from Fig. 46 that a plane which passes through the center of a sphere cuts the surface in a *great circle.*

Although the earth is somewhat oblate, for the purpose of this discussion we may consider it spherical. Every circle passing through the North and the South Poles on the earth's surface is a great circle (longitude), but with the exception of the equator, the circles of latitude are not. Straight lines drawn on the surface of the earth are always parts of great circles, and *even if two such lines are perpendicular to a third line (which, in Euclidean geometry, would imply they are parallel), they will always intersect at a pair of poles.* Thus, the elements for a geometry which will satisfy the surface of the earth are identical

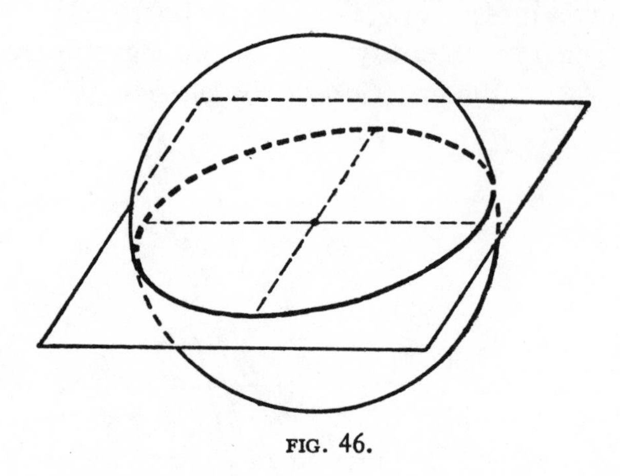

FIG. 46.

with those of Riemannian geometry. For example, a triangle drawn on the surface of the earth will have

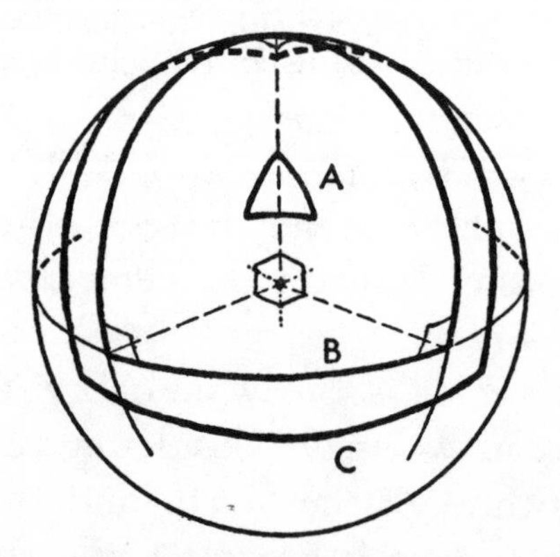

FIG. 47(a).—Triangle *A* is small compared with the sphere; thus it is nearly a plane triangle and its angle sum is near 180°.

But let it grow into triangle *B*, the sides of which lie on three perpendicular great circles, and the angle sum = 90° + 90° + 90° = 270°.

In the still larger triangle *C*, the angles of which are all obtuse, the sum is greater than 270°.

angles totaling more than 180 degrees, and the larger the triangle, the greater the sum of the angles.

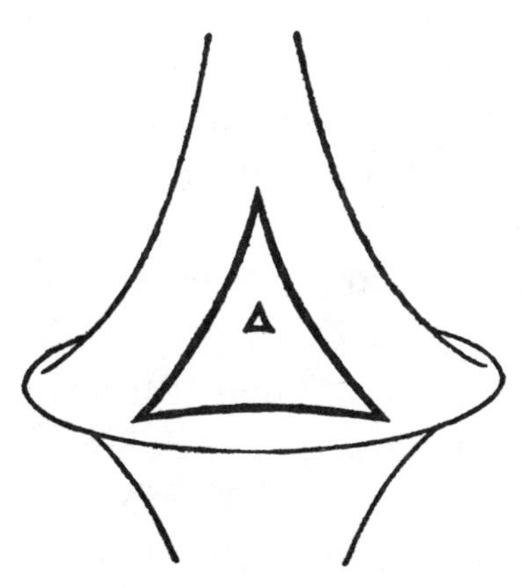

FIG. 47(b).—This is the reverse of what happens on a sphere, Fig. 47(a). On a pseudosphere, the larger the triangle, the smaller the sum of the angles.

Furthermore, two straight lines drawn on the earth's surface, if sufficiently extended, will always enclose an area. It is convenient at this point to recall the important distinction noted by Riemann that a surface may be finite but unbounded, so that straight lines drawn upon the surface of the earth *can* be infinitely extended, although the surface is evidently not infinite, but merely unbounded. The Riemannian properties of the sphere are amusingly set out by the following riddle:

A group of sportsmen, having pitched camp, set forth to go bear hunting. They walk 15 miles due south, then 15 miles due east, where they sight a bear. Bagging their game, they return to camp and find that altogether they have traveled 45 miles. What was the color of the bear?

*

Our brief discussion of non-Euclidean geometry is bound to raise in the mind of the reader many questions outside our province, but the literature, even the popular literature, is so extensive that anyone sufficiently interested and curious need not go begging for answers.

Yet it is perhaps proper that we should consider one very natural question which might take this form: "On a sphere, two straight lines, even though parallel at one place, are certain (if sufficiently extended) to intersect, and may enclose an area. Why, then, call such lines 'straight'? Are they not really curved?"

At the outset it is obvious that whether a line is straight or not depends on the definition of "straight." In mathematics, it has been found convenient to formulate such a definition only with reference to the particular surface under consideration. One way of defining a straight line is to say that it is the shortest distance between two points. On the other hand, everyone knows, from the many references in recent times to aeronautical exploits, that the shortest route between two points on the surface of the earth can be covered by following the arc of the great circle lying between them. Conveniently enough, through each two points on the surface of a sphere there does pass a great circle.

The great circle, then, on the sphere, corresponds to the straight line in the plane—it is the shortest distance between two points. Suitable curves may be found for other types of surfaces, for instance, the pseudosphere, or a saddle-shaped surface which will fulfill the same role. Generalizing this notion, a curve which is the shortest distance between two points (analogue of the straight line in the plane) on *any* kind of a surface is called a *geodesic* of that surface. When we sought entities that would satisfy the geometry of Lobachevsky, and that of Riemann, we were really looking for surfaces, the geodesics of which would obey the parallel postulates of these geometries.

In the plane, if we adopt Euclid's hypothesis, a pair of geodesics meet in one point, unless they are parallel, in which case they do not meet at all. On a sphere, a pair of geodesics (arcs of great circles), even if parallel, always meet in two points, and therefore the sphere obeys the geometry of Riemann. On a pseudosphere, obeying Lobachevsky's geometry, parallel geodesics may approach one another *asymptotically*, but never intersect.

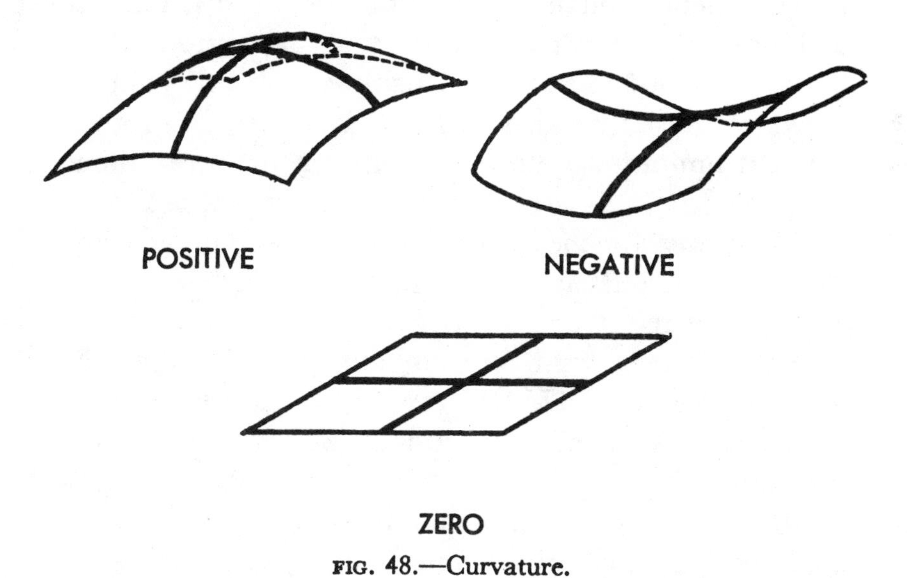

FIG. 48.—Curvature.

The geodesics of a surface are determined by its *curvature*. Curvature is not easy to explain, although we all have an intuitive notion of what it means. A plane has a curvature of 0. A surface like that of a sphere or an ellipsoid is one of positive curvature, whereas the saddle-shaped surface or the pseudosphere is said to be of *negative* curvature. We can imagine more complicated surfaces, parts of which may have a positive, parts a negative, and parts a 0 curvature. The geodesics of a surface, as well as its most appropriate geometry, depend upon such curvature—positive, negative, or 0. Whence the geometry of a surface of constant negative curvature is Lobachevskian, that of a surface of constant positive curvature Riemannian, and that of a surface of 0 curvature Euclidean.

*

All that has been said about non-Euclidean geometry, while evident enough when we talk of *geometry*, tends to become obscure when applied to everyday surroundings. We are inclined to pity the inhabitants of a two-dimensional world, as much for their ignorance as for their physical limitations. They cannot even dream of doing things which to us are perfectly commonplace. Yet we tend to show the same intellectual limitations in picturing our world to ourselves. Indeed, we go further, for we deliberately reject our own experience. Our experience is that space is finite but unbounded, and that the straight lines we are able to draw on the surface on which we live can never really be straight, but must be curved. (Of course the earth's curvature differs from 0.) But we continue to confuse infinity and unboundedness, to reject the latter which constitutes our actual spatial knowledge and to embrace the former for religious and aesthetic reasons. And, although every intelligent person knows the earth's surface is curved, and every navigator practices great-circle sailing, most of us behave like Seventh-Day Adventists in reasoning that our straight lines are drawn in a plane of 0 curvature—or, in effect, in a world that is flat. From this it is only a step to the belief that Euclid's fifth postulate is sacred and any substitute is "against nature." A little curvature, even more than a little learning, has its disadvantages.

Although we know a good deal more about the surface we inhabit than about the physical space in which we live, there is hardly any choice between the absurdities of our beliefs about either one. The geometry of Euclid, which considers surfaces of 0 curvature, in the strictest sense (disregarding convenience in computation) does not suit the surface on which we live as well as that of Riemann. Unmistakably, our geometries, though suggested by our sense perceptions, are not dependent upon them.

The geometries we have discussed are only three of an infinite number of possible ones. Any geometry, whatever its postulates (provided they lead to no contradictions), will be just as "true" as the geometry of Euclid. For every surface, however complex its curvature, there is a peculiarly suited geometry. It is true we start our geometries as purely logical structures, but, as in other branches of mathematics, we find that Nature has anticipated us, and that a surface often waits upon our inventiveness. For that reason, the non-Euclidean mathematics has found enormously important fields of application in the weird topsy-turvy of modern physics.

While we have considered the applications of two-dimensional non-Euclidean geometries to familiar surfaces, the mathematical physicist studies the application of higher-dimensional non-Euclidean geometries to higher-dimensional space manifolds. In attempting to discover experimentally what space we actually live in, scientists have obtained results which lead them to believe that space is curved rather than straight. Having emancipated ourselves from the primitive idea that we live on a *plane* surface, curved space should not be so hard to take.

There is a final point: If we consider the geometries of Euclid, Lobachevsky and Riemann as applied, and not as pure, mathematics, if we ask which one is most suitable to the space immediately surrounding us and the surface on which we live, what shall our answer be? Experiment and measurement alone can answer that question. It turns out that Euclid's geometry is the most *convenient*, and the one, in consequence, which we shall continue to use to build our bridges, tunnels, skyscrapers, and highways. The geometries of Lobachevsky, or Riemann, properly handled, would do just as well.[13] Our skyscrapers would stand it, and so would our bridges, tunnels, and highways; our engineers might not. The geometry of Euclid is easier to teach, fits in more readily with misguided common sense, but above everything, is easier to use. And we are concerned, after all, in such matters with living, and not with logic.

Yet our vistas have widened and our vision is clearer. Mathematics has helped us to transcend those sense impressions which we now say "deceive us never, while lying ever."

FOOTNOTES

1. St. Augustine, *Confessions.*—P. 112.
2. An illustration of pure mathematics: *

   Consider the following propositions, which are the axioms for a special kind of geometry.

   Axiom 1. If *A* and *B* are distinct points on a plane, there is at least one line containing both *A* and *B*.

   Axiom 2. If *A* and *B* are distinct points on a plane, there is not more than one line containing both *A* and *B*.

* Morris Raphael Cohen and Ernest Nagel, *An Introduction to Logic and Scientific Method* (New York: Harcourt Brace, 1936), pp. 133-139.

Axiom 3. Any two lines on a plane have at least one point of the plane in common.

Axiom 4. There is at least one line on a plane.

Axiom 5. Every line contains at least three points of the plane.

Axiom 6. All the points of a plane do not belong to the same line.

Axiom 7. No line contains more than three points of the plane.

These axioms seem clearly to be about points and lines on a plane. In fact, if we omit the seventh one, they are the assumptions made by Veblen and Young for "projective geometry" on a plane in their standard treatise on that subject. It is unnecessary for the reader to know anything about projective geometry in order to understand the discussion that follows. But what are points, lines and planes? The reader may think he "knows" what they are. He may "draw" points and lines with pencil and ruler, and perhaps convince himself that the axioms state truly the properties and relations of these geometric things. This is extremely doubtful, for the properties of marks on paper may diverge noticeably from those postulated. But in any case the question whether these actual marks do or do not conform is one of *applied* and not of *pure* mathematics. The axioms themselves, it should be noted, do not indicate what points, lines, and so on "really" are. For the purpose of discovering the implications of these axioms, it is unessential to know what we shall understand by points, lines, and planes. These axioms imply several theorems, not in virtue of the visual representation which the reader may give them, but in virtue of their logical form. Points, lines, and planes may be any entities whatsoever, undetermined in every way except by the relations stated in the axioms.

Let us, therefore, suppress every explicit reference to points, lines, and planes, and thereby eliminate all appeal to spatial intuition in deriving several theorems from the axioms. Suppose, then, that instead of the word "plane," we employ the letter *S;* and instead of the word "point," we use the phrase "*element of S*." Obviously, if the plane (*S*) is viewed as a collection of points (elements of *S*), a line may be viewed as a class of points (elements) which is a subclass of the points of the plane (*S*). We shall therefore substitute for the word "line" the expression "*L-class*." Our original set of axioms then reads as follows:

Axiom 1′. If *A* and *B* are distinct elements of *S*, there is at least one *L-class* containing both *A* and *B*.

Axiom 2′. If *A* and *B* are distinct elements of *S*, there is not more than one *L-class* containing both *A* and *B*.

Axiom 3′. Any two *L-classes* have at least one element of S in common.

Axiom 4′. There exists at least one *L-class* in S.

Axiom 5′. Every *L-class* contains at least three elements of S.

Axiom 6′. All the elements of S do not belong to the same *L-class*.

Axiom 7′. No *L-class* contains more than three elements of S.

In this set of assumptions no explicit reference is made to any specific subject matter. The only notions we require to state them are of a completely general character. The ideas of a "class," "subclass," "elements of a class," the relation of "belonging to a class" and the converse relation of a "class containing elements," the notion of "number," are part of the fundamental equipment of logic. If, therefore, we succeed in discovering the implications of these axioms, it cannot be because of the properties of space as such. (As a matter of fact, none of these axioms can be regarded as propositions; none of them is in itself either true or false. For the symbols S, *L-class*, A, B, and so on are variables. Each of the variables denotes any one of a class of possible entities, the only restriction placed upon it being that it must "satisfy," or conform to, the formal relations stated in the axioms. But until the symbols are assigned specific values, the axioms are propositional functions, and not propositions.)

Our "assumptions," therefore, consist in relations considered to hold between undefined terms. But the reader will note that although no terms are *explicitly* defined, an *implicit* definition of them is made. They may denote anything whatsoever, provided that what they denote conforms to the stated relations between themselves. This procedure characterizes modern mathematical technique. In Euclid, for example, *explicit* definitions are given of points, lines, angles, and so on. In a modern treatment of geometry, these elements are defined *implicitly* through the axioms. As we shall see, this latter procedure makes it possible to give a variety of interpretations to the undefined elements, and so to exhibit an identity of structure in different concrete settings. . . .

STRUCTURAL IDENTITY OR ISOMORPHISM

We want to show now that an abstract set such as the one discussed in the previous section may have more than one concrete representation, and that these different representations, though extremely unlike in material content, will be identical in logical structure.

Let us suppose there is a banking firm with seven partners. In order to assure themselves of expert information concerning various securities, they decide to form seven committees, each of which will study a special field. They agree, moreover, that each partner will act as chairman of one committee, and that every partner will serve on three, and only three, committees. The following is the schedule of committees and their members, the first member being chairman:

Domestic railroads	Adams	Brown	Smith
Municipal bonds	Brown	Murphy	Ellis
Federal bonds	Murphy	Smith	Jones
South American securities	Smith	Ellis	Gordon
Domestic steel industry	Ellis	Jones	Adams
Continental securities	Jones	Gordon	Brown
Public utilities	Gordon	Adams	Murphy

An examination of this schedule shows that it "satisfies" the seven axioms if the class S is interpreted as the banking firm, its elements as the partners, and the *L-classes* as the various committees. . . .

Paraphrasing:

One further interpretation illustrates the same seven formal relations.

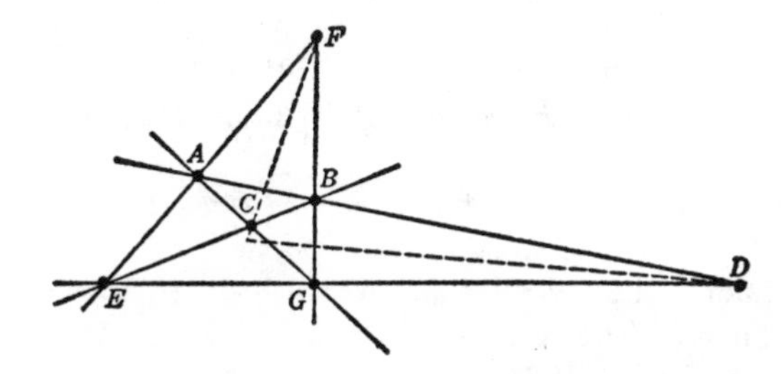

In the diagram there are seven points lying by threes on seven lines, one of which is "bent." If each point represents an element of S and each set of three points lying on a line an *L-class*, then all the seven assumptions are satisfied. Thus, for example, the three-termed relation between Adams, Brown, and Smith, by virtue of which they are on the same committee, holds for the points A, B, D, in virtue of which they lie on the same line. And in general, what may be deduced for A from the assumptions holds for Mr. Adams, what may be deduced for *B* holds for Mr. Brown, and so on.—P. 114.

3. Forsyth, *Geometry of Four Dimensions*.—P. 119.
4. It should be emphasized that a manifold, as usually defined, is stripped of every attribute, except that it is a class. Accordingly, it is easy to think of many familiar kinds of manifolds which have nothing to do with either space or geometry. A three-dimensional manifold would be a class of elements, each of which would require exactly three numbers to identify it—to distinguish it from every other element in the class. Think of a cylinder containing a quantity of three gases which have been thoroughly mixed so that the volume of gas or any portion of it, is uniquely determined by three numbers, x, y, and z, each of which represents the percentage of the three respective gases in the mixture. Or, one further instance: A group of people might be thought of as a manifold. If we find that five numbers are necessary and sufficient to individualize each one, say x equals age, y equals bank balance, z equals telephone number, u equals height, v equals weight, then they constitute a five-dimensional manifold. Other examples of manifolds can be devised: (a) four-dimensional: particles of air, 3 dimensions to fix them in space, 1 to fix their density; (b) four-dimensional: all conceivable spheres in space, 3 dimensions to fix their centers, 1 to determine their radii.—P. 119.
5. Nöbeling, "Die vierte Dimension und der krumme Raum," in *Krise und Neuaufbau*, Leipzig: Deuticke, 1933.—P. 123.
6. Eddington, *Space, Time, and Gravitation*.—P. 131.
7. Lindsay and Margenau, *Foundation of Physics*.—P. 132.
8. *Op. cit.*—P. 133.
9. Young, *Fundamental Concepts of Algebra and Geometry*, New York: Macmillan, 1911.—P. 135.
10. Morris Raphael Cohen, *Reason and Nature*.—P. 137.
11. Cohen and Nagel, *Introduction to Logic and Scientific Method*, New York: Harcourt Brace, 1934.—P. 142.
12. The diagram illustrates somewhat in detail what is meant. A perpendicular is drawn to the line G on the pseudosphere; through the point O two parallels are to be drawn to the line G. Mark off the distance S on G, terminating in Q. At Q erect a perpendicular to G. If, then, about the point O with S as radius we draw a circle, this circle will cut QT at S_1 and S_2. These two points, together with the point O, determine the two parallels to G, P_1 and P_2. All of the lines through O making an angle smaller than Θ do not intersect G, although they are not parallel

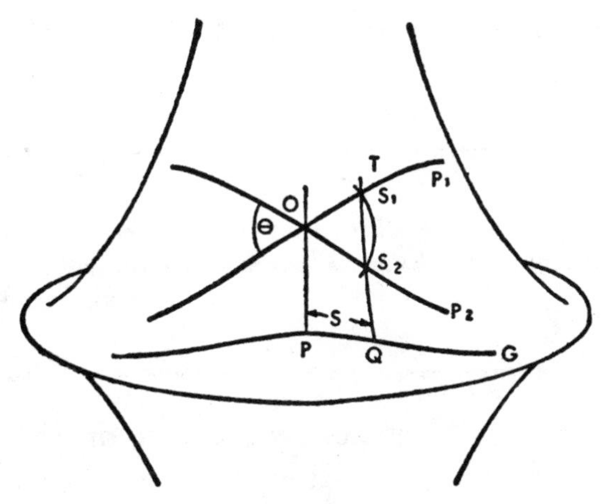

FIG. 49.

to it. This diagram is found in Colerus, *Vom Punkt zur vierten Dimension*, Vienna: Zsolnay, 1935.—P. 143.

13. These geometries are indispensable in the physics of the atom and the stars, in regions of space which are not a part of our immediate experience.—P. 150.

Pastimes of Past and Present Times

Work consists of whatever a body is obliged *to do, and play consists of whatever a body is not obliged to do.*

—MARK TWAIN

IT HAS been said, "It is not by amusing oneself that one learns," [1] and in reply: "It is *only* by amusing oneself that one can learn." Wherever the truth may lie, somewhere between those extremes, it is undeniable that mathematical recreations furnish a challenge to imagination and a powerful stimulus to mathematical activity. The theory of equations, of probability, the infinitesimal calculus, the theory of point sets, of topology—all are fruits grown from seeds sown in the fertile soil of creative imagination—all have grown out of problems first expressed in puzzle form.

Puzzles and paradoxes have been popular since antiquity, and in amusing themselves with these playthings men sharpened their wits and whetted their ingenuity. But it was not for amusement alone that Kepler, Pascal, Fermat, Leibniz, Euler, Lagrange, Hamilton, Cayley, and many others devoted so much time to puzzles. Researches in recreational mathematics sprang from the same desire to know, were guided by the same principles, and required the exercise of the same faculties as the researches leading to the most profound discoveries in mathematics and mathematical physics. Accordingly, no branch of intellectual activity is a more appropriate subject for discussion than puzzles and paradoxes.

*

The field is enormous. Puzzles have been in the making since Egyptian times and probably before. From the cryptical utterances of the oracle of Delphi, through the time of Charlemagne, down to the golden age of the crossword, paradoxes and puzzles, like the creatures of the earth, have assumed every shape and form and have multiplied. We can examine only a few of the dominating species, those which have survived in one shape or another and continue to thrive in streamlined form.

Most of the famous puzzles invented before the 17th century may be found in the first great puzzle book, *Les problèmes plaisants et délectables, qui se font par les nombres*, by Claude-Gaspard Bachet, Sieur de Meziriac. Although it appeared in 1612, two years before Napier's work on logarithms, it is still a delightful book and a quarry of information. Many collections have appeared since then,[2] Bachet's volume alone having been enlarged to almost five times its original size.

All we can hope to do is to follow the illustrious example of Mark Twain in a similar predicament. He attempted to reduce all jokes to a dozen primitive or elementary forms (mother-in-law, farmer's daughter, etc.). We shall attempt to present a few of the typical puzzles that illustrate the basic ideas from which all are evolved. We shall restrict our interest to puzzles and problems, reserving for another chapter some of the more celebrated logical and mathematical paradoxes. Although it may not always be easy to distinguish between a puzzle and a paradox, for our purposes it is sufficient to consider a puzzle as an ingenious game or problem, and a paradox as an apparently fallacious and self-contradictory proof or statement.

*

Puzzles often seem difficult because they are not easy to interpret in precise terms. In attempting the solution of a problem, the method of trial and error is not only more natural, but generally easier than the mathematical attack. It is common experience that often the most formidable algebraic equations are easier to solve than problems formulated in words. Such problems must first be translated into symbols, and the symbols placed into the proper equations before the problems can be solved.

When Flaubert was a very young man, he wrote a letter to his sister, Carolyn, in which he said: "Since you are now studying geometry and trigonometry, I will give you a problem. A ship sails the ocean. It left Boston with a cargo of wool. It grosses 200 tons. It is bound for Le Havre. The mainmast is broken, the cabin boy is on deck, there are 12 passengers aboard, the wind is blowing East-North-East, the clock points to a quarter past three in the afternoon. It is the month of May—How old is the captain?" Flaubert was not only teasing, he was uttering a complaint shared by that large and respectable company "not good at puzzles," that the average puzzle both confuses and overwhelms with superfluous words.[3] For that reason, the following puzzles have been stripped of all inessential elements so as to exhibit their underlying mathematical structure. And we understand by the term "mathematical structure" not necessarily something expressed by numbers, angles, or lines, but the essential internal relationship between the component elements of the puzzle. For, at bottom, that is all that mathematical analysis can reveal, all that mathematics itself signifies.

*

Among the oldest problems are those which involve ferrying people and their belongings across a river under somewhat trying conditions. Alcuin, the friend of Charlemagne, suggested a problem which has since been restated and complicated in many ways. A traveler comes to a riverbank with his possessions: a wolf, a goat, and a head of cabbage. The only available boat is very small and can carry no more than the traveler and *one* of his possessions. Unfortunately, if left together, the goat will eat the cabbage and the wolf dine on the goat. How shall the traveler transport his belongings to the other side of the river, keeping his vegetables and animals intact? [4] The solution may be attempted with the aid of a match box, representing the boat, and four slips of paper for its occupants.

A more elaborate version of this problem was suggested in the sixteenth century by Tartaglia. Three beautiful brides with their jealous husbands also come to a river. The small boat which is to take them across holds only two people. To avoid any compromising situations, the crossings are to be so arranged that no woman shall be left with a man unless her husband is present. Eleven passages are required. Five passages would be required for two couples, but with four or more couples the crossing under the conditions stated would be impossible.

Similar problems involve shunting. In Fig. 50 there is a locomotive, L, and 2 freight cars, W_1 and W_2. The common portion of the rails of the two sidings on which W_1 and W_2 are standing, DA, is long enough to hold

W_1 *or* W_2, but not both, nor the locomotive L. Thus, a car on DA can be shunted to either siding. The engineer's job is to switch the positions of W_1 and W_2. How can this be done? Although this problem presents no particular

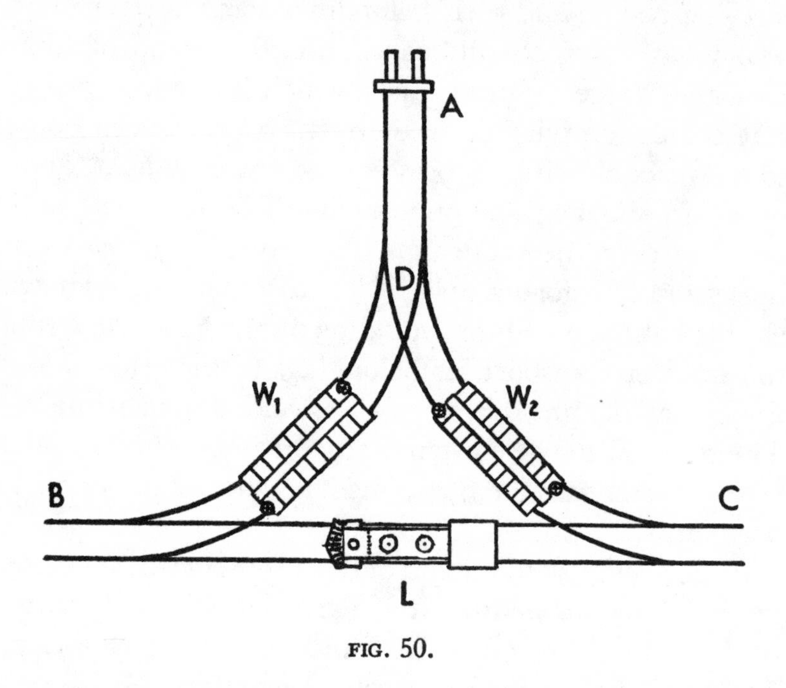

FIG. 50.

difficulties, the same theme in more complex form may demand of the engineer mathematical talents of a high order.

*

Simeon Poisson's family tried to make him everything from a surgeon to a lawyer, the last on the theory that he was fit for nothing better. One or two of these professions he tackled with singular ineptitude, but at last he found his métier. It was on a journey that someone posed him a problem similar to the one below. Solving it immediately, he realized his true calling and thereafter devoted himself to mathematics, becoming one of the greatest mathematicians of the nineteenth century.[5]

Two friends who have an eight-quart jug of wine wish to share it evenly. They also have two empty jars, one

FIG. 51.—Solution to the problem of the 3 jars.

holding five quarts, the other three. The diagram illustrates how they were able to divide their wine into two portions of four quarts each.[6]

This brings to mind another "pouring problem," perhaps not exactly germane at this point, but a good exercise in logical rigor and liquid refreshment.

THE INTERNATIONAL BEER-DRINKING PUZZLE

In a certain town lying on the border between Mexico and the United States a peculiar currency situation exists. In Mexico a U. S. dollar is worth only ninety cents of their money, while in the United States the value of the Mexican dollar is only ninety cents of our money. One day a cowhand strolls into a Mexican cantina and orders a ten-cent beer. He pays for it with a Mexican dollar, receiving for change an American dollar, worth just ninety cents there. After drinking his beer, he strolls over the border to an American saloon and orders another. This he pays for with an American dollar receiving a Mexican bill for change. He takes this back across the border and repeats the process, drinking beer merrily all day, and ends up as rich as he started, with a dollar.

The question: Who paid for the beer?

The moral: Visit sunny Mexico on your vacation.

*

The mystifying nature of so many arithmetic tricks lies, as we have indicated, in their structure, not their content. With a strainer to sift out the essential ideas hidden between dozens of useless ones, every man could be his own magician. A silly little riddle, oft repeated among mathematicians, comes to mind. "How shall one catch the lions in the desert?" it is asked. Since there is so much sand and so few lions, simply take a strainer, strain out the sand, and there are the lions! Such a strainer, then, or perhaps a scalpel, is needed to get at the rudiments. When the verbiage has been swept away the puzzle skeletons succumb to simple arithmetic of algebra. The parlor tricks of guessing numbers which others have selected, or cards which someone has chosen seem almost as wonderful as instances of "extra sensory perception." But after we have learned to separate the lions from the sand, caging them is comparatively simple.

Card tricks are usually arithmetic puzzles in disguise. Generally, they are amenable to mathematical analysis, and are not, as is commonly believed, performed by sleight of hand. One important principle, easily overlooked, is that "cutting a pack of cards never alters the relative positions of the cards, provided that, if necessary, we regard the top card as following immediately after the bottom card in the pack." [7] Once this is understood, many tricks cease to be baffling.

Seven poker players have invested in a new deck of cards. In keeping with tradition the cards are cut, not shuffled, on the first deal. The dealer, pretending to cheat, takes his second and fourth cards from the bottom of the deck. This lapse is noticed by everyone, as intended. However, when the other players pick up their cards, they are reluctant to demand a new deal, each one finding that he has a full house. But still fearful that the dealer has fixed up a better hand for himself, they insist that he discard his 5 cards and take the next 5 from the top of the deck. Feigning indignation he acquiesces—and wins with a straight flush. Try it. Ninety-nine times out of one hundred you will succeed in cheating your friends—but then, you can't cheat an honest man.

Frequently, the arithmetic tricks of guessing a number selected by another depend on the "scale of notation."

When a number is expressed in the decimal system, such as 3976, what is actually meant is

$$(3 \times 10^3) + (9 \times 10^2) + (7 \times 10^1) + (6 \times 10^0).$$

The table [8] further illustrates other numbers written to the base 10.

EXAMPLE		10^0		10^1		10^2		10^3		10^4
469	=	9×10^0	+	6×10^1	+	4×10^2				
	=	9	+	60	+	400				
7901	=	1×10^0	+	0×10^1	+	9×10^2	+	7×10^3		
	=	1	+	0	+	900	+	7000		
30,000	=	0×10^0	+	0×10^1	+	0×10^2	+	0×10^3	+	3×10^4
	=	0	+	0	+	0	+	0	+	30,000
21148	=	8×10^0	+	4×10^1	+	1×10^2	+	1×10^3	+	2×10^4
	=	8	+	40	+	100	+	1000	+	20,000

Among the wide variety of problems which arise from the use of the decimal system, the following are of some interest:

A useful device for checking multiplication goes by the copybook title of "casting out nines."

Consider 1234 × 5678 = 7006652. Add the digits of the multiplier, multiplicand, and product, thus obtaining 10, 26 and 26 respectively. Since each of these numbers is greater than 9, add the digits of the individual sums once more,* obtaining 1, 8, and 8. (If, after the first repetition a sum greater than 9 remains, the digits may be added once again.) Now, take the product of the integers corresponding to the multiplier, and the multi-

* Thus 10 = 1 + 0 = 1
26 = 2 + 6 = 8, etc.

plicand, i.e., 1×8, and compare this with the integer corresponding to the sum of the digits of the product, which is also 8. Since they are the same, the result of the original multiplication is correct.

Using the same rule, let us test whether the product of 31256 and 8427 is 263395312. Again the sum of the digits of the multiplicand, multiplier, and the product are respectively 17, 21 and 34; repeating, the sum of these digits is 8, 3, and 7. The product of the first two equals 24 which has 6 for the sum of its digits. But the sum of the digits of the product equals 7. Thus, we have two different remainders, 6 and 7, whence the multiplication must be incorrect.

Closely connected with the rule of casting out nines is the following trick, which reveals a remarkable property common to all numbers.

Take any number and rearrange its digits in any order you please to form another number. The difference between the first number and the second is always divisible by 9.[9]

Another type of problem dependent on the decimal scale of notation involves finding numbers which may be obtained by multiplying their *reversals* by integers. Among such numbers with 4 digits, 8712 equals 4 times 2178, and 9801 equals 9 times 1089.

The *binary* or *dyadic* notation (using the base 2) is hardly a new concept, having been referred to in a Chinese book believed to have been written about 3000 B.C. Forty-six centuries later, Leibniz rediscovered the wonders of the binary scale and marveled at it as though it were a new invention—somewhat like the twentieth-century city dweller, who, upon seeing a sundial for the first time, and having it explained, remarked with awe:

"What will they think of next?" In its use of only two symbols, Leibniz saw in the dyadic system something of great religious and mystic significance: God could be represented by unity, and nothingness by zero, and since God had created all forms out of nothingness, zero and one combined could be made to express the entire universe. Anxious to impart this gem of wisdom to the heathens, Leibniz communicated it to the Jesuit Grimaldi, president of the Tribunal of Mathematics in China, in the hope that he could thus show the Emperor of China the error of his ways in clinging to Buddhism instead of adopting a God who could create a universe out of nothing.

Whereas the decimal scale requires ten sumbols: 0, 1, 2, 3, 4, . . . , 9, the binary scale uses only two: 0 and 1. Below are the first 32 integers given in the binary scale.

DECIMAL		BINARY	DECIMAL		BINARY
1	=	1	17	=	10001
2	=	10	18	=	10010
3	=	11	19	=	10011
4	$= 2^2 =$	100	20	=	10100
5	=	101	21	=	10101
6	=	110	22	=	10110
7	=	111	23	=	10111
8	$= 2^3 =$	1000	24	=	11000
9	=	1001	25	=	11001
10	=	1010	26	=	11010
11	=	1011	27	=	11011
12	=	1100	28	=	11100
13	=	1101	29	=	11101
14	=	1110	30	=	11110
15	=	1111	31	=	11111
16	$= 2^4 =$	10000	32	$= 2^5 =$	100000

Since $2^0 = 1$, it may readily be seen that *any* number can be expressed as the *sum* of powers of 2, just as any number in the decimal system can be expressed as the *sum* of powers of 10. For example, the number expressed in the decimal system as 25, is expressed in the binary system, using only the two symbols 1 and 0, by 11001.

DECIMAL		DYADIC
25	=	11001
↕		↕
$(2 \times 10^1) + (5 \times 10^0)$.		$(1 \times 2^4) + (1 \times 2^3) + (0 \times 2^2) + (0 \times 2^1) + (1 \times 2^0)$.

Because numbers can be more briefly written in the decimal scale than in the binary, it is more convenient, although in every other respect the latter is just as accurate and efficient. Even fractions have their place in the dyadic notation. The fraction $\frac{1}{3}$, for example, given by the nonterminating decimal, .33333 . . . , is represented in the binary notation by a nonterminating binary, .01010101 . . .[10]

The binary system easily makes understandable the solution of problems such as:

I. In many sections of Russia, the peasants employed until recently what appears to be a very strange method of multiplication. In substance, this was at one time in use in Germany, France, and England, and is similar to a method used by the Egyptians 2000 years before the Christian era.

It is best illustrated by an example: To multiply 45 by 64, form two columns. At the head of one put 45, at the head of the other, 64. Successively multiply one column by 2 and divide the other by the same number. When an odd number is divided by 2, discard the remaining fraction. The result will be:

	DIVIDE	MULTIPLY
	45	64
	22	128
(A)	11	256
	5	512
	2	1024
	1	2048

Take from the second column those numbers which appear opposite an odd number in the first. Add them and you obtain the desired product:

	45	64	$64 = 2^0 \times 64$
	22	128	$= 2^1 \times 64$
(B)	11	256	$256 = 2^2 \times 64$
	5	512	$512 = 2^3 \times 64$
	2	1024	$= 2^4 \times 64$
	1	2048 . . .	$2048 = 2^5 \times 64$
			$2880 = 45 \times 64$

The relation of this method to the dyadic system may be seen upon expressing 45 in the dyadic notation.

$$
\begin{aligned}
45 &= (1 \times 2^5) + (0 \times 2^4) + (1 \times 2^3) + (1 \times 2^2) + (0 \times 2^1) \\
&\quad + (1 \times 2^0) \\
&= 101101 \\
&= 32 + 0 + 8 + 4 + 0 + 1
\end{aligned}
$$

Therefore,

$$
\begin{aligned}
45 \times 64 &= (2^5 + 2^3 + 2^2 + 2^0) \times 64 \\
&= (2^5 \times 64) + (2^3 \times 64) + (2^2 \times 64) + (2^0 \times 64).
\end{aligned}
$$

Since 2^4 and 2^1 do not appear in the dyadic expression for 45, the products $(2^4 \times 64)$ and $(2^1 \times 64)$ are not included in the numbers to be added in (B). Thus, what the peasant does in multiplying 45 by 64 is to multiply 2^5, 2^3, 2^2, 2^0, successively by 64, and then take the sum.

II. Another well-known problem, already mentioned by Cardan, consists in the removal of a number of rings from a bar. The puzzle can best be analyzed by the use of the dyadic system, although the actual manipulation of the rings is at all times extremely difficult.

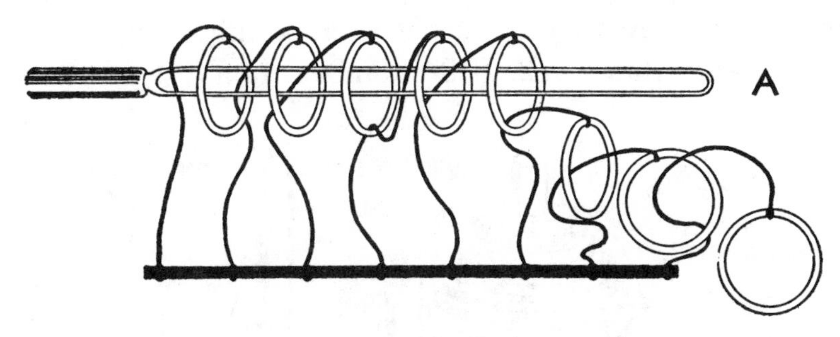

FIG. 52 —The Chinese ring puzzle.

The rings on the bar are so connected that although the end one can be removed without difficulty, any other ring can be put on or removed only when the one next to it, toward the end (*A* in the figure) is *on* the bar, and all the rest of the rings are *off*. Thus, to remove the fifth ring, the first, second, third must be off the bar, and the fourth must be on. If the position of all the rings on or off the rack are written in the dyadic notation, 1 designating a ring which is off, and 0 designating a ring which is on, the mathematical determination of the number of moves required to remove a given number of rings is not too hard. The solution without the aid of the dyadic notation, as the rings increase in number, would be wholly beyond one's imaginative powers.

III. The problem of the Tower of Hanoi is similar in principle. The game consists of a board with three pegs, as illustrated in Fig. 53.

On one of these pegs rest a number of discs of various sizes, so arranged that the largest disc is on the bottom, the next largest rests on that one, the next largest on that, and so on, up to the smallest disc which is on top. The problem is to transfer the entire set of discs to one of the other two pegs, moving only one disc at a time, and making certain that no disc is ever permitted to rest on one smaller than itself. If the removal of a disc from one

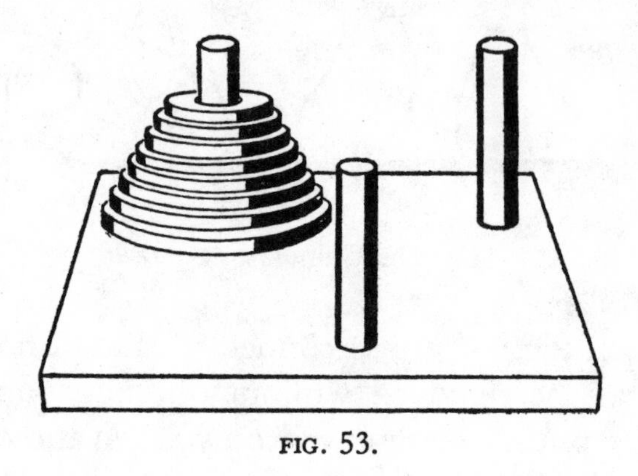

FIG. 53.

peg to another constitutes one transfer, the following table shows the number of transfers required for various numbers up to *n* discs:

TABLE FOR TRANSFERS [11]

DISCS	TRANSFERS
1	1
2	3
3	7
4	15
5	31
6	63
7	127
⋮	⋮
n	2^n-1

There is a charming story about this toy:[12]

In the great temple at Benares beneath the dome which marks the center of the world, rests a brass plate in which are fixed three diamond needles, each a cubit high and as thick as the body of a bee. On one of these needles, at the creation, God placed sixty-four discs of pure gold, the largest disc resting on the brass plate and the others getting smaller and smaller up to the top one. This is the Tower of Brahma. Day and night unceasingly, the priests transfer the discs from one diamond needle to another, according to the fixed and immutable laws of Brahma, which require that the priest on duty must not move more than one disc at a time and that he must place this disc on a needle so that there is no smaller disc below it. When the sixty-four discs shall have been thus transferred from the needle on which, at the creation, God placed them, to one of the other needles, tower, temple, and Brahmans alike will crumble into dust, and with a thunderclap, the world will vanish.

The number of transfers required to fulfill the prophecy is $2^{64} - 1$, that is 18,446,744,073,709,551,615. If the priests were to effect one transfer every second, and work 24 hours a day for each of the 365 days in a year,[13] it would take them 58,454,204,609 centuries plus slightly more than 6 years to perform the feat, assuming they never made a mistake—for one small slip would undo all their work.

IV. One other game may be mentioned in connection with the dyadic system—Nim. In this game, two players play alternately with a number of counters placed in several heaps. At his turn, a player picks up one of the heaps, or as many of the counters from it as he pleases. The player taking the last counter loses. If the number of counters in each heap is expressed in the binary scale, the game readily lends itself to mathematical analysis. A player who can bring about a certain arrangement of the number of counters in each heap may force a win.[14]

It is interesting to note that the number 2^{64}—18,446,-744,073,709,551,616—represented in the dyadic system by a number with 64 digits, appears in the solution of

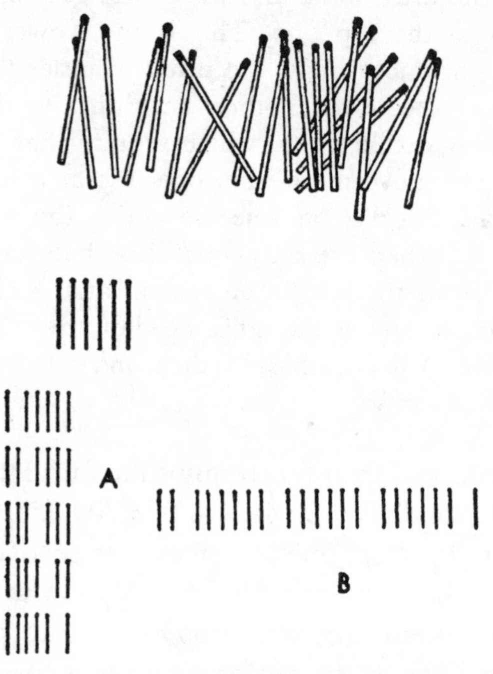

FIG. 54.—The diagram illustrates how to force a win at the game of Nim. Assume each player at his turn must pick up at least one match and may pick up as many as five. The rule is that the player picking up the last match loses. Then, for example, imagine that the original heap consists of 21 matches. In that case, the one playing first can force a win by mentally dividing the matches into groups of 1, 6, 6, 6, and 2 (as in *B*). Since he plays first, he picks up 2 matches. Then, however many his opponent picks, the first player picks up the complement of 6. This is shown in *A*: If the second player takes 1, the first player takes 5; if the second player takes 2, the first player takes 4, and so on. Each of the three groups of 6 is thus exhausted, and the second player is left with the last match.

Had there been 47 matches, say, the grouping to force a win for the first player would have been: 1, 6, 6, 6, 6, 6, 6, 6, and 4. Rules for any other variation of Nim can also be easily formulated.

a puzzle connected with the origin of the game of chess.

According to an old tale, the Grand Vizier Sissa Ben Dahir was granted a boon for having invented chess for the Indian King, Shirhâm. Since this game is played on a board with 64 squares, Sissa addressed the king: "Majesty, give me a grain of wheat to place on the first square, and two grains of wheat to place on the second square, and four grains of wheat to place on the third, and eight grains of wheat to place on the fourth, and so, Oh, King, let me cover each of the 64 squares of the board." "And is that all you wish, Sissa, you fool?" exclaimed the astonished King. "Oh, Sire," Sissa replied, "I have asked for more wheat than you have in your entire kingdom, nay, for more wheat than there is in the whole world, verily, for enough to cover the whole surface of the earth to the depth of the twentieth part of a cubit." [15] Now the number of grains of wheat which Sissa demanded is $2^{64} - 1$, exactly the same as the number of disc transfers required to fulfill the prophecy of Benares related on page 171.

Another remarkable way in which 2^{64} arises is in computing the number of each person's ancestors from the beginning of the Christian era—just about 64 generations ago. In that length of time, assuming that each person has 2 parents, 4 grandparents, 8 great-grandparents, etc., and not allowing for incestuous combinations, everyone has at least 2^{64} ancestors, or a little less than eighteen and a half quintillion lineal relations alone. A most depressing thought.

*

The Josephus problem is one of the most famous and certainly one of the most ancient. It generally relates a story about a number of people on board a ship, some of whom must be sacrificed to prevent the ship from sinking. Depending on the time that the version of the puzzle was written, the passengers were Christians and Jews, Christians and Turks, sluggards and scholars, Negroes and whites, etc. Some ingenious soul with a knowledge of mathematics always managed to preserve the favored group. He arranged everyone in a circle, and reckoning from a certain point onward, every nth person was to be thrown overboard—n being a specified integer. The arrangement of the circle by the mathematician was such that either the Christians, or the industrious scholars, or the whites,—in other words, the assumedly superior group —were saved, while the rest were thrown overboard in accordance with the Golden Rule.

Originally, this tale was told of Josephus who found himself in a cave with 40 other Jews bent on self-extinction to escape a worse fate at the hands of the Romans. Josephus decided to save his own neck. He placed everyone in a circle and made them agree that each third person, counting around and around, should be killed. Placing himself and another provident soul in the 16th and the 31st position of the circle of 41, he and his companion, being the last ones left, were conveniently able to avoid the road to martyrdom.

A later version of this problem places 15 Turks and 15 Christians on board a storm-ridden ship which is certain to sink unless half the passengers are thrown overboard. After arranging everyone in a circle, the Christians, *ad majorem Dei gloriam*, proposed that every ninth person be sacrificed.

Thus, every infidel was properly disposed of, and all true Christians saved.[16]

Among the Japanese, the Josephus problem assumed another form: Thirty children, 15 of the first marriage, and 15 of the second, agree that their father's estate is too small to be divided among all of them. So the second

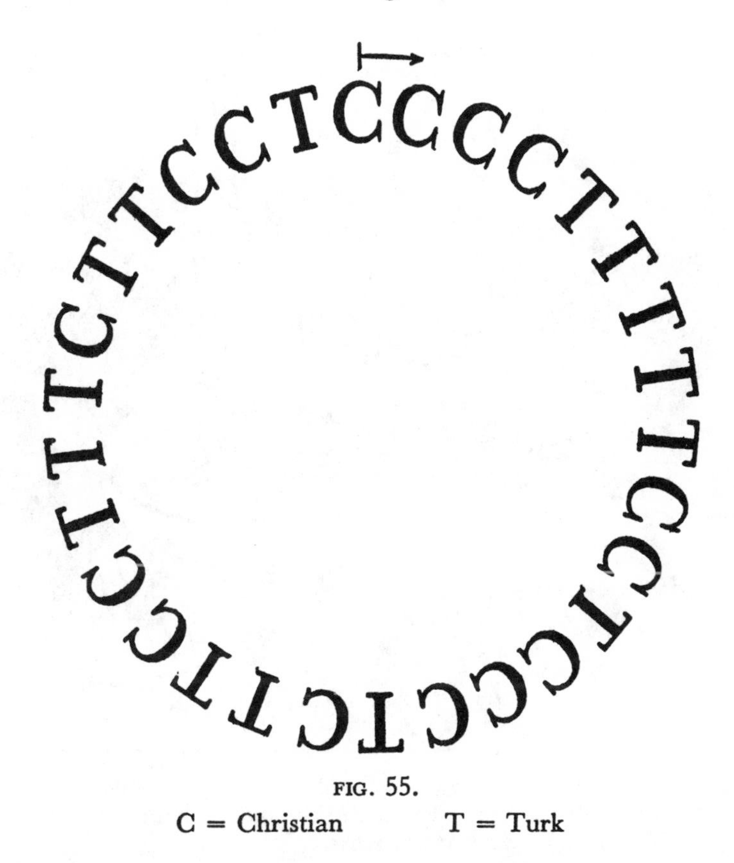

FIG. 55.

C = Christian T = Turk

wife proposes that all the children be arranged in a circle, in order to determine her husband's heir by a process of elimination. Being a prudent mathematician, as well as the proverbially wicked stepmother, she arranges the children in such a way that one of her own is certain to be chosen. After 14 of the children of the first marriage have been eliminated, the remaining child, evidently a keener mathematician than his stepmother, proposes that the counting shall start afresh in the opposite direction. Certain of her advantage, and therefore disposed to be generous, she consents, but finds to her

FIG. 56.—The Josephus problem, from Miyake Kenryu's *Shojutsu*. (From Smith and Mikami, *A History of Japanese Mathematics*.)

dismay that all 15 of her own children are eliminated, leaving the one child of the first marriage to become the heir.[17]

Elaborate mathematical solutions of more difficult and generalized versions of the Josephus problem were given by Euler, Schubert, and Tait.

No discussion of puzzles, however brief, can afford to omit mention of the best-known of the many puzzles invented by Sam Lloyd. "15 Puzzle," "Boss Puzzle," "le Jeu de Taquin," are a few of its names. For several years after its appearance in 1878, this puzzle enjoyed a popularity, particularly throughout Europe, greater than swing and contract bridge combined enjoy today. In Germany, it was played in the streets, in factories, in the royal palaces, and in the Reichstag. Employers were forced to post notices forbidding their employees to play the "15 Puzzle" during business hours under penalty of dismissal. The electorate, having no such privileges, had to watch their duly elected representatives play the "Boss Puzzle" in the Reichstag while Bismarck played the Boss. In France, the "Jeu de Taquin" was played on the boulevards of Paris and in every tiny hamlet from the Pyrenees to Normandy. A scourge of mankind was the "Jeu de Taquin," according to a contemporary French journalist, —worse than tobacco and alcohol—"responsible for untold headaches, neuralgias, and neuroses."

For a time, Europe was "15 Puzzle" mad. Tournaments were staged and huge prizes offered for the solution of apparently simple problems. But the strange thing was that no one ever won any of these prizes, and the apparently simple problems remained unsolved.

The "15 Puzzle" (figure below) consists of a square shallow box of wood or metal which holds 15 little square blocks numbered from 1 to 15. There is actually room for 16 blocks in the box so that the 15 blocks can be moved about and their places interchanged. The number of conceivable positions is 16! = 20,922,789,888,000. A problem consists of bringing about a specified arrangement of the blocks from a given initial position, which is frequently the *normal* position illustrated in Fig. 57.

Shortly after the puzzle was invented, two American mathematicians [18] proved that from any given initial order only *half* of all the conceivable positions can actu-

FIG. 57.—The 15 Puzzle (also Boss Puzzle or Jeu de Taquin) in normal position.

ally be obtained. Thus, there are always approximately ten trillion positions which the possessor of a "15 Puzzle" can bring about, and ten trillion that he *cannot.*

The fact that there are impossible positions makes it easy to understand why such generous cash prizes were offered by Lloyd and others, since the problems for which prizes were offered always entailed impossible positions. And it is heart-breaking to think of the headaches, neuralgias, and neuroses that might have been spared—to say nothing of the benefits to the Reichstag—if *The American Journal of Mathematics* had been as widely circulated as the puzzle itself. With ten trillion possible solutions there still would have been enough fun left for everyone.

In the normal position (Fig. 57), the blank space is in

the lower right-hand corner. When making a mathematical analysis of the puzzle, it is convenient to consider that a rearrangement of the blocks consists of nothing more than moving the blank space itself through a specific path, always making certain that it ends its journey in the lower right-hand corner of the box. To this end, the blank space must travel through the same number of boxes to the left as to the right and through the same number of boxes upwards as downwards. In other words, *the blank space must move through an even number of boxes.* If, starting from the normal position, the desired one can be attained while complying with this requirement, it is a *possible* position, otherwise it is *impossible.*

Based upon this principle, the method of determining whether a position is possible or impossible is very simple. In the normal position every numbered block appears in its proper numerical order, i.e., regarding the boxes, row by row, from left to right, no number precedes any number smaller than itself. To bring about a position different from the normal one, the numerical order of the blocks must be changed. Some numbers, perhaps all, will precede others smaller than themselves. Every instance of a number preceding another smaller than itself is called an *inversion.* For example, if the number 6 precedes the numbers 2, 4, and 5, this is an inversion to which we assign the value 3, because 6 precedes three numbers smaller than itself. If the *sum* of the *values* of *all the inversions* in a given position is even, the position is a possible one—that is, it can be brought about from the normal position. If the sum of the values of the inversions is odd, the position is impossible and cannot be brought about from the normal configuration.

The position illustrated in Fig. 58 can be created

from the normal position since the sum of the values of the inversions is six—an even number.

2	1	4	3
6	5	8	7
10	9	12	11
13	14	15	

FIG. 58.

But the position shown in Fig. 59 is impossible, since, as may readily be seen, the sum of the value of the inversions brought about is odd:

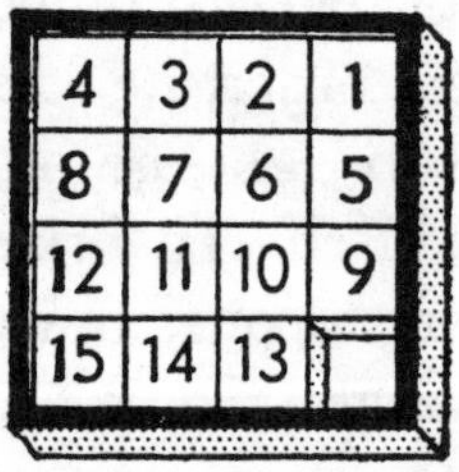

4	3	2	1
8	7	6	5
12	11	10	9
15	14	13	

FIG. 59.

Figures 60 a, b, c illustrate three other positions. Are they possible, or impossible to obtain from the normal order?

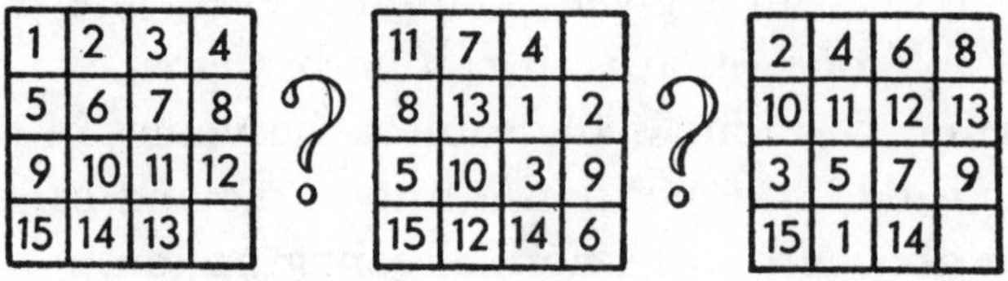

1	2	3	4
5	6	7	8
9	10	11	12
15	14	13	

?

11	7	4	
8	13	1	2
5	10	3	9
15	12	14	6

?

2	4	6	8
10	11	12	13
3	5	7	9
15	1	14	

FIGS. 60 (a, b, c).

SPIDER AND FLY PROBLEM

Most of us learned that a straight line is the shortest distance between two points. If this statement is supposed to apply to the earth on which we live, it is both useless and untrue. As we have seen in the previous chapter, the nineteenth-century mathematicians Riemann and Lobachevsky knew that the statement, if true at all, applied only to special surfaces. It does not apply to a *spherical* surface on which the shortest distance between two points is the arc of a great circle. Since the shape of the earth approximates a sphere, the shortest distance between two points anywhere on the surface of the earth is *never* a straight line, but is a portion of the arc of a great circle. (See page 146.)

Yet, for all *practical* purposes, even on the surface of the earth, the shortest distance between two points is given by a straight line. That is to say, in measuring ordinary distances with a steel tape or a yardstick, the principle is substantially correct. However, for distances beyond even a few hundred feet, allowance must be made for the curvature of the earth. When a steel rod over 600 feet in length was recently constructed in a large Detroit automobile factory, it was found that the exact measurement of its length was impossible without allowing for the earth's curvature. We indicated that the determination of a geodesic is very difficult for complicated surfaces. But we can give one puzzle showing how deceptive this problem may be for even the simplest case—the flat surface.

In a room 30 feet long, 12 feet wide, and 12 feet high, there is a spider in the center of one of the smaller walls, 1 foot from the ceiling; and there is a fly in the middle of the opposite wall, 1 foot from the floor. The spider has designs on the fly. What is the shortest possible route along which the spider may crawl to reach his

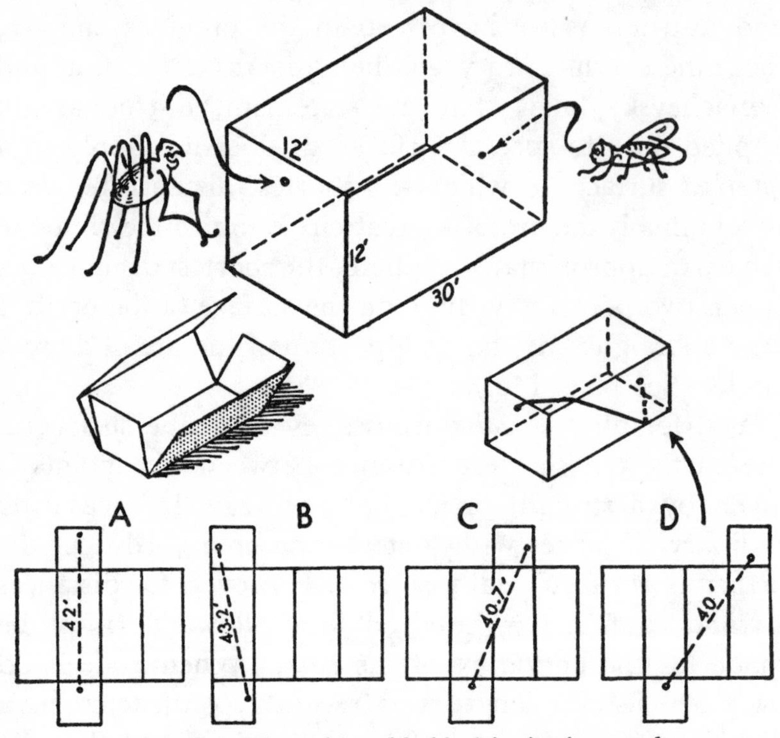

FIG. 61.—The spider, his kind invitation to the fly having been rebuffed, sets out for dinner along the shortest possible route. What path is the geodesic for the hungry spider?

prey? If he crawls straight down the wall, then in a straight line along the floor, and then straight up the other wall, or follows a similar route along the ceiling, the distance is 42 feet. Surely it is impossible to imagine a shorter route! However, by cutting a sheet of paper, which, when properly folded, will make a model of the room (see Fig. 61), and then by joining the points representing the spider and the fly by a straight line, a geodesic is obtained. The length of this geodesic is only 40 feet, in other words, 2 feet shorter than the "obvious" route of following straight lines.

There are several ways of cutting the sheet of paper, and accordingly, there are several possible routes, but that of 40 feet is the shortest; and remarkably enough, as may be seen from cut *D* in Fig. 61, this route requires the spider to pass over 5 of the 6 sides of the room.

This problem graphically reveals the point emphasized throughout—our intuitive notions about space almost invariably lead us astray.

*

RELATIONSHIPS

Ernest Legouvé,[19] the well known French dramatist, tells in his memoirs that, while taking the baths at Plombières, he proposed a question to his fellow bathers: "Is it possible for two men, wholly unrelated to each other, to have the same sister?" "No, that's impossible," replied a notary at once. An attorney who was not quite so quick in giving his answer, decided after some deliberation, that the notary was right. Thereupon, the others quickly agreed that it was impossible. "But still it is possible," Legouvé remarked, "and I will name two such men. One of them is Eugene Sue, and I am the other." In the midst of cries of astonishment and demands that he explain, he called the bath attendant and asked for the slate on which the attendant was accustomed to mark down those who had come for their baths. On it, he wrote:

(~ means married to; | means offspring of)

Mrs. Sue~Mr. Sue	Mrs. Sauvais~Mr. Sue	Mrs. Sauvais~Mr. Legouvé
|	|	|
Eugene Sue	Flore Sue	Ernest Legouvé

"Thus, you see," he concluded, "it is quite possible for two men to have the same sister, without being related to each other."

Most of the puzzles treated hitherto required four steps for their solution:

1. Sifting out the essential facts.
2. Translating these facts into the appropriate symbols.
3. Setting up the symbols in equations.
4. Solving the equations.

To solve the problems of relationship two of these steps must be modified. A simple diagram replaces the algebraic equation; inferences from the diagram replace the algebraic solution. Without the symbols and diagrams, however, the problems may become extremely confusing.

Alexander MacFarlane, a Scotch mathematician, developed an "algebra of relationships" which was published in the proceedings of the Royal Society of Edinburgh, but the problems to which he applied his calculus were easily solvable without it. McFarlane used the well-known jingle:

> Brothers and sisters have I none,
> But this man's father is my father's son,

as a guinea pig for his calculus, although the diagrammatic method gives the solution much more quickly.

An old Indian fairy tale creates an intricate series of relationships which would probably prove too much even for MacFarlane's algebra. A king, dethroned by his relatives, was forced to flee with his wife and daughter. During their flight they were attacked by robbers; while defending himself, the king was killed, although his wife and daughter managed to escape. Soon they came to a forest in which a prince of the neighboring country and his son were hunting. The prince (a widower) and his son (an eligible bachelor) noticing the footsteps of the mother and daughter decided to follow them. The father declared that he would marry the woman with the large footsteps—undoubtedly the older—and the son said he would marry the woman with the small footsteps who was surely the younger. But on their return to the castle, the father and his son discovered that the small feet belonged to the mother and the big feet to the daughter. Nevertheless, mastering their disappointment, they married as they had planned. After the marriage, the mother, daughter-in-law of her daughter, the daughter, mother-in-law of her mother, both had children—sons and daughters. The task of disentangling the resulting relationships we entrust to the reader, as well as the explanation of the following verse found on an old gravestone at Alencourt, near Paris:

> Here lies the son; here lies the mother;
> Here lies the daughter; here lies the father;
> Here lies the sister; here lies the brother;
> Here lie the wife and the husband.
> Still, there are only three people here.

*

In Albrecht Dürer's famous painting, "Melancholia," there appears a device about which more has been written than any other form of mathematical amusement. The device is a *magic square.*

A magic square consists of an array of integers in a square which, when added up by rows, diagonals, or columns, yield the same total. Magic squares date back at least to the Arabs. Great mathematicians like Euler and Cayley found them amusing and worth studying. Benjamin Franklin admitted somewhat apologetically that he had spent some time in his youth on these "trifles"—time "which," he hastened to add, "I might have employed more usefully." Mathematicians have never pretended that magic squares were anything more than amusement, however much time they spent on them, although the continual study devoted to this puzzle form may incidentally have cast some light on relations between numbers. Their chief appeal is still mystical and recreational.

There are other puzzles of considerable interest not discussed here because we treat them more fully in their proper place.[20] Among these are problems connected with the theory of probability, map-coloring, and the one-sided surfaces of Möbius.

Only one extensive group of problems remains—those connected with the theory of numbers. The modern theory of numbers, represented by a vast literature, engages the attention of every serious mathematician. It is a branch of study, many theorems of which, though exceedingly difficult to prove, can be simply stated and readily understood by everyone. Such theorems are therefore more widely known among educated laymen than theorems of far greater importance in other branches of mathematics, theorems which require technical knowledge to be understood. Every book on mathematical recreations is filled with simple or ingenious, cunning or marvelous, easy or difficult puzzles based on the behavior and properties of numbers. Space permits us to mention only one or two of those significant theorems about numbers which, in spite of their profundity, can be easily grasped.

Ever since Euclid proved [21] that the number of primes is infinite, mathematicians have been seeking for a test which would determine whether or not a given number is a prime. But no test applicable to all numbers has been found. Curiously enough, there is reason to believe that certain mathematicians of the seventeenth century, who spent a great deal of time on number theory, had means of recognizing primes unknown to us. The French mathematician Mersenne and his much greater contemporary, Fermat, had an uncanny way of determining the values of p, for which $2^p - 1$ is a prime. It has not yet been clearly determined how completely they had developed their method, or indeed, exactly what method they employed. Accordingly, it is still a source of wonder that Fermat replied without a moment's hesitation to a letter which asked whether 100895598169 was a prime, that it was the product of 898423 and 112303, and that each of these numbers was prime.[22] Without a general formula for all primes, a mathematician, even today, might spend years hunting for the correct answer.

One of the most interesting theorems of number theory is Goldbach's, which states that every even number is the sum of two primes. It is easy to understand; and there is every reason to believe that it is true, no even number having ever been found which is *not* the sum of two primes; yet, no one has succeeded in finding a proof valid for all even numbers.

But perhaps the most famous of all such propositions, believed to be true, but never proved, is "Fermat's Last Theorem." In the margin of his copy of Diophantus, Fermat wrote: "If n is a number greater than two, there are no whole numbers, a, b, c such that $a^n + b^n = c^n$.

I have found a truly wonderful proof which this margin is too small to contain." What a pity! Assuming Fermat actually had a proof, and his mathematical talents were of such a high order that it is certainly possible, he would have saved succeeding generations of mathematicians unending hours of labor if he had found room for it on the margin. Almost every great mathematician since Fermat attempted a proof, but none has ever succeeded.

Many pairs of integers are known, the sum of whose squares is also a square, thus:

$$3^2 + 4^2 = 5^2; \text{ or, } 6^2 + 8^2 = 10^3.$$

But no three integers have ever been found where the sum of the cubes of two of them is equal to the cube of the third. It was Fermat's contention that this would be true for all integers when the power to which they were raised was greater than 2. By extended calculations, it has been shown that Fermat's theorem is true for values of n up to 617. But Fermat meant it for *every* n greater than 2. Of all his great contributions to mathematics, Fermat's most celebrated legacy is a puzzle which three centuries of mathematical investigation have not solved and which skeptics believe Fermat, himself, never solved.

*

Somewhat reluctantly we must take our leave of puzzles. Reluctantly, because we have been able to catch only a glimpse of a rich and entertaining subject, and because puzzles in one sense, better than any other single branch of mathematics, reflect its always youthful, unspoiled, and inquiring spirit. When a man stops wondering and asking and playing, he is through. Puzzles are made of the things that the mathematician, no less than the child, plays with, and dreams and wonders about, for they are made of the things and circumstances of the world he lives in.

FOOTNOTES

1. Anatole France, *The Crime of Sylvestre Bonnard*.—P. 156.
2. W. W. R. Ball, *Mathematical Recreations and Essays*, 11th ed. New York: Macmillan, 1939.
   W. Lietzmann, *Lustiges und Merkwürdiges von Zahlen und Formen*, Breslau: Hirt, 1930.
   Helen Abbot Merrill, *Mathematical Excursions*, Boston: Bruce Humphries, 1934.
   W. Ahrens, *Mathematische Unterhaltungen und Spiele*, Leipzig: B. G. Teubner, 1921, vols. I and II.
   H. E. Dudeney, *Amusements in Mathematics*, London: Thomas Nelson, 1919.
   E. Lucas, *Récréations Mathématiques*, Paris: Gautier-Villars, 1883–1894, vols. I, II, III and IV.—P. 157.
3. Here is an example of a type of puzzle quite fashionable of late, which, though apparently wordy, contains no unessential facts:

   THE ARTISANS

   There are three men, John, Jack and Joe, each of whom is engaged in two occupations. Their occupations classify each of them as two of the following: chauffeur, bootlegger, musician, painter, gardener, and barber.

   From the following facts find in what two occupations each man is engaged:

   1. The chauffeur offended the musician by laughing at his long hair.
   2. Both the musician and the gardener used to go fishing with John.
   3. The painter bought a quart of gin from the bootlegger.
   4. The chauffeur courted the painter's sister.
   5. Jack owed the gardener $5.
   6. Joe beat both Jack and the painter at quoits.—P. 158.
4. There are two different ways, both of which are symbolized in the following table.—P. 159.

FIRST SOLUTION SECOND SOLUTION
W = WOLF C = CABBAGE
G = GOAT → = CROSSING

1. WGC			1. WGC		
2. WC	G→	G	2. WC	G→	G
3. WC	←	G	3. WC	←	G
4. C	W→	WG	4. W	C→	GC
5. GC	←G	W	5. WG	←G	C
6. G	→C	WC	6. G	W→	WC
7. G	←	WC	7. G	←	WC
8.	G→	WGC	8.	G→	WGC

5. At least so says his biographer, Arago. Not only was the quality of Poisson's work extremely high, but the output was enormous. Besides occupying several important official positions, he turned out over 300 works in a comparatively short lifetime, (1781–1840). "La vie, c'est le travail," said this erstwhile shadow on the Poisson household, though oddly enough, a puzzle brought him to a life dedicated to unceasing labor.—P. 161.
6. Fill the 5 quart jar from the 8 quart jar and pour 3 quarts from the 5 quart jar into the 3 quart jar. Then pour the 3 quarts back into the 8 quart jar. Pour the remaining 2 quarts from the 5 quart jar into the 3 quart jar. Now fill the 5 quart jar again. Since there are 2 quarts in the 3 quart jar, one additional quart will fill this jar. Pour enough wine from the 5 quart jar to fill the 3 quart jar. The 5 quart jar will then have 4 quarts remaining in it. Now pour the 3 quarts from the 3 quart jar into the 8 quart jar. This, together with the 1 quart remaining in the 8 quart jar, will make 4 quarts.—P. 162.
7. W. W. R. Ball, *op. cit.*—P. 163.
8. Other bases have been suggested. There is reason to believe that the Babylonians employed the base 60, and in more recent times, the use of the base 12 has been urged rather strongly.—P. 164.
9. Hall and Knight, *Higher Algebra.*—P. 165.

10. Arnold Dresden, *An Invitation to Mathematics*, New York: Henry Holt & Co., 1936.—P. 167.
11. W. Ahrens, *op. cit.*—P. 170.
12. W. W. R. Ball, *op. cit.*—P. 170.
13. (Making allowance for leap years.—Ed.)—P. 171.
14. See W. Ahrens, *op. cit.*, and Bouton, *Annals of Mathematics*, series 2, vol. III (1901–1902), pp. 35–39, for the mathematical proof of Nim.—P. 171.
15. One-twentieth of a cubit is about one inch.—P. 173.
16. The general rule for solution of all such problems may be found in P. G. Tait, *Collected Scientific Papers*, 1900.—P. 174.
17. Smith & Mikami, *A History of Japanese Mathematics*, p. 83.—P. 176.
18. Johnson & Story, *American Journal of Mathematics*, vol. 2 (1879).—P. 178.
19. Ahrens, *op. cit.*, volume 2.—P. 183.
20. There are also puzzles which though very amusing and deceptive, present no mathematical idea which has not been already considered—and such puzzles have, therefore, been omitted. We may, nevertheless, give three examples, chosen because they are so often solved incorrectly:
    (a) A glass is half-filled with wine, and another glass half-filled with water. From the first glass remove a teaspoonful of wine and pour it into the water. From the *mixture* take a teaspoonful and pour it into the wine. Is the quantity of wine in the water glass now greater or less than the quantity of water in the wine glass? To end all quarrels—they are the same.
    (b) The following puzzle troubled the delegates to a distinguished gathering of puzzle experts not long ago. A monkey hangs on one end of a rope which passes through a pulley and is balanced by a weight attached to the other end. The monkey decides to climb the rope. What happens? The astute puzzlers engaged in all sorts of futile conjectures and speculations, ranging from doubts as to whether the monkey could climb the rope, to rigorous "mathematical demonstrations" that he couldn't. (We yield to a shameful and probably superfluous urge to point out the solution—the weight rises, like the monkey!)
    (c) Imagine we have a piece of string 25,000 miles long, just long enough to exactly encircle the globe at the equator. We take the string and fit it snugly around, over oceans, deserts, and jungles. Unfortunately, when we have com-

pleted our task we find that in manufacturing the string there has been a slight mistake, for it is just a yard too long.

To overcome the error, we decide to tie the ends together and to distribute this 36 inches evenly over the entire 25,000 miles. Naturally (we imagine) this will never be noticed. How far do you think that the string will stand off from the ground at each point, merely by virtue of the fact that it is 36 inches too long?

The correct answer seems incredible, for the string will stand 6 inches from the earth over the entire 25,000 miles.

To make this seem more sensible you might ask yourself: In walking around the surface of the earth, how much further does your head travel than your feet?—P. 186.

21. Euclid's proof that there is an infinite number of primes is an elegant and concise demonstration. If P is any prime, a prime greater than P can always be found. Construct $P! + 1$. This number, obviously greater than P, is not divisible by P or any number less than P. There are only two alternatives: (1) It is not divisible at all; (2) It is divisible by a prime lying between P and $P! + 1$. But both of these alternatives prove the existence of a prime greater than P. Q.E.D.—P. 187.
22. Ball, *op. cit.*—P. 187.

Paradox Lost and Paradox Regained

How quaint the ways of paradox—
At common sense she gaily mocks.
—W. S. GILBERT

PERHAPS THE greatest paradox of all is that there are paradoxes in mathematics. We are not surprised to discover inconsistencies in the experimental sciences, which periodically undergo such revolutionary changes that although only a short time ago we believed ourselves descended from the gods, we now visit the zoo with the same friendly interest with which we call on distant relatives. Similarly, the fundamental and age-old distinction between matter and energy is vanishing, while relativity physics is shattering our basic concepts of time and space. Indeed, the testament of science is so continuously in a flux that the heresy of yesterday is the gospel of today and the fundamentalism of tomorrow. Paraphrasing Hamlet—what was once a paradox is one no longer, but may again become one. Yet, because mathematics builds on the old but does not discard it, because it is the most conservative of the sciences, because its theorems are deduced from postulates by the methods of logic, in spite of its having undergone revolutionary changes we do not suspect it of being a discipline capable of engendering paradoxes.

Nevertheless, there are three distinct types of paradoxes which do arise in mathematics. There are contradictory and absurd propositions, which arise from fallacious reasoning. There are theorems which seem strange and incredible, but which, because they are logically unassailable, must be accepted even though they transcend intuition and imagination. The third and most important class consists of those logical paradoxes which arise in connection with the theory of aggregates, and which have resulted in a re-examination of the foundations of mathematics. These logical paradoxes have created confusion and consternation among logicians and mathematicians and have raised problems concerning the nature of mathematics and logic which have not yet found a satisfactory solution.

PARADOXES—STRANGE BUT TRUE

This section will be devoted to apparently contradictory and absurd propositions which are nevertheless true.[1] Earlier, we examined the paradoxes of Zeno. Most of these were explained by means of infinite series and the transfinite mathematics of Cantor. There are yet others involving motion, but unlike Zeno's puzzles, they do not consist of logical demonstrations that motion

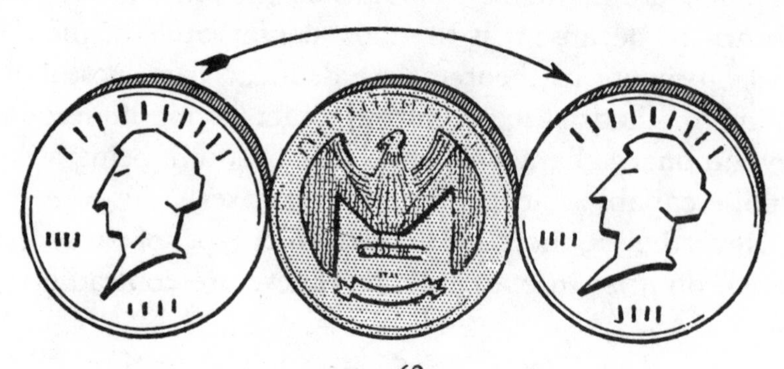

FIG. 62.

is impossible. However, they graphically illustrate how false our ideas about motion may be; how easily, for example, one may be deceived by the path of a moving object.

In Fig. 62, there are two identical coins. If we roll the coin at the left along half the circumference of the other, following the path indicated by the arrow, we might suspect that its final position, when it reaches the extreme right, should be with the head inverted and not in an upright position. That is to say, after we revolved the coin through a semicircle (half of its circumference), the head on the face of the coin, having started from an upright position, should now be upside down. If, however, we perform the experiment, we shall see that the final position will be as illustrated in Fig. 62, just as though the coin had been revolved once completely about its own circumference.

The following enigma is similar. The circle in Fig. 63 has made one complete revolution in rolling from *A* to *B*. The distance *AB* is therefore equal in length to the

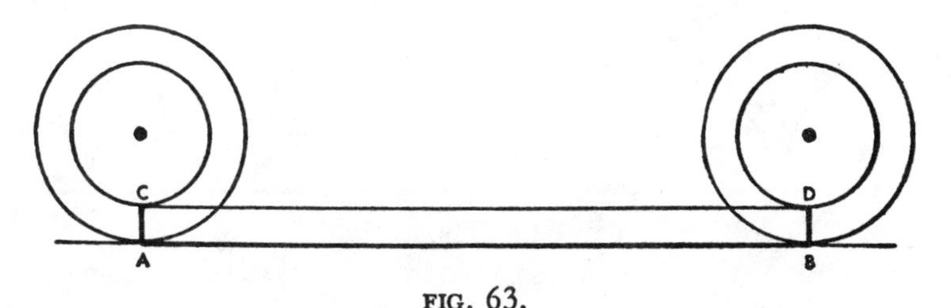

FIG. 63.

circumference of the circle. The smaller circle inside the larger one has also made one complete revolution in traversing the distance *CD*. Since the distance *CD* is equal to the distance *AB* and each distance is apparently equal to the circumference of the circle which has been unrolled upon it, we are confronted with the evident absurdity that the circumference of the small circle is equal to the circumference of the large circle.

In order to explain these paradoxes, and several others of a similar nature, we must turn our attention for a moment to a famous curve—the *cycloid.* (See Fig. 64.)

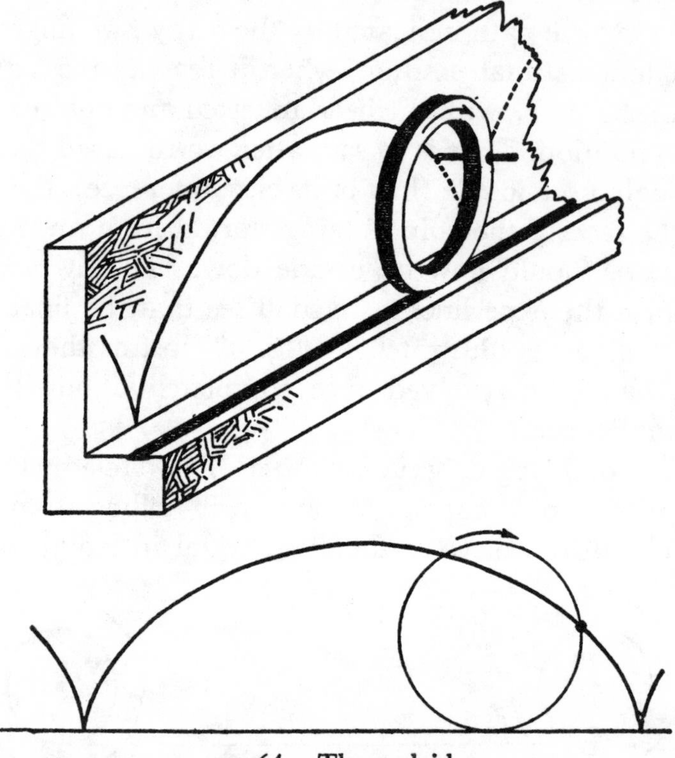

FIG. 64.—The cycloid.

The cycloid is the path traced by a fixed point on the circumference of a wheel as it rolls without slipping upon a fixed straight line.

In Fig. 65, as the wheel rolls on the line MN, the points A and B describe a cycloid. After the wheel has made half a revolution, the point A_1 is at A_3, and B_1 is at B_3. At this juncture, there is nothing to indicate that the point

A and the point B have not traveled throughout at the same speed, since it is evident that they have covered the same distance. But, if we examine the intermediary points A_2 and B_2 which show the respective positions of A and B after a *quarter*-turn of the wheel, it is clear that in the same time A has traveled a greater distance than B. This difference is compensated for in the second quarter turn in which B, traveling from B_2 to B_3, covers the same distance that A covered moving from A_1 to A_2; it is obvious that the distance along the curve from B_2 to B_3 is equal in length to the distance from A_1 to A_2. Hence in one-half revolution, both A and B have traversed exactly the same distance.

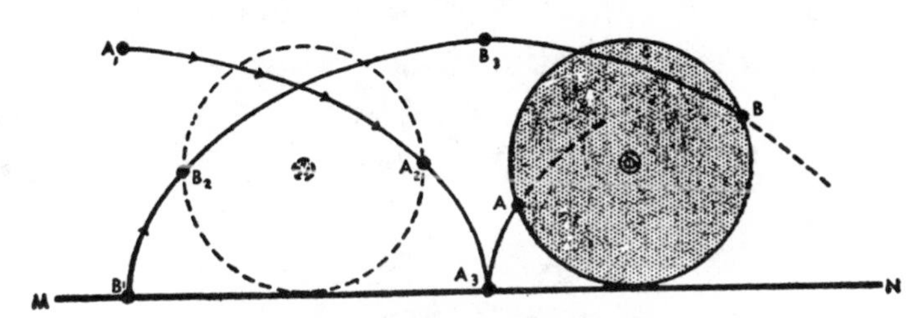

FIG. 65.—When the rolling wheel is in the dotted position it has completed one-quarter of a turn, and A has traveled from A_1 to A_2, but B only from B_1 to B_2. The shaded circle indicates the wheel has completed three-quarters of a revolution.

This strange behavior of the cycloid explains the fact that when a wheel is in motion, the part furthest from the ground, at any instant, actually moves along horizontally faster than the part in contact with the ground. It can be shown that as the point of a wheel in contact with the road starts moving up, it travels more and more quickly, reaching its maximum horizontal speed when its position is furthest from the ground.

Another interesting property of the cycloid was discovered by Galileo. It was pointed out in the chapter on Pie that the area of a circle could only be expressed with the aid of π, the transcendental number. Since the numerical value of π can only be approximated (although as closely as we please, by taking as many terms of the infinite series as we wish), the area of a circle can also only be expressed as an approximation. Remarkably enough, however, with the aid of a cycloid, we may *construct* an area *exactly* equal to the area of a given circle. Based upon the fact that the length of a cycloid, from cusp to cusp, is equal to four times the length of the diameter of the generating circle, it may be shown that the area bound by the portion of the cycloid between the two cusps and the straight line joining the cusps, is equal to three times the area of the circle. From which it follows that the enclosed space (shaded in Fig. 66) on *either* side of the circle in the center is *exactly* equal to the area of the circle itself.

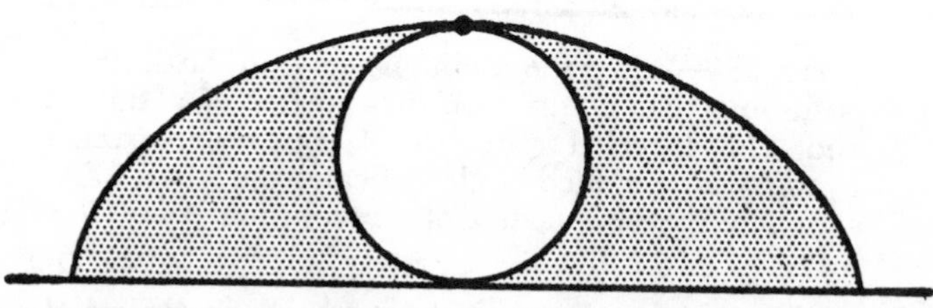

FIG. 66.—When the rolling circle is in the indicated position, the shaded areas on each side are exactly equal to the area of the circle.

The paradox resulting from the pseudo-proof that the circumference of a small circle is equal to that of a larger circle can be explained with the aid of another member of the cycloid family—the prolate cycloid (Fig. 67).

An inner point of a wheel which rolls on a straight line

describes the prolate cycloid. Thus, a point on the circumference of a smaller circle concentric with a larger one will generate this curve. The small circle in Fig. 63 makes only one complete revolution in moving from *C* to *D* and a point on the circumference of this circle will describe a prolate cycloid. However, by comparing the prolate cycloid with the cycloid, we observe that the

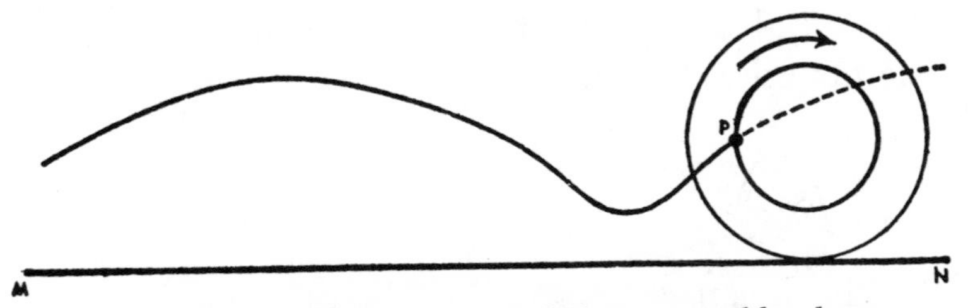

FIG. 67.—The prolate cycloid is generated by the point *P* on the smaller circle as the larger circle rolls along the line *MN*.

small circle would not cover the distance *CD* merely by making one revolution as the large circle does. Part of the distance is covered by the circle while it is unrolling, but simultaneously, it is being carried forward by the large circle as this moves from *A* to *B*. This may be seen even more clearly if we regard the *center* of the large circle in Fig. 63. The center of a circle, being a mathematical point and having no dimensions, does not revolve at all, but is carried the entire distance from *A* to *B* by the wheel.

With regard to the problems arising from a wheel rolling on a straight line, we have discussed the trajectory (path) of a point on the circumference of the wheel and found this path to be a cycloid, and we have considered the curve traced by a point on the inside of the wheel and discovered the prolate cycloid. In addition, it is interesting to mention the path traced by a point *outside* of the circumference of a wheel, such as the outermost point of the flanged wheels used on railway trains. Such a point is not actually in contact with the rail upon which the wheel is revolving. The curve generated is called a curtate cycloid (Fig. 68) and explains the curious paradox that, at any instant of time, a railroad train never moves entirely in the direction in which the engine is pulling. There are always parts of the train which are traveling in the opposite direction!*

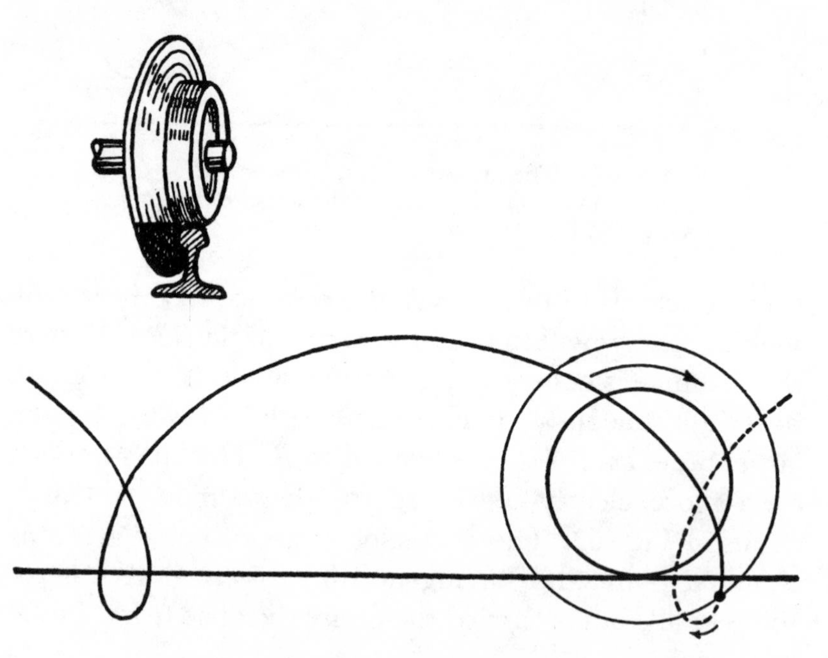

FIG. 68.—The Curtate Cycloid.

A point on the flange of a moving railway wheel generates this curve. The part of a railway train which moves *backward* when the train moves forward is the shaded portion of the wheel.

Among the innovations in mathematics of the last quarter-century, none overshadows in importance the development of the theory of point sets and the theory of functions of a real variable. Based entirely upon the new methods of mathematical analysis, a greater rigor and generality in geometry was achieved than could have been imagined had science been developed entirely by intuitive means. It was found that all conventional geometrical ideas could be redefined with increased accuracy by drawing upon the theory of aggregates and the powerful new tools of analysis. In rubber-sheet geometry, as we shall see, curves are defined in such a way as to eliminate every naïve appeal to intuition and experience. A simple closed curve is defined as a set of points possessing the property that it divides the plane into exactly two regions: an *inside* and an *outside*, where *inside* and *outside* are precisely formulated by analytical methods without reference to our customary notions of space. By just such means, figures far more complex than had ever before been studied were developed and investigated. Indeed, although analytical geometry is limited to contours which can be described by algebraic equations whose variables are the co-ordinates of the points of the configuration, the new analysis made possible the study of forms which *cannot* be described by *any* algebraic equation. Some of these we shall encounter in the section on Pathological Curves.

Extended studies were also undertaken of certain classes of points—like the points in space—and the notion of dimensionality was freshly re-examined. In connection with this study, one of the great accomplishments of recent years has been to assign to each configuration a number: 0, 1, 2, or 3, to denote its dimensionality. The established belief had been that this was a simple and obvious matter which did not require mathematical analysis and could be solved intuitively. Thus, a point would be said to have zero dimension, a line or a curve—one dimension, a plane or a surface—two dimensions, and a solid—three dimensions. It must be conceded that the problem of determining whether an object has 0, 1, 2, or 3 dimensions does not look very formidable. However, one remarkable paradox which was uncovered is sufficient in itself to show that this is not the case at all and that our intuitive ideas about dimensionality, as well as area, are not only lacking in precision, but are often wholly misleading.

The paradox appeared in trying to ascertain whether a number (called a measure) could be uniquely assigned to every figure in the plane so that the following three conditions would be satisfied: (1) the word "congruent" being used in the same sense in which it was learned in elementary geometry,[2] two congruent figures were to have the same measure; (2) if a figure were divided into two parts, the sum of the measures assigned to each of the two parts was to be exactly equal to the measure assigned to the original figure; (3) as a model for determining the method of assigning a measure to each figure in the plane, it was agreed that the measure 1 should be assigned to the square whose side has a length of one unit.

What is this concept of measure? From the foregoing, it would seem to follow that the *measure* to be assigned to each figure in the plane is nothing more than the *area* of that figure. In other words, the problem is to ascertain whether the *area* of *every figure* in the plane, regardless of its complexity, can be uniquely determined. It need hardly be pointed out that this was intended as a general and theoretical exercise and not as the vast and obviously impossible undertaking of actually measuring every conceivable figure. The problem was to be considered solved if a theoretical proof were given that *every figure* could be assigned a unique measure. But it should be noted that the principal aim was to keep this investigation free from the traditional concepts of classical geometry—the notion of area understood in the old way was taboo, and the customary methods of determining it specifically excluded; the approach was to be *analytic* (by means of point sets), rather than geometric. Adhering to just such restrictions, it was proved that no matter how complex a figure is, no matter how many times the boundary crosses and recrosses itself, a unique measure can be assigned to it.

Then came the débacle. For the amazing fact was uncovered that the same problem, when extended to surfaces, was not only unsolvable but led to the most stunning paradoxes. Indeed, the very same methods which had been so fruitful in the investigations in the plane, when applied to the surface of a sphere proved inadequate to determine a unique measure.

Does this really mean that the area of the surface of a sphere cannot be uniquely determined? Does not the conventional formula $4\pi r^2$ give correctly the area of the surface of a sphere? Unfortunately, we cannot undertake to answer these questions in detail, for to do so would carry us far afield and require much technical knowledge. We admit that the area of a spherical surface as determined by the old classical methods is $4\pi r^2$. But the old methods were lacking in generality; they were found to be inadequate to determine the area of complex figures; furthermore, we already gave warning that the naïve concept of area was deliberately to be omitted from the measuring attempt. While the advance in function theory and the new methods of analysis did overcome some of these difficulties, they also introduced new problems closely connected with the infinite, and as mathematicians have long realized, the presence of that concept is by no means an unmixed blessing. Though it has enabled mathematics to make great strides forward, these have always been in the shadow of uncertainty. One may continue to employ such formulae as $4\pi r^2$, for the very good reason that they work; but if one wishes to keep pace with the bold and restless mathematical spirit, one is faced with the comfortless alternatives of abandoning logic to preserve the classical concepts, or of accepting the paradoxical results of the new analysis and casting horse sense to the winds.

The conditions for assigning a measure to a surface are similar to the conditions for assigning a measure to figures in the plane: (1) The same measure shall be assigned to congruent surfaces; (2) The sum of the measures assigned to each of two component parts of a surface shall be equal to the measure assigned to the original surface; (3) If S denotes the entire surface of a a sphere of radius r, the measure assigned to S shall be $4\pi r^2$.

The German mathematician Hausdorff showed that this problem is insoluble, that a measure *cannot* uniquely be assigned to the portions of the surface of a sphere so that the above conditions will be satisfied. He showed that if the surface of a sphere were divided into three separate and distinct parts: A, B, C, so that A is congruent to B and B is congruent to C, a strange paradox arises which is strongly reminiscent of, and indeed, related to some of the paradoxes of transfinite arithmetic. For Hausdorff proved that not only is A congruent to C (as might be expected), *but also* that A is congruent to $B + C$. What are the implications of this startling result?

If a measure is assigned to A, the same measure must be assigned to B and to C, because A is congruent to B, B is congruent to C and A is congruent to C. But, on the other hand, since A is congruent to $B + C$, the measure assigned to A would also have to be equal to the sum of the measures assigned to B and C. Obviously, such a relationship could only hold if the measures assigned to A, B, and C were all equal to 0. But that is impossible by condition (3), according to which the sum of the measures assigned to the parts of the surface of a sphere must be equal to $4\pi r^2$. How then is it possible to assign a measure?

From a slightly different viewpoint, we see that if A, B, and C are congruent to each other and together make up the surface of the entire sphere, the measure of any one of them must be the measure of one-third of the surface of the entire sphere. But if A is not only congruent to B and C, but also to $B + C$ (as Hausdorff has shown), the measure assigned to A and the measure assigned to $B + C$ must each be equal to half the surface of the sphere. Thus, whichever way we look at it, assigning measures to portions of the surface of a sphere involves us in a hopeless contradiction.

Two distinguished Polish mathematicians, Banach and Tarski, have extended the implications of Hausdorff's paradoxical theorem to three-dimensional space, with results so astounding and unbelievable that their like may be found nowhere else in the whole of mathematics. And the conclusions, though rigorous and unimpeachable, are almost as incredible for the mathematician as for the layman.

Imagine two bodies in three-dimensional space: one very large, like the sun; the other very small, like a pea. Denote the sun by S and the pea by S'. Remember now that we are referring not to the surfaces of these two spherical objects, but to the *entire solid spheres of both the sun and the pea*. The theorem of Banach and Tarski holds that the following operation can theoretically be carried out:

Divide the sun S into a great many small parts. Each part is to be separate and distinct and the totality of the parts is to be finite in number. They may then be designated by $s_1, s_2, s_3, \ldots s_n$, and together these small parts will make up the entire sphere S. Similarly, S'—the pea—is to be divided into an equal number of mutually exclusive parts—$s'_1, s'_2, s'_3, \ldots s'_n$, which together will make up the pea. Then the proposition goes on to say that if the sun and the pea have been cut up in a suitable manner, so that the little portion s_1 of the sun is congruent to the little portion s'_1 of the pea, s_2 congruent to s'_2, s_3 congruent to s'_3, up to s_n congruent to s'_n, this process will exhaust not only all the little portions of the pea, *but all the tiny portions of the sun as well.*

In other words, the sun and the pea may both be divided into a finite number of disjoint parts *so that every single part of one is congruent to a unique part of the other, and so that after each small portion of the pea has been matched with a small portion of the sun, no portion of the sun will be left over.**

* We recognize this, of course, to be a simple one-to-one correspondence between the elements of one set which make up the sun, and the elements of another set which make up the pea. The paradox lies in the fact that each element is matched with one which is completely congruent to it (at the risk of repeating, congruent means identical in size and shape) and that there are enough elements in the

To express this giant bombshell in terms of a small firecracker: *There is a way of dividing a sphere as large as the sun into separate parts, so that no two parts will have any points in common, and yet without compressing or distorting any part, the whole sun may at one time be fitted snugly into one's vest pocket.* Furthermore the pea may have its component parts so rearranged that without expansion or distortion, no two parts having any points in common, *they will fill the entire universe solidly, no vacant space remaining either in the interior of the pea, or in the universe.*

Surely no fairy tale, no fantasy of the Arabian nights, no fevered dream can match this theorem of hard, mathematical logic. Although the theorems of Hausdorff, Banach, and Tarski cannot, at the present time, be put to any practical use, not even by those who hope to learn how to pack their overflowing belongings into a week-end grip, they stand as a magnificent challenge to imagination and as a tribute to mathematical conception.[3]

*

As distinguished from the paradoxes just considered, there are those which are more properly referred to as mathematical fallacies. They arise in both arithmetic and geometry and are to be found sometimes, although not often, even in the higher branches of mathematics as, for instance, in the calculus or in infinite series. Most mathematical fallacies are too trivial to deserve attention; nevertheless, the subject is entitled to some consideration because, apart from its amusing aspect, it shows how a chain of mathematical reasoning may be entirely vitiated by one fallacious step.

set making up the pea to match exactly the elements which make up the sun.

ARITHMETIC FALLACIES

I. A proof that 1 is equal to 2 is familiar to most of us. Such a proof may be extended to show that *any* two numbers or expressions are equal. The error common to all such frauds lies in dividing by zero, an operation strictly forbidden. For the fundamental rules of arithmetic demand that every arithmetic process (addition, subtraction, multiplication, division, evolution, involution) should yield a unique result. Obviously, this requirement is essential, for the operations of arithmetic would have little value, or meaning, if the results were ambiguous. If $1 + 1$ were equal to 2 *or* 3; if 4×7 were equal to 28 *or* 82; if $7 \div 2$ were equal to 3 *or* $3\frac{1}{2}$, mathematics would be the Mad Hatter of the sciences. Like fortunetelling or phrenology, it would be a suitable subject to exploit at a boardwalk concession at Coney Island.

Since the results of the operation of division are to be unique, division by 0 *must* be excluded, for the result of this operation is anything that you may desire. In general, division is so defined that if a, b, and c are three numbers, $a \div b = c$, *only when* $c \times b = a$. From this definition, what is the result of $5 \div 0$? It cannot be any number from zero to infinity, for no number when multiplied by 0 will be equal to 5. Thus $5 \div 0$ is meaningless. And even $5 \div 0 = 5 \div 0$ is a meaningless expression.

Of course, fallacies resulting from division by 0 are rarely presented in so simple a form that they may be detected at a glance. The following example illustrates how paradoxes arise whenever we divide by an expression, the value of which is 0:

Assume $A + B = C$, and assume $A = 3$ and $B = 2$.

Multiply both sides of the equation $A + B = C$ by $(A + B)$.

We obtain $A^2 + 2AB + B^2 = C(A + B)$.

Rearranging the terms, we have
$$A^2 + AB - AC = -AB - B^2 + BC.$$

Factoring out $(A + B - C)$, we have
$$A(A + B - C) = -B(+A + B - C).$$

Dividing both sides by $(A + B - C)$, that is, dividing by zero, we get $A = -B$, or $A + B = 0$, which is evidently absurd.

II. In extracting square roots, it is necessary to remember the algebraic rule that the square root of a positive number is equal to *both* a negative and a positive number. Thus, the square root of 4 is -2 as well as $+2$ (which may be written $\sqrt{4} = \pm 2$), and the square root of 100 is equal to $+ 10$ and $- 10$ (or, $\sqrt{100} = \pm 10$). Failure to observe this rule may generate the following contradiction: [4]

(a) $(n + 1)^2 = n^2 + 2n + 1$

(b) $(n + 1)^2 - (2n + 1) = n^2$

(c) Subtracting $n(2n + 1)$ from both sides and factoring, we have

(d) $(n + 1)^2 - (n + 1)(2n + 1) = n^2 - n(2n + 1)$

(e) Adding $\frac{1}{4}(2n + 1)^2$ to both sides of (d) yields
$$(n + 1)^2 - (n + 1)(2n + 1) + \tfrac{1}{4}(2n + 1)^2 = n^2 - n(2n + 1) + \tfrac{1}{4}(2n + 1)^2.$$

This may be written:

(f) $[(n + 1) - \frac{1}{2}(2n + 1)]^2 = [n - \frac{1}{2}(2n + 1)]^2$.

Taking square roots of both sides,

(g) $n + 1 - \frac{1}{2}(2n + 1) = n - \frac{1}{2}(2n + 1)$

and, therefore,

(h) $n = n + 1.$

III. The following arithmetic fallacy the reader may disentangle for himself: [5]

(1) $\sqrt{a} \times \sqrt{b} = \sqrt{a \times b}$ ...... true
(2) $\sqrt{-1} \times \sqrt{-1} = \sqrt{(-1) \times (-1)}$ ...... true
(3) Therefore, $(\sqrt{-1})^2 = \sqrt{1}$; i.e., $-1 = 1$ ......?

IV. A paradox which cannot be solved by the use of elementary mathematics is the following: Assume that $\log(-1) = x$. Then, by the law of logs,

$$\log(-1)^2 = 2 \times \log(-1) = 2x.$$

But, on the other hand, $\log(-1)^2 = \log(1)$, which is equal to 0. Therefore, $2 \cdot x = 0$. Therefore, $\log(-1) = 0$, which is obviously not the case. The explanation lies in the fact that the function that represents the log of a negative, or complex, number is not *single-valued*, but is *many-valued*. That is to say, if we were to make the usual functional table for the logarithm of negative and complex numbers, there would be an *infinitude* of values corresponding to each number.[6]

V. The infinite in mathematics is always unruly unless it is properly treated. Instances of this were found in the development of the theory of aggregates and further examples will be seen in the logical paradoxes. One instance is appropriate here.

Just as transfinite arithmetic has its own laws differing from those of finite arithmetic, special rules are required for operating with *infinite series.* Ignorance of these rules, or failure to observe them brings about inconsistencies. For instance, consider the series equivalent to the natural logarithm of 2:

$$\text{Log } 2 = 1 - \tfrac{1}{2} + \tfrac{1}{3} - \tfrac{1}{4} + \tfrac{1}{5} - \tfrac{1}{6} \ldots$$

If we rearrange these terms as we would be prompted to do in finite arithmetic, we obtain:

$$\text{Log } 2 = (1 + \tfrac{1}{3} + \tfrac{1}{5} + \tfrac{1}{7} \ldots) - (\tfrac{1}{2} + \tfrac{1}{4} + \tfrac{1}{6} + \tfrac{1}{8} \ldots)$$

Thus,

$$\begin{aligned} \text{Log } 2 &= \{(1 + \tfrac{1}{3} + \tfrac{1}{5} + \tfrac{1}{7} \ldots) + (\tfrac{1}{2} + \tfrac{1}{4} + \tfrac{1}{6} + \tfrac{1}{8})\} \\ &\qquad -2(\tfrac{1}{2} + \tfrac{1}{4} + \tfrac{1}{6} + \tfrac{1}{8} \ldots) \\ &= \{1 + \tfrac{1}{2} + \tfrac{1}{3} + \tfrac{1}{4} + \tfrac{1}{5} + \ldots\} \\ &\qquad - \{1 + \tfrac{1}{2} + \tfrac{1}{3} + \tfrac{1}{4} + \tfrac{1}{5} + \ldots\} \\ &= 0 \end{aligned}$$

Therefore, $\log 2 = 0$.

On the other hand,

$$\log 2 = 1 - \tfrac{1}{2} + \tfrac{1}{3} - \tfrac{1}{4} + \tfrac{1}{5} - \tfrac{1}{6} \ldots = 0.69315,$$

an answer that can be obtained from any logarithmic table.

Rearranging the terms in a slightly different way:

$$\begin{aligned} \log 2 &= 1 + \tfrac{1}{3} - \tfrac{1}{2} + \tfrac{1}{5} + \tfrac{1}{7} - \tfrac{1}{4} + \tfrac{1}{9} + \tfrac{1}{11} - \tfrac{1}{6} \ldots \\ &= \tfrac{3}{2} \times 0.69315 \text{ or, in other words,} \end{aligned}$$

$$\log 2 = \tfrac{3}{2} \times \log 2.$$

A famous series which had troubled Liebniz is the beguilingly simple: $+1 - 1 + 1 - 1 + 1 - 1 + 1 \ldots$ By pairing the terms differently, a variety of results is obtained; for example: $(1-1) + (1-1) + (1-1) + \ldots = 0$, but $1 - (1-1) + (1-1) \ldots = 1$

GEOMETRIC FALLACIES

Optical illusions concerning geometric figures account for many deceptions. We confine our attention to fallacies which do not arise from physiological limitations,[7] but from errors in mathematical argument. A well-known geometric "proof" is that every triangle is isosceles. It assumes that the line bisecting an angle of the triangle and the line which is the perpendicular bisector of the side opposite this angle intersect at a point inside the triangle.

The following is a similarly fallacious proof, namely, that a right angle is equal to an angle greater than a right angle.[8]

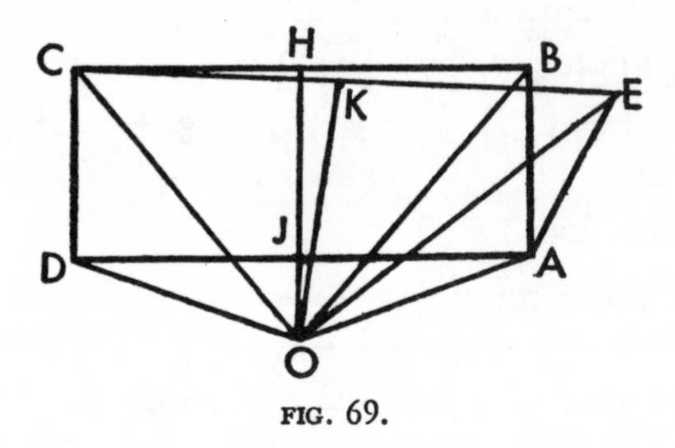

FIG. 69.

In Fig. 69, *ABCD* is a rectangle. If *H* is the midpoint of *CB*, through *H* draw a line at right angles to *CB*, which will bisect *DA* at *J* and be perpendicular to it. From *A* draw the line *AE* outside of the rectangle and equal to *AB* and *DC*. Connect *C* and *E*, and let *K* be the midpoint of this line. Through *K* construct a perpendicular to *CE*. *CB* and *CE* not being parallel, the lines through *H* and *K* will meet at a point *O*. Join *OA*, *OE*, *OB*, *OD* and *OC*. It will be made clear that the triangle *ODC* and *OAE* are equal in all respects. Since *KO* is the perpendicular bisector of *CE* and thus any point on *KO* is equidistant from *C* and *E*, *OC* is equal to *OE*. Similarly, since *HO* is the perpendicular bisector of *CB* and *DA*, *OD* equals *OA*.

As *AE* was constructed to equal *DC*, the three sides of the triangle *ODC* are equal respectively to the three sides of the triangle *OAE*. Hence, the two triangles are equal, and therefore, the angle *ODC* is equal to the angle *OAE*. But angle *ODA* is equal to angle *OAD*, because side *AO* is equal to side *OD* in the triangle *OAD* and the base angles of the isosceles triangle are equal. Therefore, the angle *JDC*, which is equal to the difference of *ODC* and *ODJ*, equals *JAE*, which is the difference between *OAE* and *OAJ*. But the angle *JDC* is a right angle, whereas the angle *JAE* is greater than a right angle, and hence the result is contradictory. Can you find the flaw? Hint: Try drawing the figure exactly.

LOGICAL PARADOXES

Like folk tales and legends, the logical paradoxes had their forerunners in ancient times. Having occupied themselves with philosophy and with the foundations of logic, the Greeks formulated some of the logical conundrums which, in recent times, have returned to plague mathematicians and philosophers. The Sophists made a specialty of posers to bewilder and confuse their opponents in debate, but most of them rested on sloppy thinking and dialectical tricks. Aristotle demolished them when he laid down the foundations of classical logic—a science which has outworn and outlasted all the philosophical systems of antiquity, and which, for the most part, is perfectly valid today.

But there were troublesome riddles that stubbornly resisted unraveling.[9] Most of them are caused by what is known as "the vicious circle fallacy," which is "due to neglecting the fundamental principle that what involves the whole of a given totality cannot itself be a member of the totality."[10] Simple instances of this are those pontifical phrases, familiar to everyone, which seem to have a great deal of meaning, but actually have none, such as "never say never," or "every rule has exceptions," or, "every generality is false." We shall consider a few of the more advanced logical paradoxes involving the same basic fallacy, and then discuss their importance from the mathematician's point of view.

(A) Poaching on the hunting preserves of a powerful prince was punishable by death, but the prince further decreed that anyone caught poaching was to be given the privilege of deciding whether he should be hanged or beheaded. The culprit was permitted to make a statement—if it were false, he was to be hanged; if it were true, he was to be beheaded. One logical rogue availed himself of this dubious prerogative—to be hanged if he didn't and to be beheaded if he did—by stating: "I shall be hanged." Here was a dilemma not anticipated. For, as the poacher put it, "If you now hang me, you break the laws made by the prince, for my statement is true, and I ought to be beheaded; but if you behead me, you are also breaking the laws, for then what I said was false and I should, therefore, be hanged." As in Frank Stockton's story of the lady and the tiger, the ending is up to you. However, the poacher probably fared no worse at the hands of the executioner than he would have at the hands of a philosopher, for until this century philosophers had little time to waste on such childish riddles—especially those they could not solve.

(B) The village barber shaves everyone in the village who does not shave himself. But this principle soon involves him in a dialectical plight analogous to that of the executioner. Shall he shave himself? If he does, then he is shaving someone who shaves himself and breaks his own rule. If he does not, besides remaining unshaven, he also breaks his rule by failing to shave a person in the village who does not shave himself.

(C) Consider the fact that every integer may be expressed in the English language without the use of symbols. Thus, (a) 1400 may be written as one thousand, four hundred, or (b) 1769823 as one million, seven hundred and sixty-nine thousand, eight hundred and twenty-three. It is evident that certain numbers require more syllables than others; in general, the larger the integer, the more syllables needed to express it. Thus, (a) requires 6 syllables, and (b) 21. Now, it may be established that certain numbers will require 19 syllables or less, while others will require more than 19 syllables. Furthermore, it is not difficult to show that among those integers requiring exactly 19 syllables to be expressed in the English language, there must be a smallest one. Now, "it is easy to see"[11] that "*The least integer not nameable in fewer than nineteen syllables*" is a phrase which must denote the specific number, 111777. But the italicized expression above is itself an unambiguous means of denoting the smallest integer expressible in nineteen syllables in the English language. Yet, the italicized statement has only eighteen syllables! Thus, we have a contradiction, for the least integer expressible in nineteen syllables can be expressed in eighteen syllables.

(D) The simplest form of the logical paradox which arises from the indiscriminate use of the word *all* may be seen in Fig. 70.

What is to be said about the statement numbered 3? 1 and 2 are false, but 3 is both a wolf dressed like a sheep and a sheep dressed like a wolf. It is neither the one thing nor the other: It is neither false nor true.

An elaboration appears in the famous paradox of Russell about the class of all classes not members of them-

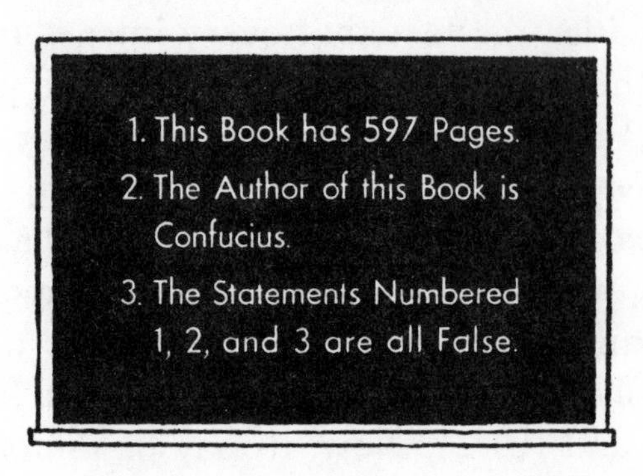

FIG. 70.

selves. The thread of the argument is somewhat elusive and will repay careful attention:

(E) Using the word class in the customary sense, we can say that there are classes made up of tables, books, peoples, numbers, functions, ideas, etc. The class, for instance, of all the Presidents of the United States has for its members every person, living or dead, who was ever President of the United States. Everything in the world other than a person who was or is a President of the United States, including the *concept* of the class itself is *not* a member of this class. This then, is an example of a class which is *not* a member of itself. Likewise, the class of all members of the Gestapo, or German secret police, which contains some, but not all, of the scoundrels in Germany; or the class of all geometric figures in a plane bounded by straight lines; or the class of all integers from one to four thousand inclusive, have for members, the things described, but the classes are not members of themselves.

Now, if we consider a class as a *concept*, then the *class* of *all concepts* in the world is itself a concept, and thus is a class which is a member of itself. Again, the class of all ideas brought to the attention of the reader in this book is a class which contains itself as a member, since in mentioning this class, it is an idea which we bring to the attention of the reader. Bearing this distinction in mind, we may divide all classes into two types: Those which are members of themselves and those which are *not* members of themselves. Indeed, we may form a class which is composed of *all those classes which* are *not* members of themselves (note the dangerous use of the word "all"). The question is presented: Is this class (composed of those classes which are not members of themselves) a member of itself, or not? Either an affirmative or a negative answer involves us in a hopeless contradiction. If the class in question *is* a member of itself, it ought not be by definition, for it should contain only those classes which are not members of themselves. But if it is *not* a member of itself, it ought to be a member of itself, for the same reason.

It cannot be too strongly emphasized that the logical paradoxes are not idle or foolish tricks. They were not included in this volume to make the reader laugh, unless it be at the limitations of logic. The paradoxes are like the fables of La Fontaine which were dressed up to look like innocent stories about fox and grapes, pebbles and frogs. For just as all ethical and moral concepts were skillfully woven into their fabric, so all of logic and mathematics, of philosophy and speculative thought, is interwoven with the fate of these little jokes.

Modern mathematics, in attempting to avoid the paradoxes of the theory of aggregates, was squarely faced with the alternatives of adopting annihilating skepticism in regard to all mathematical reasoning, or of reconsidering and reconstructing the foundations of mathematics as well as logic. It should be clear that if paradoxes can arise from apparently legitimate reasoning about the theory of aggregates, they may arise *anywhere* in mathematics. Thus, even if mathematics could be reduced to logic, as Frege and Russell had hoped, what purpose would be served if logic itself were insecure? In proposing their "Theory of Types" Whitehead and Russell, in the *Principia Mathematica*, succeeded in avoiding the contradictions by a formal device. Propositions which were *grammatically* correct but contradictory, were branded as meaningless. Furthermore, a principle was formulated which specifically states what form a proposition must take to be meaningful; but this solved only half the difficulty, for although the contradictions could be recognized, the arguments leading to the contradictions could not be invalidated without affecting certain accepted portions of mathematics. To overcome this difficulty, Whitehead and Russell postulated the *axiom of reducibility* which, however, is too technical to be considered here. But the fact remains that the axiom is not acceptable to the great majority of mathematicians and that the logical paradoxes, having divided mathematicians into factions unalterably opposed to each other, have still to be disposed of.[12]

*

It has been emphasized throughout that the mathematician strives always to put his theorems in the most general form. In this respect, the aims of the mathematician and the logician are identical—to formulate propositions and theorems of the form: if *A* is true, *B* is true, where *A* and *B* embrace much more than merely cabbages and kings. But if this is a high aim, it is also dangerous, in the same way that the concept of the infinite is dangerous. When the mathematician says that such and such a proposition is true of one thing, it may be interesting, and it is surely safe. But when he tries to extend his proposition to *everything*, though it is much more interesting, it is also much more dangerous. In the transition from *one* to *all*, from the specific to the general, mathematics has made its greatest progress, and suffered its most serious setbacks, of which the logical paradoxes constitute the most important part. For, if mathematics is to advance securely and confidently it must first set its affairs in order at home.

FOOTNOTES

1. Strictly speaking, mathematical propositions are neither true nor false; they are merely implied by the axioms and postulates which we assume. If we accept these premises and employ legitimate logical arguments, we obtain legitimate propositions. The postulates are not characterized by being true or false; we simply agree to abide by them. But we have used the word *true* without any of its philosophical implications to refer unambiguously to propositions logically deduced from commonly accepted axioms. —P. 194.
2. Two point sets (configurations) are called congruent if, to every pair of points P, Q of one set, there uniquely corresponds a pair of points P', Q' of the other set, such that the distance between P' and Q' equals the distance between P and Q.—P. 202.
3. In the version given of the theorems of Hausdorff, Banach, and Tarski, we have made liberal use of the lucid explanation given by Karl Menger in his lecture: "Is the Squaring of the Circle Solvable?" in *Alte Probleme—Neue Lösungen*, Vienna: Deuticke, 1934.—P. 207.
4. Lietzmann, *Lustiges und Merkwürdiges von Zahlen und Formen*, Breslau: Ferd. Hirt, 1930.—P. 209.
5. Ball, *op. cit.*—P. 210.
6. Weismann, *Einführung in das mathematische Denken*, Vienna, 1937. —P. 210.

7. The following optical illusions, while not properly part of a book on mathematics, may be of some interest—at least to the imagination.—P. 212.

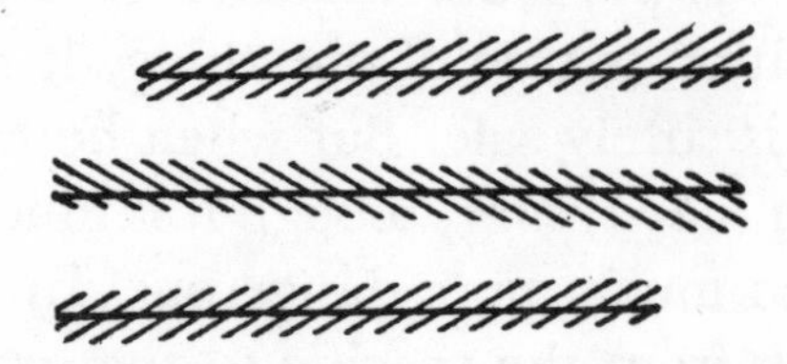

FIG. 71.—Are the three horizontal lines parallel?

FIG. 72.—The white square is of course larger than the black. Or is it smaller?

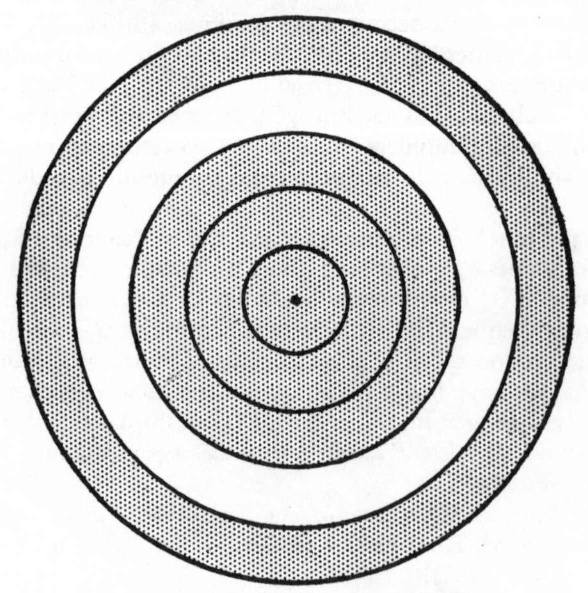

FIG. 73.—The two shaded regions have the same area.

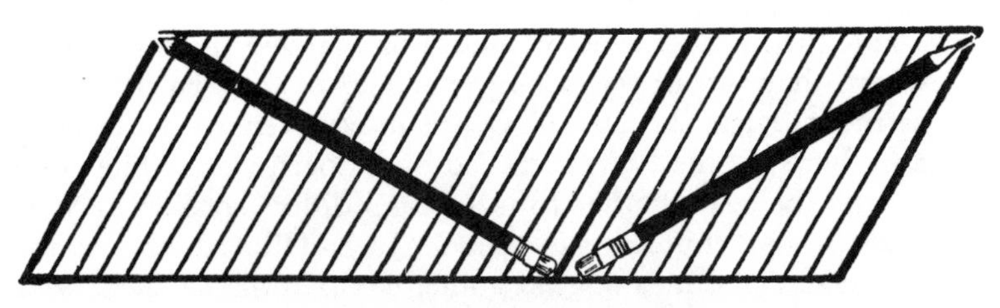

FIG. 74.—Which of the two pencils is longer? Measure them and find out.

FIG. 75.—What do you see? Now look again.

8. Ball, *op. cit.*—P. 212.
9. For instance, the riddle of the Epimenides concerning the Cretan who says that all Cretans are liars (Chapter II).—P. 213.
10. Ramsay, Frank Plumpton. Articles on "Mathematics," and "Logic," *Encyclopedia Britannica*, 13th edition.—P. 214.
11. This expression may, perhaps, be taken in the sense in which Laplace employed it. When he wrote his monumental *Mécanique Céleste*, he made abundant use of the expression, "It is easy to see" often prefixing it to a mathematical formula which he had arrived at only after months of labor. The result was that scientists who read his work almost invariably recognized the expression as a danger signal that there was very rough going ahead.—P. 215.
12. As was pointed out in the chapter on the googol, there are the followers of Russell who are satisfied with the theory of types and the axiom of reducibility; there are the Intuitionists, led by Brouwer and Weyl, who reject the axiom and whose skepticism

about the infinite in mathematics has carried them to the point where they would reject large portions of modern mathematics as meaningless, because they are interwoven with the infinite; and there are the Formalists, led by Hilbert, who, while opposed to the beliefs of the Intuitionists, differ considerably from Russell and the Logistic school. It is Hilbert who considers mathematics a meaningless game, comparable to chess, and he has created a subject of metamathematics which has for its program the discussion of this meaningless game and its axioms. —P. 218.

Chance and Chanceability

There once was a brainy baboon
Who always breathed down a bassoon,
For he said, "It appears
That in billions of years
I shall certainly hit on a tune."

—SIR ARTHUR EDDINGTON

HOLMES HAD been seated for some hours in silence with his long, thin back curved over a chemical vessel in which he was brewing a particularly malodorous product. His head was sunk upon his breast, and he looked from my point of view like a strange, lank bird, with dull gray plumage and a black topknot.

"So Watson," said he, suddenly, "you do not propose to invest in South African securities?"

I gave a start of astonishment. Accustomed as I was to Holmes' curious faculties, this sudden intrusion into my most intimate thoughts was utterly inexplicable.

"How on earth do you know that?" I asked.

He wheeled round upon his stool with a steaming test tube in his hand, and a gleam of amusement in his deep-set eyes.

"Now, Watson, confess yourself utterly taken aback," said he.

"I am."

"I ought to make you sign a paper to that effect."

"Why?"

"Because in five minutes you will say that it is all so absurdly simple."

"I am sure that I shall say nothing of the kind "

"You see, my dear Watson"—he propped his test tube in

the rack, and began to lecture with the air of a professor addressing his class—"It's not really difficult to construct a series of inferences, each dependent upon its predecessor and each simple in itself. If, after doing so, one simply knocks out all the central inferences and presents one's audience with the starting point and the conclusion, one may produce a startling, though possibly a meretricious, effect. Now, it was not really difficult, by an inspection of the groove between your left forefinger and thumb, to feel sure that you did *not* propose to invest your small capital in the gold fields."

"I see no connection."

"Very likely not; but I can quickly show you a close connection. Here are the missing links of the very simple chain. 1. You had chalk between your left finger and thumb when you returned from the club last night. 2. You put chalk there when you play billiards, to steady the cue. 3. You never play billiards except with Thurston. 4. You told me, four weeks ago, that Thurston had an option on some South African property which would expire in a month, and which he desired you to share with him. 5. Your check book is locked in my drawer and you have not asked for the key. 6. You do not propose to invest your money in this manner."

"How absurdly simple!" I cried.

"Quite so!" said he, a little nettled. "Every problem becomes very childish when once it is explained to you. . . . " [1]

This excerpt from the adventures of Mr. Sherlock Holmes, distinguished consulting detective, is an excellent caricature of *reasoning by probable inference*. Such a method of reasoning, while it resembles the formal procedure of the syllogism, is more loose-jointed and less confined to an exact framework. Accordingly, it is better suited to daily thinking.

Reasoning of the type: *

* Cohen and Nagel, *op. cit.*

A. No fossil can be crossed in love.
An oyster can be crossed in love.
Therefore, oysters are not fossils.

B. No ducks waltz.
No officers ever decline to waltz.
All my poultry are ducks.
Therefore, my poultry are not officers.

carries with it great compulsion. It is clear, exact and precise, securing for our thoughts the maximum of formal validity. Just as in mathematics, certain fundamental assumptions are made and we deduce conclusions from them. But most of our thinking is non-mathematical, most of our beliefs are not certain, only probable. As Locke once wrote, "In the greatest part of our concernment God has afforded us only the twilight, as I may so say, of Probability, suitable, I presume, to that state of Mediocrity and Probationership He has been pleased to place us in here."

It is then the relation of probability, not certainty, that obtains between most of our premises and conclusions. We are *certain* that a coin will fall after being tossed. We are equally certain that a black ball cannot be drawn from an urn containing only white ones. But most of our beliefs fall short of certainty, though they may range from very weak to very strong. Thus, we are nearly certain that an ordinary penny will not fall heads 100 times in succession. Or we may faintly believe that we will win the grand prize in the next sweepstakes.

Perhaps it is possible to explain this attitude. Some things in the world happen in conformity with natural laws, which (unless we believe in miracles) operate inexorably. Thus, because of gravitation, pennies when tossed in the air will fall. The sun will rise tomorrow because the planets follow regular courses. All men are mortal because death is a biological necessity—and so on.

But about most of the phenomena which surround us we know very little. We know neither the laws they obey, nor indeed, whether they obey any laws. One given to pointing morals about man's limitations would not have to go beyond trivial instances for startling confirmation. We are able to predict the motions of planets millions of miles off in space, but no one can predict the outcome of tossing a penny or throwing a pair of dice. Events in this category, and countless others, we ascribe to chance. But chance is merely a euphemism for ignorance. To say an event is determined by chance is to say we do not know how it is determined.

Nevertheless, even within the realm of chance we sense a certain regularity, a certain symmetry—an order within disorder—and so even about events which we ascribe to chance we form various degrees of rational belief. The theory of probability considers what are paradoxically called "the laws of chance." Part of its critical analysis is an attempt to formulate rules about *when* and *how* mathematics may be employed to measure the relation of probability. However, the intrinsic meaning of probability must be made clear before it is possible to turn to a consideration of its rules.

*

Though most of our judgments are based upon probability rather than certainty, careful thought is rarely given to the mechanics of this method of reasoning. In the laboratory, in business, as jurors, or at the bridge table, judgments are formed by probable inference. Few have the powers of a Sherlock Holmes, or can point to such successful deductions. Nevertheless, in almost all out daily thinking, everyone is called upon to play the part of amateur detective, logician, and mathematician.

When it is cloudy and warm, we say "It will *probably* rain." The meteorologist may require better evidence before venturing a prediction. He will want to know about barometric pressure, isobars, and tables of precipitation. But the average man makes his prediction with much less to go on. Money quickly, abundantly, and mysteriously earned during prohibition (it was judged, without consulting Bradstreet's) was *probably* the fruit of bootlegging. And the man who gets a few kicks under the bridge table infers that he is *probably* playing the wrong suit, whether he is a businessman or a scientist.

And so do we reason about matters ranging from the most trivial to the most important, making frequent use of words and expressions such as: "probable," "the probability is," or "the chances are" without, however, having a precise idea of what is meant by probability. Yet, this is not for want of definitions. It is true that practical scientists have generally left the job of defining and interpreting probability to the philosophers, mindful, perhaps, of the Gallic aphorism that science is continually making progress because it is never certain of its results. But while scientists have been satisfied to enlarge upon the uses of mathematical probability and to perfect its methods, philosophers and mathematicians have repeatedly attempted to define it.

Out of many conflicting opinions and theories three principal interpretations have crystallized.

*

The subjective view of probability, though now somewhat outmoded, at one time (particularly during the last century) held a very respectable position. One of its chief adherents and expositors was Augustus De Morgan, the celebrated logician and mathematician. He thought that probability referred to a *state of mind*, to the *degree* of certainty or of uncertainty which characterizes our beliefs. This is not entirely an erroneous view; the principal difficulties which it entails, as we shall see, arise when we attempt to justify a calculus of probability upon such foundations.

A proposition is either true or false,[2] but our knowledge is for the most part so limited as to make it impossible to be rationally certain of either its truth or falsity. To form a rational *belief*, we must have some pertinent knowledge. Occasionally, such knowledge may be sufficient to justify our certainty that the proposition is true or false. Thus we are certain that Socrates was not an American citizen; and we are equally certain that Hitler should have remained a house painter. On the other hand, between the extremes of certainty there is a rainbow of shadings of belief corresponding to the degree of our knowledge.

In a sense, it is undoubtedly true that our rational beliefs are subjective. Still, if we are convinced of the objective truth or falsity of all propositions, we cannot, if we wish to be rational, permit ourselves to be guided by mere intensity of belief. As a matter of principle, faulty conclusions based on limited knowledge and correct reasoning, are infinitely preferable to correct results obtained by faulty reasoning. It is only thus that we faintly approach the life of reason.

Moreover, we feel that if the relation of probability is to be treated mathematically, it must furnish us with better material for measurement than mere strength of belief. In most instances a numerical magnitude cannot be assigned to the relation of probability, yet it can only be considered by the mathematician when it is measurable and countable. If probability is to serve in describing certain aspects of the world in terms of fractions, it must be expressible as a number. When a thing cannot happen, its probability is 0; if it is certain to happen its probability is 1. Every probability between these extremes is expressible by a fraction between zero and one. But to form these fractions entails measurement and counting, and how is the mathematician to measure "intensity of belief"? At best this is a problem for the psychologist. Even if an instrument could be devised to measure intensity of belief, its value would be little more than that of the lie detector, that gem of jurisprudence. People differ widely in their beliefs based upon the same set of facts. What is perfectly evident to one man is thoroughly unconvincing to another; and our beliefs often vaguely conceived and loosely drawn are too interwoven with our emotions and our prejudices to justify considering one without the other.

One of the difficulties arising out of the subjective view of probability results from the *principle of insufficient reason.* This principle, the logical basis upon which the calculus of probability must rest according to the subjective view, holds that *if we are wholly ignorant of the different ways an event can occur and therefore have no reasonable ground for preference, it is as likely to occur one way as another.* Since first enunciated by James Bernoulli, this principle has been exhaustively analyzed by mathematicians. As the principle rests on ignorance, it would seem to follow that the calculus of probability was most effective when used by those who had an "equally balanced ignorance." However well men approximate to this ideal, philosophers and mathematicians hold themselves in higher esteem, and so the principle has fallen on lean days.

Nevertheless, it contains an element of truth, and no consistent calculus of probability can be developed without in some measure being dependent on it. Mainly, it has value as a negative criterion, in the sense that it cannot be said that two events are equally probable if there is ground for preferring one to the other.

When the principle of insufficient reason is used without great caution, it gives rise to contradictions. Two examples: Take the case of an ape, who is given a number of cards, each with an English word written upon it. Is it equally probable that any way he arranges the cards will, or will not, produce a meaningful English sentence? By the principle of insufficient reason this would seem to follow, although it is evidently absurd. Or, having no evidence relevant to whether Mars is inhabited or not, we could conclude that the probability is $\frac{1}{2}$ that it is exclusively inhabited by Nazis, and we could just as well conclude that the probability of each of the propositions, "Mars is exclusively inhabited by jackasses" and "Mars is exclusively inhabited by termites," is also $\frac{1}{2}$. But this confronts us with the impossible case of *three exclusive alternatives* all as likely as not.[3]

*

A much more workable and widely held theory which avoids some of these difficulties is the *relative frequency*, or *statistical* interpretation. In a large measure, this view is responsible for the advance in applying probability, not only to physics and astronomy, but also to biology, to the social sciences, and to business. The statistical interpretation comes close to the view expressed by Aristotle that the probable is that which usually happens.

Probability is considered to be the relative frequency with which an event occurs in a certain class of events. Thus, the probability of an event is expressed as a definite mathematical ratio which is hypothetically assigned. The hypothesis may be verified either rationally; by showing, for example, from our knowledge of mechanical causes, that a penny, or a pair of dice *must* fall in a certain way; or experimentally, by showing that the penny, or the pair of dice, *do*, in fact, fall in that way.

Suppose a penny is tossed in a random manner. Having no special information, there is no reason to predict how the coin will fall, either head or tail. If it is tossed a great many times and the ratio of heads to tails recorded, let us assume the following frequencies are obtained:

TOSSES	RESULTS
15	6 heads; 9 tails
20	9 heads; 11 tails
30	16 heads; 14 tails
40	21 heads; 19 tails
80	41 heads; 39 tails
150	74 heads; 76 tails

We notice that the *ratio* of heads to the total number of tosses, as these increase, approaches more and more closely to the fraction $\frac{1}{2}$. This represents the relative frequency of the class of heads in the larger class of tosses. We then advance to a general prediction from a large number of particular instances, and assume the future will be consistent with the past.

However, consider for the moment: What justification is there for such a step? Having performed our experiment and determined the relative frequency, we now say that the probability of getting a head is $\frac{1}{2}$. Evidently, that statement is a hypothesis. Further experiments may serve to strengthen our belief in that hypothesis or cause us to either modify or abandon it. The assumption (based on our experiment) is that in a great number of cases, heads will appear as often as tails. If the results do not corroborate the hypothesis, we conclude that the coin is perhaps heavier on one side than on the other. But it is important to remember that since the proof is not logical, but only experimental, it is never complete, it is subject always to further experiments. A logical proof is only possible if every cause that affects an event is known. Obviously, such an occasion cannot arise outside of mathematics itself. Thus the verification of an hypothesis by experiment can only show that in actual practise, the relative frequency *approaches* the predicted probability—that our assumptions are borne out by experience.

It is appropriate to point out how the logical or deductive method of proof differs from the experimental one. "The process of induction, which is basic in all experimental sciences, is forever banned from rigorous mathematics . . ."[4] In order to prove a proposition in mathematics, even a vast number of instances of its validity would not be sufficient, whereas one exception will suffice to disprove it. The propositions of mathematics are true only if they lead to no contradictions. But outside of mathematics, in all other human activities, such a restriction would have a paralyzing effect. Scientific procedure rests on the same convenient rule of thumb as that which guides us in practical affairs: A hypothesis is valuable if it leads to correct results more often than not; experimental verifications are quite final—until the next day's experiments upset them.

*

"The Adventure of the Dancing Men," from which was selected the incident at the beginning of this chapter, may serve again to illustrate how the statistical method serves probable inference.

Holmes is confronted with a cryptogram composed of several messages; (see Fig. 76):

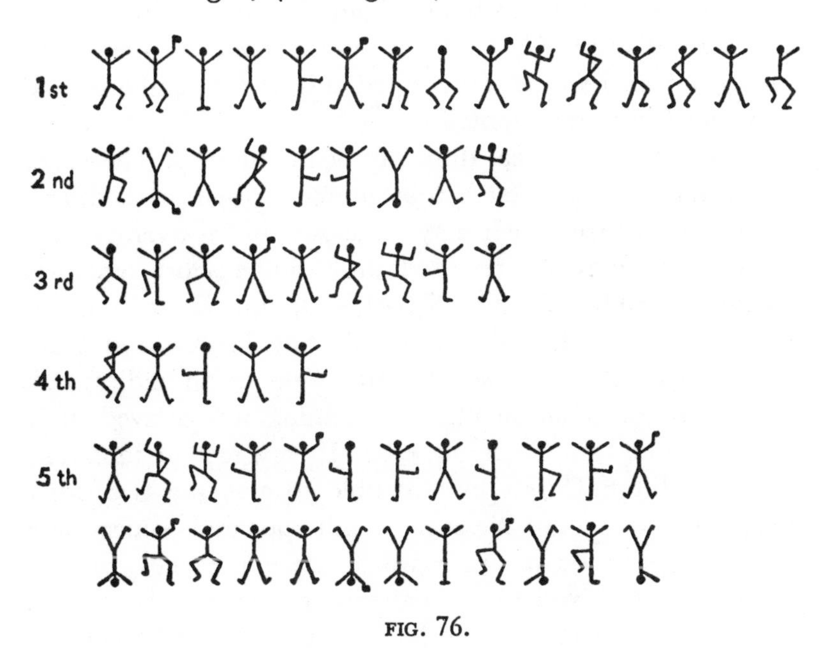

FIG. 76.

The solution of most cryptograms depends to a large extent upon certain statistical knowledge as well as upon shrewd inferences. Holmes derived his method of solution from a method already referred to by Edgar Allan Poe in *The Gold Bug*.

> Having once recognized, however, that the symbols stood for letters, and having applied the rules which guide us in all forms of secret writings, the solution was easy enough. The first message submitted to me was so short that it was impossible for me to say with confidence, that the symbol stood for E. As you are aware, E is the most common letter in the English alphabet, and it predominates to so marked an extent that

even in a short sentence one would expect to find it most often. Out of fifteen symbols in the first message, four were the same, so it was reasonable to set this down as E. It is true that in some cases the figure was bearing a flag and in some cases not, but it was probable, from the ways in which the flags were distributed, that they were used to break the sentence up into words. I accepted this as a hypothesis, and noted that E was represented by .

But now came the real difficulty of the inquiry. The order of the English letters after E is by no means well marked, and any preponderance which may be shown in an average of a printed sheet may be reversed in a single short sentence. Speaking roughly, T,A,O,I,N,S,H,R,D, and L are the numerical order in which letters occur, but T,A,O, and I are nearly abreast of each other, and it would be an endless task to try each combination until a meaning was arrived at. I therefore waited for fresh material. In my second interview with Mr. Hilton Cubitt he was able to give me two other short sentences and one message, which appeared—since there was no flag—to be a single word. Here are the symbols. Now, in the single word I have already got the two E's second and fourth in a word of five letters. It might be "sever" or "lever," or "never." There can be no question that the latter as a reply to an appeal is far the most probable, and the circumstances pointed to its being a reply written by the lady. Accepting it as correct, we are now able to say that the symbols stand respectively for N,V, and R.

Even now I was in considerable difficulty, but a happy thought put me in possession of several other letters. It occurred to me that if these appeals came, as I expected, from someone who had been intimate with the lady in her early life, a combination which contained two E's with three letters between might very well stand for the name "ELSIE." On examination I found that such a combination formed the termination of the message which was three times repeated. It was certainly some appeal to "Elsie." In this way I had got my L,S, and I. But what appeal could it be? There were only four letters in the word which preceded "Elsie," and it ended in E. Surely the word must be "COME." I tried all other four letters ending in E, but could find none to fit the case. So now I was in possession of C,O, and M, and I was in a position to attack the first message once more, dividing it into words and putting dots for each symbol which was still unknown. So treated, it worked out in this fashion:

.M .ERE ..E SL.NE.

Now the first letter *can* only be A which is a most useful discovery, since it occurs no fewer than three times in this short sentence, and the H is also apparent in the second word. Now it becomes:

AM HERE A.E SLANE.

Or filling in the obvious vacancies in the name:

AM HERE ABE SLANEY

*

In spite of the brilliant successes achieved by the statistical method, it is open to serious objections. While some of the difficulties can be remedied without greatly impairing its usefulness, others are not so easily disposed of.

The concept of the limit, which plays such an important role in many branches of mathematics, is also used in statistics, although its use here can hardly be defended, for this concept arises properly only in connection with infinite processes. The statistician uses it in saying that frequencies approach a limiting ratio, but the statistician, and also the physicist, do not deal with infinity—rather with phenomena which, however vast and complex, are *finite* and *limited.* Because an experiment yields the same result a thousand times is no proof that the results to follow will be consistent. Even Scheherezade may tell an unpleasing tale on the thousand and second night.

Relative frequencies can hardly be said to approach a mathematical limit. The limiting concept as it is used in the theory of relative frequency bears roughly the same relation to the mathematical concept of limit as reasoning by probable inference bears to the syllogism.

Reference is often made to the probability of past events, although such probability in terms of the relative frequency view has apparently no meaning. "It is improbable that John Wilkes Booth escaped the federal soldiers after the assassination of Lincoln"; or "Henry VIII was probably not so much interested in reform when he broke with the Pope as in getting rid of Catherine of Aragon." How shall such statements be evaluated if probability is the relative frequency of an event within a class of events? Indeed, whether the event be past or future, *what is meant by the probability of any single event?*

Whatever interpretation of probability is advanced, this problem is particularly troublesome. Perhaps sad necessity accounts for the best accredited opinion that probability has no meaning whatsoever when applied to a single event, either past or future.

According to the statistical interpretation, probability can refer to a single event only in relation to a class of similar events. But this often makes for confusion. Everyone would agree that the following reasoning is absurd: In a certain community records of births for the past 10 years indicate a ratio of 51 females to 50 males. The first 35 children born in a particular month are all girls. Mr. Jones, expectant father, is therefore quite certain that the odds are heavily in his favor that his wife will present him with a boy, because of the "law of averages." *

* Not to leave the reader in suspense we can tell him that Jones is just as well off as though he were starting from scratch.

On the other hand, it is a very common misapprehension of the very same kind to which we still cling intuitively that if X throws five sevens in a row at dice, the chance of his tossing another seven the next throw is much less than his chance of throwing some other particular number. We find it hard to believe that the mathematical probability, the chance of a future event, where the events are independent, is unaffected by what has already happened.

In our daily lives we instinctively and deliberately reject this principle. When logic says "You must," we often reply "Not this time." Charles S. Peirce, the famous pragmatist, illustrates the point exceedingly well: "If a man had to choose between drawing a card from a pack containing 25 red cards and a black one, or from a pack containing 25 black cards and a red one; and if the drawing of a red card were destined to transport him to eternal felicity and that of a black one to consign him to everlasting woe, it would be foolish to deny that he ought to prefer the pack containing the larger portion of red cards, although from the nature of the risk, it could not be repeated. It is not easy to reconcile this with our analysis of the conception of chance. But suppose he should choose the red pack and should draw the black card. What consolation would he have? He might say that he had acted in accordance with reason, but that would only show that his reason was absolutely worthless. And if he should choose the red card, how could he regard it as anything but a happy accident? He could not say that if he had drawn from the other pack he might have drawn the wrong one, because an hypothetical proposition such as: 'If A, then B' means nothing with reference to a single case."[5]

Finally, a brief allusion to an interpretation of probability, accredited chiefly to Peirce, which seems to avoid some of the difficulties inherent in the interpretations already examined.[6]

Peirce holds that probability refers not to *events* but to *propositions*. With some modifications, his view is adhered to by John Maynard Keynes in his remarkable *Treatise on Probability*. According to Peirce, probability has nothing to do with either intensity of belief or with statistical frequencies. "Instead of talking about such an event as 'heads,' the *truth frequency theory* discusses *propositions* such as: This coin will fall head uppermost on one toss." The probability of the truth of this proposition must be the same as the relative frequency with which the event "head" occurs in a series of tosses.

This interpretation of probability is better able to take care of single events. The statement, "It will probably rain tomorrow" means that the propositions about the state of the weather, temperature, barometric pressure and so on, more often than not imply propositions of the type: "It will probably rain tomorrow." In other words, if from our knowledge of the weather we conclude this latter proposition, we will be right more often than wrong.

Before passing to a consideration of a few of the theorems of the calculus of probability, there is one further caution. Everything said thus far points to one fact unmistakably: *No proposition has any probable truth except in relation to other knowledge.* To say that a proposition is probable, when the knowledge on which it is based is either obscure or nonexistent, is absurd. To be sure, we often make elliptical statements about probability, where it is clearly understood to what body of knowledge we refer. This is just as permissible as to say that San

Francisco is 3000 miles away, it being evident that what is meant is "San Francisco is 3000 miles away from New York." As already emphasized, it is more laudable to adhere to a statement which turns out to be wrong, so long as the evidence from which we reach our conclusion is the best available, than to advance a true proposition on the basis of faulty reasoning or incorrect facts. Herodotus says: "There is nothing more profitable for a man than to take good counsel with himself; for even if the event turns out contrary to one's hopes, still one's decision was right even though fortune has made it of no effect; whereas if a man acts contrary to good counsel, although, being lucky, he gets what he had no right to expect, his decision was not any the less fallacious."

THE CALCULUS OF CHANCE

In moderation, gambling possesses undeniable virtues. Yet it presents a curious spectacle replete with contradictions. While indulgence in its pleasures has always lain beyond the pale for fear of Hell's fires, the great laboratories and respectable insurance palaces stand as monuments to a science originally born of the dice cups.

The Chevalier de Méré, euphemistically called a "gaming philosopher" of the seventeenth century, desired some information about the division of stakes at games of dice. He directed his inquiries to one of the ablest mathematicians of all times—the gentle and devoutly religious Blaise Pascal. Pascal, in turn, wrote to an even more celebrated mathematician, the Parliamentary Town Councilor of Toulouse, Pierre de Fermat, and in the correspondence that ensued, the theory of probability first saw the light of day.

Pascal could not forbear from a mild rebuke of De Méré, not because he was a gambler, but for the more serious reason that De Méré was not a mathematician: "*Car, il a très bon esprit,*" (he wrote to Fermat) "*mais il n'est pas géomètre; c'est comme vous savez un grand défaut.*" Indeed, the Chevalier deserved worse, for the answer to his question evidently interfered with his business so that he took the occasion to write a diatribe on the worthlessness of all science, in particular arithmetic. And that was the fate of the first brain trust.

Interest in probability grew, encouraged by the researches of such eminent mathematicians as Leibniz, James Bernoulli, De Moivre, Euler, the Marquis de Condorcet, and above all, Laplace. The latter's epochal work on the analytic theory of probability brought the calculus to the point where Clerk Maxwell could say that it is "mathematics for practical men," and Jevons could wax quite lyrical (quoting without acknowledgment from Bishop Butler) that the mathematics of probability is "the very guide of life and hardly can we take a step or make a decision without correctly or incorrectly making an estimation of probability." And these opinions were uttered even before the calculus had achieved its most brilliant successes in physics and genetics as well as in more practical spheres.[7] It was indeed remarkable, as Laplace wrote, that "a science which began with the considerations of play has risen to the most important objects of human knowledge."

*

In developing a calculus of probability it is necessary to make certain ideal assumptions. Particularly since a great many things to which we would like to apply it are *not measurable*, we must be doubly careful that the axioms and postulates which we formulate are precise, so that their range of application may be readily judged. We have already referred to the fact that the mathematical probability of an event lies between 0 and 1. The probability of an impossible event is 0, that of a certain event, 1.

We must now define what is meant by "equiprobable" (equally probable). This is a rather difficult task; for our purposes we can shorten the road by employing a rough definition.

Two contingent events will be considered equiprobable if, either in the absence of any evidence or after considering all the relevant evidence, one event cannot be expected in preference to the other.

Perhaps the reader detects an incongruity. Had he not been cautioned that no probability can be estimated where there is no appropriate or relevant knowledge? Yet here it is said that two propositions, or events, can be equally probable, even if we have no knowledge about them whatsoever. But therein lies the clue! A little knowledge is dangerous. None at all is much more satisfactory. For our purposes we invoke the principle of insufficient reason, according to which, in the absence of any knowledge about two events, they are considered equally likely. The reader must bear in mind that our definition is rough —very rough. And also, that it is possible to know that two quantities are equal without knowing what they are. Thus, one may know from a general knowledge of games that in chess both sides start with equal forces without knowing what these are, or anything else about the game.

If we assume, then, that a penny is symmetrical, it is *equiprobable* that it will fall heads or tails, since there is no more reason to anticipate one result than the other.

If there are a number of equiprobable ways in which

an event can happen and a number of equiprobable ways in which it cannot happen, the probability of the occurrence of the event is the ratio of the number of ways in which the event can happen to the total number of ways in which it can and cannot happen. The coin may fall heads or tails. The probability of its falling heads is thus the ratio $\frac{H}{H + T} = \frac{1}{2}$. In general, if we call the ways in which an event can happen, *favorable*, and the ways in which it cannot happen, *unfavorable*, the probability of an event is the fraction $\frac{F}{F + U}$.

That branch of mathematics which considers permutations and combinations is concerned with the number of different ways in which an event *can* happen. It is the study of *mathematical possibility*, and furnishes an ideal framework for the *mathematics of probability*.

The typical problems of permutations and combinations have a dry and dreary look. At first it is hard to believe that information gained in solving problems of this type can be of much service in other studies: "Four travelers arrive at a town where there are 5 inns. In how many ways can they take up their quarters, each at a different hotel?" Nor does it seem that a theory which is used to determine in how many different ways the letters of the word *Mississippi* [8] may be arranged, would be useful in determining either the physics of the atom or in fixing insurance rates. Nevertheless, the theorems of combinatorial analysis are the basis for the calculus of probability. We have to know how to calculate the total number of different ways an event *can* happen before aspiring to predict how it is *likely* to happen.

Our overworked penny again furnishes an example.

A penny is tossed three times in succession. The possible results are:

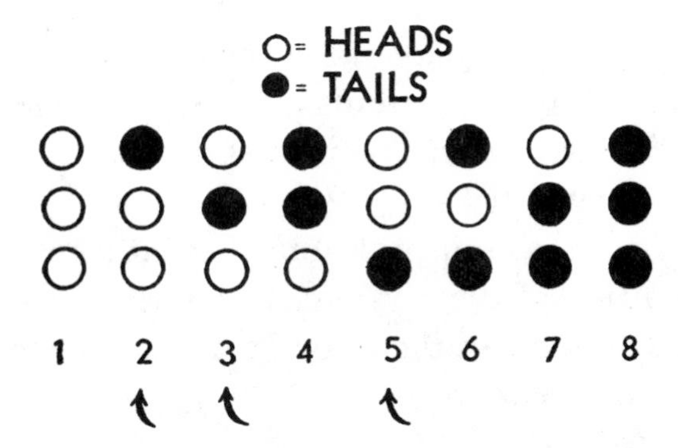

FIG. 77.—The possible results of tossing a penny three times. The arrows indicate the cases of two heads and one tail.

These eight possible results answer all the questions which might be asked in permutations and combinations. But, further, any others that arise in the calculus of probability can also be answered by referring to the diagram. Thus, the *probability* of getting 3 heads, the ratio $\frac{F}{F + U}$, is $\frac{1}{8}$. The probability of getting 2 heads and 1 tail is the ratio of cases 2, 3, 5 to all the possible cases, i.e., $\frac{3}{8}$.

Now it is plain that the enumeration of all possible cases becomes both tedious and unwieldy as these increase in number. For that reason the calculus contains many theorems taken from combinatorial analysis which make direct enumeration unnecessary.

MUTUALLY EXCLUSIVE EVENTS

I. Since there are four aces in a deck, the probability of drawing an ace from 52 cards is $\frac{4}{52} = \frac{1}{13}$. But what is the probability of drawing *either* an ace *or* a king from a deck of cards in one draw? This is the probability of *mutually exclusive*, or *alternative*, events; if one of the two events occurs, the other cannot. A theorem in the calculus states that the probability of the occurrence of one of several mutually exclusive events is the *sum of the probabilities of each of the single events*. The probability of getting an ace *or* a king is therefore, $\frac{1}{13} + \frac{1}{13} = \frac{2}{13}$.

What is the probability of obtaining either a 6 or a 7 in throwing a pair of dice? We may enumerate the number of cases favorable to either 6 or 7 and then check our results with the theorem.

FIRST DIE		SECOND DIE	FIRST DIE		SECOND DIE
1		5	1		6
2		4	2		5
3	VI	3	3	VII	4
4		2	4		3
5		1	5		2
			6		1

There are 36 *possible combinations* of the dice and 11 are favorable to the event; therefore, the probability of obtaining either a 6 or a 7 is $\frac{11}{36}$.

Had we used the theorem, we would have taken the sum of the separate probabilities, i.e., $\frac{5}{36} + \frac{6}{36}$, and of course, obtained the same result.

INDEPENDENT EVENTS

II. Two events are said to be *independent* of each other if the happening of one is in no way connected with the happening of the other. A penny is tossed twice in succession. What is the probability of getting 2 heads in a row? The appropriate theorem states that the *probability of the joint occurrence of two independent events is the product of the separate probabilities of each of the events*. The probability of getting 2 heads in succession is, therefore, $\frac{1}{2} \times \frac{1}{2} = \frac{1}{4}$. And, as we saw above, by direct enumeration, the probability of getting three heads in a row is $\frac{1}{8}$. Checking this against the theorem gives $\frac{1}{2} \times \frac{1}{2} \times \frac{1}{2} = \frac{1}{8}$.

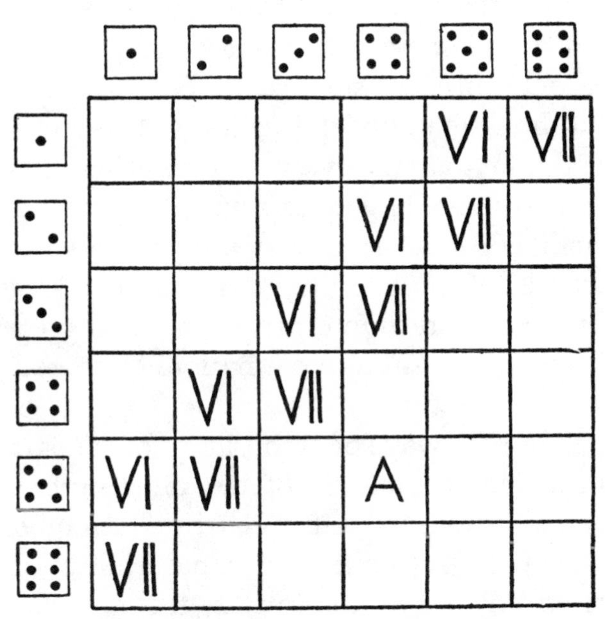

FIG. 78.—Each square represents an equiprobable result. For instance, the square marked *A* represents getting a 4 with one die and 5 with the other. Of the thirty-six possibilities, five result in a 6, and six result in a 7.

Consider now a problem slightly different in form:

In tossing a coin twice in succession, what is the probability of getting *at least* one head? This problem may be solved easily without enumeration, by ascertaining the probability of the desired event *not* happening and subtracting this fraction from 1. Since the probability of getting *two tails* in succession, which is the sole alternative to getting at least one head, is $\frac{1}{4}$, the probability of at least one head is $1 - \frac{1}{4} = \frac{3}{4}$.

D'Alembert, in his article on probability in the famous *Encyclopédie*, revealed that he did not understand the theorem of multiplying independent probabilities. He doubted that the last-named probability was $\frac{3}{4}$, reasoning that if a head appeared at the first throw, the game was finished and there was no need for a second. Enumerating only three possible cases: H, TH, and TT, he arrived at the probability of $\frac{2}{3}$. What he failed to consider was that the first alternative was in itself no more likely than the alternative of getting a tail.

Although D'Alembert consistently misunderstood the fundamentals of probability, some of his ideas foreshadowed the statistical interpretation. He suggested that by making experiments approximations of desired probabilities could be estimated.

Long before the wave of enthusiasm for statistics swept over Europe in the middle of the nineteenth century, experiments of the kind suggested by D'Alembert were carried out. The eighteenth-century naturalist, Count Buffon, carried on many experiments, the most famous of which is his "Needle Problem." A plain surface is ruled by parallel lines (as in Fig. 79), the distance between the lines being equal to H. Taking a needle whose length, L, is less than H, Buffon dropped it, permitting it to fall each time on the ruled surface. He considered the toss favorable when the needle fell across a line, unfavorable when it rested between two lines. His amazing discovery was that the *ratio* of successes to failures was an expression in which π appears. Indeed, if L is equal to H, the probability of a success is $\frac{2}{\pi}$. The larger the number of trials, the more closely did the result approximate the value of π, even to three decimal places.

Elaborate experiments were performed in 1901 by an Italian mathematician, Lazzerini, who made 3,408 tosses, giving a value for π equal to 3.1415929, an error of only 0.0000003. One could scarcely expect to find a better ex-

FIG. 79.—Count Buffon's needle problem.

ample of the interrelatedness of all mathematics. Thus far we have seen π in three guises: as the ratio of the circumference of a circle to its diameter; as the limit of infinite series; and as a measure of probability.

COMPOUND PROBABILITY

III. The theorem dealing with the probability of independent events may sometimes be usefully extended to deal with cases where the probabilities are not actually independent.

A bag contains one white ball (W) and two black balls (B); the probability of drawing a black ball is $\frac{2}{3}$; a white ball $\frac{1}{3}$. Assume two successive drawings from the same bag, with the ball replaced after each drawing. Now the probability of drawing two W's in succession is $\frac{1}{3} \times \frac{1}{3} = \frac{1}{9}$, of drawing two B's in succession $\frac{2}{3} \times \frac{2}{3} = \frac{4}{9}$. However, if after each drawing the balls are *not* replaced, the drawings are no longer *independent* but *dependent* on each other. After each drawing the new probability must be calculated to form the correct *compound* probability. After one ball has been drawn, the probability of drawing two B's in succession, with no replacements is $\frac{2}{3} \times \frac{1}{2} = \frac{1}{3}$. That the probability of the second drawing depends on the outcome of the first is also shown by the fact that the probability of drawing two W's in a row is 0, if no replacement is made, whereas it is $\frac{1}{9}$ if the W is replaced on being drawn the first time.

IV. Thus far we have considered the probability of events that are mutually exclusive, dependent, and independent. If these factors are varied and combined, new, interesting methods result.

A bag contains 6 W's and 6 B's. If one ball is drawn, two events are equiprobable—either W or B. This may be denoted by

(a) $$(1)\ W,\ (2)\ B = 2^1$$

The possible results in two drawings are:

(b) (1) WW, (2) WB, (3) BW, (4) $BB = 2^2$

In three drawings there are eight possible results:

(c)

$$\left.\begin{array}{llll} (1)\ WWW & (3)\ WBB & (5)\ BWB & (7)\ BBW \\ (2)\ WBW & (4)\ WWB & (6)\ BWW & (8)\ BBB \end{array}\right\} = 2^3$$

In four drawings there are sixteen:

(d)

$$\left.\begin{array}{llll} (1)\ WWWW & (5)\ BWWW & (9)\ BWBW & (13)\ BBBW \\ (2)\ WWWB & (6)\ WWBB & (10)\ WBWB & (14)\ BBWB \\ (3)\ WBWW & (7)\ WBBW & (11)\ BWWB & (15)\ BWBB \\ (4)\ WWBW & (8)\ BBWW & (12)\ WBBB & (16)\ BBBB \end{array}\right\} = 2^4$$

In general, then, in n drawings there are 2^n possible results.

But this information is the clue to a most valuable method! Let us avail ourselves of an important theorem from another branch of mathematics—the Binomial Theorem.

Let W denote the *drawing* of a white ball, and B the drawing of a black. Expanding the expression $(W + B)^2$ in accordance with the binomial theorem, we obtain

$$W^2 + 2WB + B^2.$$

Now this algebraic expression pictures compactly what was already explicitly set forth in (b) above, namely: every possible result of two drawings from a bag containing the same number of black and white balls. Thus,[9]

$$\begin{array}{ll} (1)\ WW & = W^2 \\ \left.\begin{array}{l}(2)\ WB \\ (3)\ BW\end{array}\right\} & = 2WB \\ (4)\ BB & = B^2 \end{array}$$

Three drawings from such a bag is represented by the expression

$$W^3 + 3W^2B + 3WB^2 + B^3$$

for again

$$\begin{array}{lll} (1) & WWW & = W^3 \\ (2) & WWB \\ (3) & WBW & = 3W^2B \\ (4) & BWW \\ (5) & BBW \\ (6) & BWB & = 3WB^2 \\ (7) & WBB \\ (8) & BBB & = B^3 \end{array}$$

There are, therefore, eight possible results, one way of getting three Whites, three ways of getting two Whites and a Black, three ways of getting two Blacks and a White, and one way of getting three Blacks.

The respective probabilities are $\frac{1}{8}$, $\frac{3}{8}$, $\frac{3}{8}$, and $\frac{1}{8}$.

For n successive drawings the general binomial theorem gives:[10]

$$(W+B)^n = W^n + nW^{n-1}B + \frac{n(n-1)}{2!}W^{n-2}B^2 + \frac{n(n-1)(n-2)}{3!}W^{n-3}B^3 + \ldots + B^n.$$

One further illustration may be considered of the application of the binomial theorem: A bag contains 3 Whites and 2 Blacks. After each drawing the ball is replaced. What is the probability of 3 W's and 2 B's in 5 drawings?

Now, for each drawing the probability of a $W = \frac{3}{5}$, of a $B = \frac{2}{5}$. Expanding:

$$(W+B)^5 = W^5 + 5W^4B + 10W^3B^2 + 10W^2B^3 + 5WB^4 + B^5.$$

The result, the probability of which we are seeking, is W^3B^2 since this represents 3 W's and 2 B's. There are ten such possible results since the coefficient of the term W^3B^2 is 10. The desired probability, which is compound, must, therefore, be

$$10 \times (\tfrac{3}{5})^3 \times (\tfrac{2}{5})^2 = \tfrac{216}{625}.$$

*

It should be even more evident now how limited are the instances when the calculus of probability is applicable. In none of the several examples which appeared on page 227, however properly they may have illustrated the concept of probability, is our mathematical apparatus of any use. Indeed, the calculus of probability, like all other mathematical disciplines, cannot be regarded as a source of information about the physical world. Furthermore, speaking mathematically, it may be possible to define what is meant by equiprobable, but it is without doubt impossible to find two events in the physical world which actually are equiprobable.

Equiprobability in the physical world is purely a hypothesis. We may exercise the greatest care and use the most accurate of scientific instruments to determine whether or not a penny is symmetrical. Even if we are satisfied that it is, and that our evidence on that point is conclusive, our knowledge, or rather our ignorance, about the vast number of other causes which affect the fall of the penny is so abysmal that the fact of the penny's symmetry is a mere detail. Thus, the statement "head and tail are equiprobable" is at best an assumption.

Yet the calculus of probability is only helpful after we have made such an assumption—an assumption which, like all hypotheses in science, must justify its existence by its usefulness and which we must be prepared to modify or reject when experience fails to corroborate it.

By following such a bold procedure, the mathematics of probability has been remarkably successful in science and in commerce. In the eighteenth and nineteenth centuries, when science and philosophy were almost entirely under the spell of mechanistic ideas, it was enthusiastically supposed that the calculus of probability would supplement every "ignorance and weakness of the human mind." The calculus would help to illuminate those regions of knowledge in which the beacon of science did not yet burn too brightly.

It is readily understandable that a convenient and dogmatic philosophy of materialism was popular in a world which had witnessed the parade of scientific achievements from Kepler and Galileo to Newton and Laplace. The materialistic concept is based on a naïve faith in the all-pervading regularity and the recurrent order of natural phenomena, from the behavior of atoms to our own behavior on arising in the morning. Men hoped, and the history of science until recently encouraged them to believe, that science would explain all miracles and disclose all secrets, that the future was contained in and would therefore resemble the past, and that consequently the experiences of the past would help in predicting the future.

As a leading exponent of this view, Laplace had far greater hopes for the limits of knowledge than the modest twilight of mediocrity in which Locke felt the human mind would forever have to grope.

"We ought then," Laplace wrote, "to regard the present state of the universe as the effect of its anterior state and as the cause of the one which is to follow. Given for one instant an intelligence which could comprehend all the forces by which nature is animated and the respective situation of the beings who compose it—an intelligence sufficiently vast to submit these data to analysis—it would embrace in the same formula the movements of the greatest bodies of the universe and those of the lightest atom; for it, nothing would be uncertain and the future, as the past, would be present to its eyes."[11]

When Napoleon asked Laplace where in his monumental *Mécanique céleste* there was any reference to the Deity, he is said to have replied, "Sire, I have no need of that hypothesis." On hearing Napoleon recount this story, Lagrange remarked "That, Sire, is a wonderful hypothesis." Modern physics, indeed all of modern science, is as humble as Lagrange, and as agnostic as Laplace. Professing no God, it attributes to itself neither divine omniscience, nor the possibility of achieving it.

*

It was expected then, in the eighteenth and nineteenth centuries, that a Utopia in morals and politics as well as in the physical sciences was not far off. If exact natural laws in these spheres had not yet been uncovered, it was not doubted that they existed. In the meantime the calculus of probability would meet the deficiency. Though social phenomena had not yet been mastered in detail, as the motions of many of the planets had been, it was certain they would exhibit the same regularities when studied on the grand scale. Probability was to be a temporary expedient, an ordnance map which scientists would fill in in due time.

Hopes were high, and among those who expected the most was the Marquis de Condorcet. The theory of probability, he thought, might be applied effectively to the judgments of tribunals in order to minimize the danger of erroneous decisions. To that end he proposed that a large increase in the number of judges on any tribunal would assure a great many independent opinions, which, when combined, would safeguard the truth by neutralizing opposingly extreme and prejudicial views. Unfortunately, Condorcet failed to take numerous other factors into consideration. Not the least of these was the logic of the guillotine. For it was to this, ironically and tragically enough, that the judgment of a revolutionary tribunal, composed of many judges, all of whom held the same extreme views, eventually consigned him.

In the less heated atmosphere of the nineteenth century some of Condorcet's views were vindicated—if not in morals and politics, then in science and industry. The statistical view of nature changed the map of science in both the nineteenth and twentieth centuries as much, perhaps, as the inventions and the discoveries of the laboratory. Indeed (and this point cannot be emphasized too strongly), the statistical view has so permeated and penetrated modern scientific thinking and method that it has gone far beyond anything that even Condorcet could have imagined. But the fundamental materialism of his time which accompanied this faith in probability has largely vanished.

Instead of serving as an expedient, as a substitute for natural laws as yet unrevealed, statistical inference has come in time to supplant them almost completely. This signifies a change in the interpretation of physical reality comparable in intellectual importance to the Renaissance. With this in mind contemporary physicists often refer to the Renaissance of Modern Physics.

*

In his great work on the *Analytical Theory of Heat*, Fourier stated the principle which best exemplifies what we have already referred to as the classical view of physics—in fact of all natural laws. "Primary causes are unknown to us, but are subject to simple and constant laws, which may be discovered by observation, the study of them being the object of natural philosophy." And he went on to add: "Profound study of nature is the most fertile source of mathematical discoveries . . . There cannot be a language more simple, more free from errors and obscurities, that is to say, more worthy to express the relations of natural things . . . It brings together phenomena the most diverse, and discovers the hidden analogies which unite them."

The scientist of the present day, particularly the physicist would be in complete agreement with the latter part of this quotation. He would agree that mathematics is the ideal language in which to express the results of his observations and even the uncertainties of his predictions. He would, however, differ with Fourier sharply when he says that the laws governing natural phenomena are "*simple* and *constant*."

Instead of holding to the opinion that nature obeys perfect and certain laws, which it is the job of the scientist to discover and explain, the physicist is now content to make hypotheses and to perform experiments, to carry on a kind of scientific bookkeeping, with the aid of which he strikes a balance from time to time. That balance bears no relation to eternal verities. It refers only to current assets and liabilities. Instead of pinning his faith on uncovering in all natural phenomena a general all-pervading, regular, and recurrent order, he is content to hope that there is occasional method in the madness of the physical world, that in the large, if not in the small, there is some semblance of a scheme.

The old materialistic dogmatism seemed to foreclose further metaphysical speculations about the nature of reality and was "comfortable and complete." It had the "compelling power of the old logic." The outlines of the world were hard and fast, and the mysteries of the universe, its apparent uncertainties, were confessions of our own incompetence, our own limitations. When we said that the fall of a penny was determined by chance, "we regarded this confession of uncertainty as due to our own ignorance, and not the uncertainties of nature."

But the new physics and the new logic have changed our outlook as profoundly as they have changed our basic distinction between matter and energy. "We start prejudiced against probability, grudging it as a makeshift, and in favor of causality," and we end convinced that the outlines of the world are "not hard, but fuzzy," and that our most exact scientific laws are merely approximations good enough for our crude senses. Thus, in place of the syllogism and the rules of formal logic our ideas about the physical universe must be gauged entirely by the rules of probable inference. We translate "Socrates is a man; all men are mortal, therefore Socrates is mortal," as a statement about the world of fact, into, "Socrates will probably die, because so far as we know all men before him have died." "The uncertainties of the world we now ascribe not to the uncertainties of our thoughts, but rather to the character of the world around us. It is a more sensible, more mature and more comprehensible view." [12]

Here we recall the moving words of Charles Peirce: "All human affairs rest upon probabilities, and the same thing is true everywhere. If man were immortal, he could be perfectly sure of seeing the day when everything in which he had trusted should betray his trust, and, in short, of coming eventually to hopeless misery. He would break down, at last, as every good fortune, as every dynasty, as every civilization does. In place of this we have death.

"But what, without death, would happen to every man, with death must happen to some man . . . It seems to me that we are driven to this, that logicality inexorably requires that our interests shall *not* be limited. They must not stop at our own fate, but must embrace the whole community."

APPENDIX

A discussion of the theory of probability can ill afford to omit some applications. They are, however, generally quite technical, but the more persevering reader will surely find these few, chosen at random, of interest.

KINETIC THEORY OF GASES AND PROBABILITY CURVE OF ERROR

The law of gases was arrived at experimentally by the English physicist and chemist Robert Boyle (1627–1691), whose most important work bears the title *The Sceptical Chymist: or Chymico–Physical Doubts and Paradoxes, touching the experiments whereby vulgar Spagirists are wont to endeavour to evince their Salt, Sulphur and Mercury to be the true Principles of Things.* His law of gases states that the pressure of a gas is inversely proportional to its volume. Thus: Pressure $\times$ Volume $=$ Constant. But any volume of gas is made up of vast numbers of moving molecules, each of which has a velocity proportional to its energy. Naturally, molecular collisions occur in great numbers at every instant. It has been estimated that in "ordinary air each molecule collides with some other molecule about 3000 million times every second and travels an average distance of about $\frac{1}{160000}$ inch between successive collisions." *

Assuming these collisions occur with perfect elasticity,

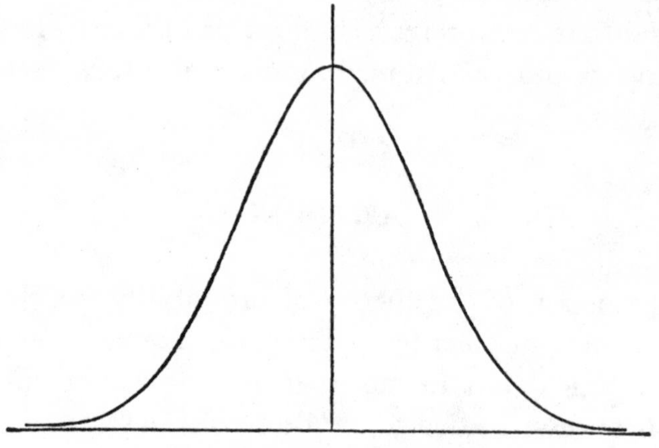

FIG. 80.—Normal probability curve.

i.e. no energy is lost, it can be inferred, based on the ideas of change, that at any instant there will be some molecules moving in all directions and with all velocities. Mathematically, it was shown first by Clausius, and later by Maxwell and Boltzmann, that $P = \frac{1}{3}\, nmV^2$ where P is pressure, n, the number of molecules in unit volume, m, the mass of each, and V^2, the average value of the square of the velocity.

To the problem of the distribution of velocities among the molecules, Maxwell applied Gauss' law of error (of importance in many branches of inquiry) derived from the theory of probability.

* Sir James Jeans, *The Universe Around Us* (New York: Macmillan, 1929).

The normal curve of error (see Fig. 80) may be obtained by the binomial expansion $(\frac{1}{2} + \frac{1}{2})^n$ as $n \rightarrow \infty$ This curve shows that in ordinary observation, small errors occur with larger frequency than great ones.

"The (kinetic) theory shows that molecules subject to chance collisions may be divided into groups, each group moving within a certain range of velocity in a manner illustrated in the diagram." * (See Fig. 81.) The resemblance of this curve to the normal curve of error is apparent.

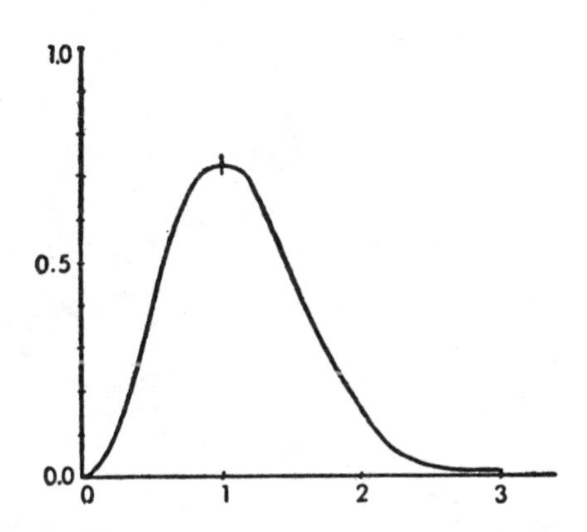

FIG. 81.—Velocity of molecules of a gas.

"The horizontal ordinate measures the velocity and the vertical ordinate, the number of molecules which move with it. The most probable velocity is taken as unity. It will be seen that the number of molecules moving with a velocity only three times the most probable velocity is almost negligible. Similar curves may be drawn to illustrate the distribution of shots on a target, or errors in a physical measurement, of men arranged in groups ac-

* Sir William Dampier, *A History of Science and its Relations with Philosophy and Religion* (New York: Macmillan, 1936).

cording to height or weight, length of life, or ability as measured by examination . . . ." *

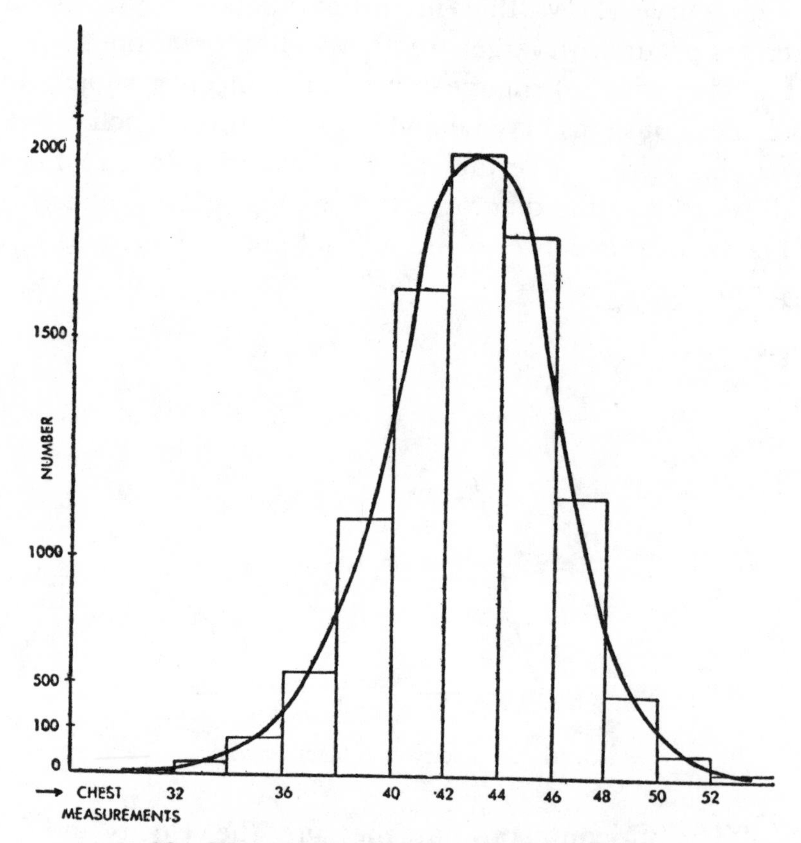

FIG. 82.—This distribution curve tells the chest measurement of Scottish soldiers. Incidentally, it also serves to describe phenomena as diverse as the following:

1. Age distribution of pensioners of a large concern.
2. Runs at roulette.
3. Scattering of bullets about a target.

STATISTICS IN ANTHROPOLOGY

The Belgian astronomer, L. A. J. Quételet (1796–1874) showed that the theory of probability could be applied

* Sir William Dampier, *op. cit.*

to human problems. Thus, the same distribution is found for "runs" at roulette, or in the distribution of bullets around the center of a target, as in the chest measurements of Scottish soldiers, or in the velocities of molecules in a gas.*

STATISTICS AND PAST EVENTS†

One of the most ancient problems in probability is concerned with the gradual diminution of the probability of a past event, as the length of the tradition increases by which it is established. Perhaps the most famous solution of it is that propounded by Craig in his *Theologiae Christianae Principia Mathematica*, published in 1699. He proves that suspicions of any history vary in the duplicate ratio of the times taken from the beginning of the history in a manner which has been described as a kind of parody of Newton's *Principia*. "Craig," says Todhunter, "concluded that faith in the Gospel, so far as it depended on oral tradition, expired about the year 880, and that, so far as it depended on written tradition, it would expire in the year 3150. Peterson, by adopting a different law of diminution, concluded that faith would expire in 1789!"

In the *Budget of Paradoxes* De Morgan quotes Lee, the Cambridge orientalist, to the effect that Mohammedan writers, in reply to the argument that the Koran has not the evidence derived from Christian miracles, contend that, as evidence of Christian miracles daily grows weaker, a time must at last arrive when it will fail of affording assurance that they were miracles at all: whence the necessity of another prophet and other miracles.

* *Ibid.*
† John Maynard Keynes, *A Treatise on Probability* (New York and London: Macmillan, 1921), chapter XVI, p. 184.

STATISTICS OF AIR-RAID CASUALTIES

Professor J. B. S. Haldane in a communication to *Nature* (October 29, 1938) discussed the mathematics of air-raid protection. A more bitter commentary on contemporary society would be hard to find, though it is coldly dispassionate and purely scientific in tone and purpose. It reads in part:

In view of the discussion which is occurring on this subject, it seems desirable to have some quantitative measure of the degree of protection afforded by a given shelter. In order to limit the problem we may consider only risks of death, and further confine ourselves to high-explosive bombs. Incendiaries have proved a negligible danger to life in Spain, and gas is also negligible except for babies and those whose respirators do not fit.

Consider a given type of bomb, say a 250 kilo. bomb, which is commonly used on central areas of Spanish cities, and a man in a given situation, whether in the street or in a shelter. Let n be the expected number of bombs falling in his neighborhood (say 1 square kilometer) during a war, the distribution of bombs over this area being supposed even, since aim is poor when cities are bombed. Let p be the probability that a single bomb falling at the point (x,y) in this area will kill him. Then the probability that he will be killed in the course of the war is $P = \int n/A \; p dx dy$, integration being taken over the whole neighborhood of area A.

The values of n and p will, of course, be different for each type of bomb, and the different expressions so obtained must be summed. Further, the man will be in different places during the war, and thus another summation is necessary. Finally, P must be summed for the whole nation.

The policy of evacuation is intended to reduce the value of n, even though it may increase that of p, as when a child is evacuated from a fairly solid house into a flimsy hut. The policy of dispersal within a dangerous area does not, of course, reduce either n or p. It merely ensures that no single bomb will kill a large number of people, while increasing the probability that any given bomb will kill at least one. It is likely to save a few lives by equalizing the numbers of wounded to be treated in different hospitals; and the psychological effect of having 20 killed in each of 10 areas may perhaps be less than that of 200 killed in one area. But, as it may actually increase the mean value of p by encouraging people to stay in a number of flimsy buildings rather than one strong one, it is at least as likely to increase the total casualties as to diminish them. The argument that a number of people must not be concentrated in one place in order that a single bomb should not kill hundreds is clearly fallacious when applied to a war in which the total casualties will be large. It is, however, true that a small group of key men each of whom can replace another should not be grouped together.

FOOTNOTES

1. A. Conan Doyle, *The Return of Sherlock Holmes*, "The Adventure of the Dancing Men."—P. 224.
2. It may also be that certain sentence structures which *look like* propositions are *neither* true nor false, but meaningless. There are, for example, propositional functions like "x is a y," or wholly meaningless statements like "A snark is a boojum." But neither of these need concern us at this point.—P. 228.
3. The following paradox which arises from the principle of insufficient reason is quoted by Keynes from the German mathematician von Kries (Keynes, *A Treatise on Probability*, London: Macmillan, 1921). Suppose that we know the specific volume of a substance to lie between 1 and 3; but we have no information as to the exact value. The principle of indifference would justify us in placing the specific volume between 1 and 2; or between 2 and 3 with equal likelihood. The specific density of a substance is the reciprocal of the specific volume; if the specific volume is V, the specific density is $1/V$, so that we know that the specific density must lie between 1 and $\frac{1}{3}$. Again, by the principle of insufficient reason, it is as likely to lie between 1 and $\frac{2}{3}$ as between

$\frac{2}{3}$ and $\frac{1}{3}$; but the specific volume being a function of the specific density, if the latter lies between 1 and $\frac{2}{3}$, the former lies between 1 and $1\frac{1}{2}$, and if the latter lies between $\frac{2}{3}$ and $\frac{1}{3}$, the former lies between $1\frac{1}{2}$ and 3. Whence it follows that the specific volume is as likely to lie between 1 and $1\frac{1}{2}$ as between $1\frac{1}{2}$ and 3, which is contrary to our first assumption that it is as likely to lie between 1 and 2 as between 2 and 3.—P. 230.

4. Dantzig, *Number, the Language of Science*, p. 67.—P. 232.
5. Charles S. Peirce, *Chance, Love, and Logic*.—P. 237.
6. For a lucid and admirably refreshing discussion of this and other problems of probability see Cohen and Nagel, *An Introduction to Logic and Scientific Method*, New York: Harcourt Brace, 1936.—P. 238.
7. See Appendix to this chapter.—P. 240.
8. As a matter of interest, there are 34,650 ways of arranging the letters of the word "Mississippi."—P. 242.
9. The reader should not be disturbed by the fact that WB and BW are represented simply by $2WB$. $2WB$ simply means two drawings in each of which there is one Black ball and one White, regardless of the order in which they appear.—P. 249.
10. Without troubling to remember the general formula, by the use of the famous triangle of Pascal, one can at once ascertain the coefficients of any binomial expansion:

$$\begin{array}{c} 1 \\ 1\ 2\ 1 \\ 1\ 3\ 3\ 1 \\ 1\ 4\ 6\ 4\ 1 \\ 1\ 5\ 10\ 10\ 5\ 1 \\ 1\ 6\ 15\ 20\ 15\ 6\ 1 \\ 1\ 7\ 21\ 35\ 35\ 21\ 7\ 1 \\ 1\ 8\ 28\ 56\ 70\ 56\ 28\ 8\ 1 \end{array}$$

By examining this array, the reader may determine for himself how each new line is formed.—P. 250.

11. Laplace, *Essai philosophique sur la probabilité*.—P. 253.
12. Quoted from C. G. Darwin, *Presidential Address to the British Association*, 1938.—P. 256.

Rubber-Sheet Geometry

It s-t-r-e-t-c-h-e-s.
—POPULAR ADVERTISEMENT

ONCE UPON a time seven bridges crossed the river Pregel as it twisted through the little German university town of Königsberg. Four of them led from opposite banks to the small island, Kneiphof. One bridge connected Kneiphof with another island, the other two joined this with the mainland. These seven bridges of the eighteenth century furnished the material for one of the celebrated problems of mathematics.

Seemingly trivial problems have given rise to the development of several mathematical theories. Probability rattled out of the dice cups of the young noblemen of France; Rubber-Sheet Geometry was brewed in the *gemütliche* air of the taverns of Königsberg. The simple German folk were not gamblers, but they did enjoy their walks. Over their beer steins they inquired: "How can a Sunday afternoon stroller plan his walk so as to cross each of our seven bridges without recrossing any of them?"

Repeated trials led to the belief that this was impossible, but a mathematical proof is based neither on beliefs nor trials.

Far away in St. Petersburg, the great Euler shivered in the midst of honors and emoluments, as mathematician at the court of Catherine the Great. To Euler, homesick and weary of pomp and circumstance, there came in some strange fashion news of this problem from his fatherland. He solved it with his customary acumen. *Topology*, or *Analysis Situs* was founded when he presented his solution to the problem of the Königsberg bridges before the

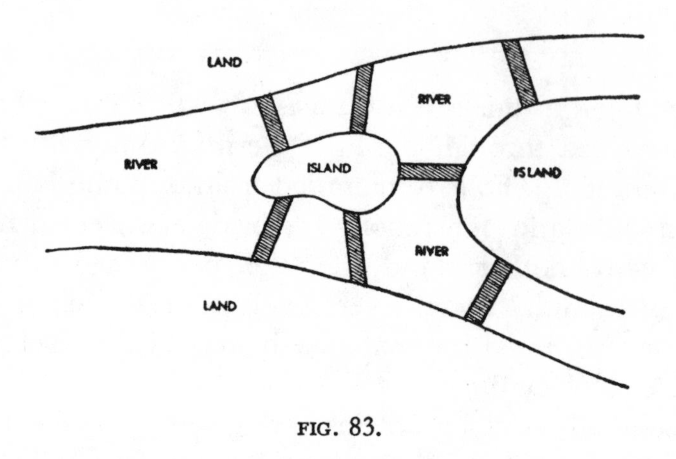

FIG. 83.

Russian Academy at St. Petersburg in 1735. This celebrated memoir proved that the journey across the seven bridges, as demanded in the problem, was impossible.

Euler simplified the problem by replacing the land (in Fig. 83) by points, and the bridges by lines connecting these points. Once this simplification has been effected, can Fig. 84 be drawn with one continuous sweep of the pencil, without lifting it from the paper? For this is the equivalent of physically traversing the seven bridges in one journey. Mathematically, the problem reduces to one of *traversing a graph*. A "graph," as the term is used here, is simply a configuration consisting of a finite number of points called vertices and a number of arcs. The vertices are the end points of the arcs, and no two arcs have a common point, except, perhaps, a common vertex.

A vertex is odd or even, according as the number of arcs forming the vertex is odd or even.

A graph is *traversed* by passing through all the arcs exactly once. Euler discovered that this can be done, starting and finishing at the same point, if the graph contains only even vertices. Further, he discovered that if the graph contains at most two *odd* vertices, it may also be traversed, but it is not possible to return to the starting point. In general, if the graph contains $2n$ odd vertices, where n is any integer, it will require exactly n distinct journeys to traverse it.[1]

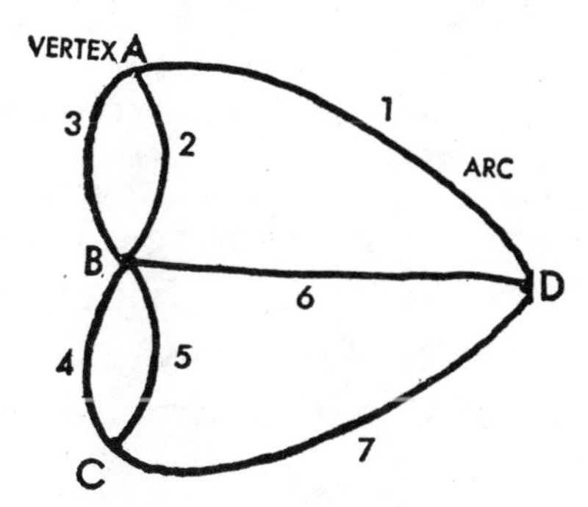

FIG. 84.—A graph with four vertices and seven arcs, illustrating the Königsberg bridges.

Figure 84 is the graph of the seven bridges of Königsberg. Since all four vertices are *odd*, that is, each one is the end point of an odd number of arcs, $2n = 2 \times 2$, and, therefore, two journeys are required to traverse the graph —a single journey will not suffice.

If, as in Fig. 85, an additional arc is drawn from *A* to *C*, representing another bridge, and the arc *BD* is removed, all the vertices become even; *A*, *B*, and *C* of order 4, and *D* of order 2, and the graph can be traversed in a single journey. If the arc *BD* is *not* removed, the stroller may take his walk, cross all the bridges only once, but will find that he cannot finish at the point where he started. Thus, if he starts at *D*, he will finish at *B*, and vice versa. (Note: he must start his walk at an odd vertex.)

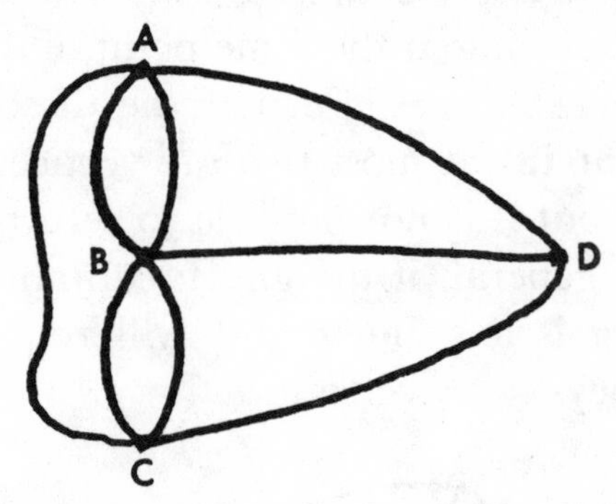

FIG. 85.—A graph with four vertices and eight arcs.

The problem of the seven bridges is representative of a group of problems, some dating back to antiquity. They exemplify the difficulty of mentally grasping the true geometric properties of all but the simplest figures.

FIG. 86.

In the history of magic and superstition, the figure shown above (Fig. 86) has played an important part as a talisman against all forms of misfortune. Known to the Mohammedans and the Hindus, to the Pythagoreans and the Cabalists, it was sometimes carved on babies' cribs to fend off evil, while in more practical countries it was painted on animals' stalls. It is possible to traverse this star, returning to the starting point, with a single pencil stroke.

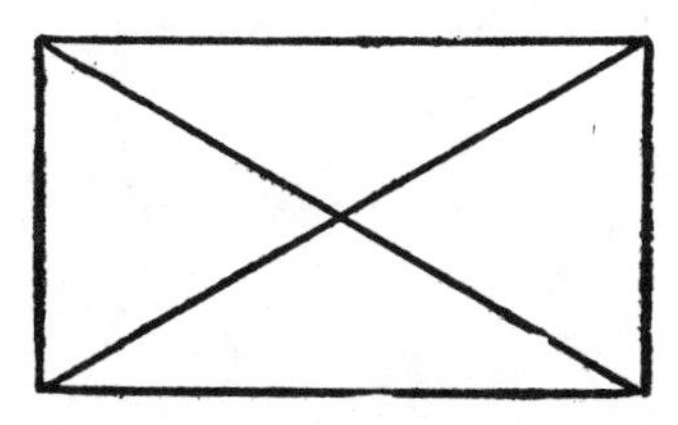

FIG. 87.

Euler's rule explains why the figure in Fig. 87 cannot be traversed with a single stroke, for there are 5 vertices, 4 of which are the terminal points of three arcs, in other words, of an odd order, and thus *two* journeys are required.

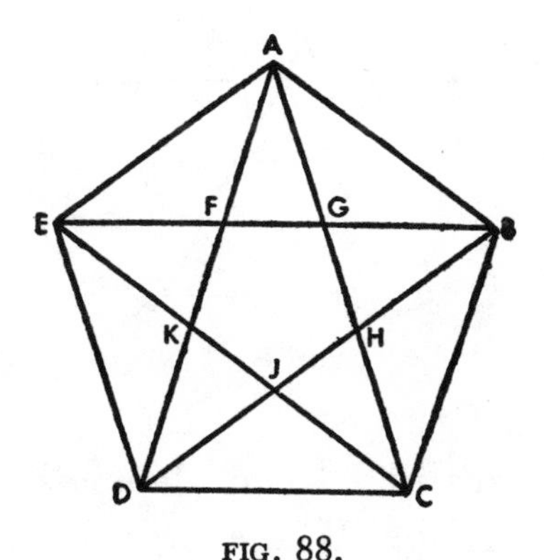

FIG. 88.

The pentagon in Fig. 88, far more complicated in appearance, can be traversed in a single journey. Starting at the point *A*, the journey would successively pass through points *ABCDEFGBHJDKFAGHCJKEA*.

Even Fig. 89 yields to a single journey, for example: *ABCcc'CDEee'EFAaa'AbBDdEfFBb'Cd'DFf'A.*

*

In grappling with the problem of the seven bridges, Euler did much more than merely solve a puzzle. He recognized the existence of certain fundamental properties

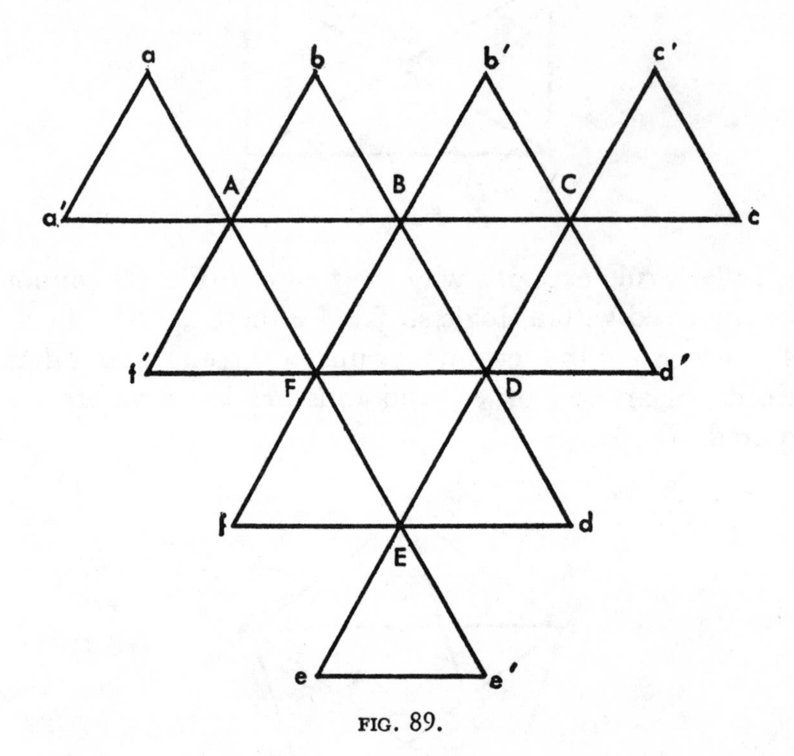

FIG. 89.

of geometric figures in no way dependent upon, or related to, size or shape. These properties are functions solely of the general position of the lines and points of a figure. For example, on a line *ABC*, the fact that the point *B* lies *between* the points *A* and *C* is just as important as the fact that the line *ABC* is straight or curved, or has a certain length. Again (Fig. 90), when an interior point of a triangle is connected with a point outside, the line joining them must cut one side of the triangle—a fact which is just as important as that the angles of a triangle equal 180°. It is the study of such properties, properties which remain unaffected when the figure is *distorted*, which constitute the science of *topology*. Topology is a geometry of place, of position (which accounts for the name *Analysis Situs*), as distinguished from the metric geometries of Euclid, Lobachevsky, Riemann, etc., which treat of lengths and angles.

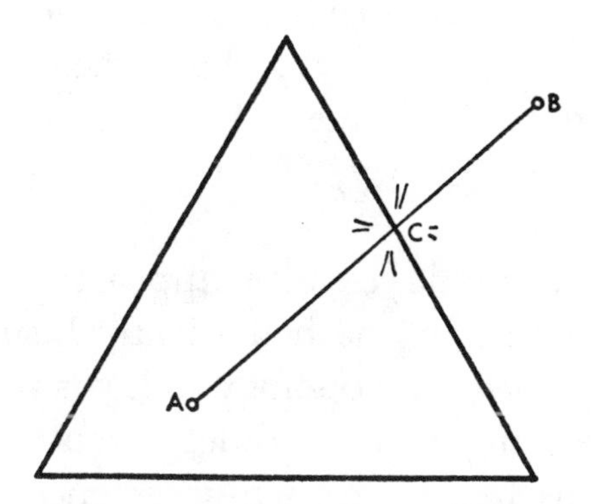

FIG. 90.—The line joining the interior point *A* to the exterior point *B cuts* the triangle at *C*. No matter how the line is drawn, it must cut the triangle at some point.

In topology we never ask "How long?" "How far?" "How big?"; but we do ask, "Where?" "Between what?" "Inside or outside?" A traveler on a strange road wouldn't ask "How far is the Jones farm?" if he didn't know the direction, for the answer, "Seven miles from here," would not help him. He is more likely to inquire, "How do I get to the Jones farm?" Then, an answer like, "Follow this road till you come to a fork, then turn to your right," will tell him just what he wants to know. Because this answer says nothing about distances and does not describe whether the path is straight or curved, it may seem nonmathematical, yet it bears the same relation to the first answer that topology bears to metric geometry.

Topology is a *non-quantitative* geometry. Its propositions would be as true of figures made of rubber as of the ordinary rigid figures encountered in metric geometry. For that reason it has been picturesquely named *Rubber-Sheet Geometry*.

*

Geometry was a very fashionable subject in the nineteenth century. The eighteenth century had been devoted to the calculus and to analysis. The nineteenth belonged in large part to the geometers. It was inevitable that topology, then in its infancy, should receive its share of attention.

The first systematic treatise appeared in 1847, the work of the German mathematician Listing, entitled *Vorstudien zur Topologie*. Topology today is still concerned with the same thing as when Euler invented it, although its language, as befits a grown-up science, has become more abstruse. It is now defined as the study of properties of spaces, or their configurations, invariant under continuous one-to-one transformations; it remains the study of the position and relation of the parts of a figure to each other without regard to shape or size. Indeed, although topology was weaned on bridges, it now feeds on pretzels and doughnuts as well as upon other more curious and less digestible objects.

*

Even a glance at one or two of the theorems of this bizarre branch of mathematics requires the introduction of a new terminology.

Poincaré pointed out that the propositions of topology

have a unique feature: "They would remain true if the figures were copied by an inexpert draftsman who should grossly change all the proportions and replace the straight lines by lines more or less sinuous." [2] In mathematical language, the theorems are not altered by any continuous point-to-point transformation. Figure 91 is an example of a plane triangle drawn by an expert draftsman, Fig. 92 its distorted surrealist twin. Nevertheless, topologically, 92 is a perfect copy of 91. The straight lines are curved, the angles changed and distorted and the lengths of the sides altered; but there remain geometric properties common to both figures. These properties which have been unaffected by the distortion are *invariants*.[3]

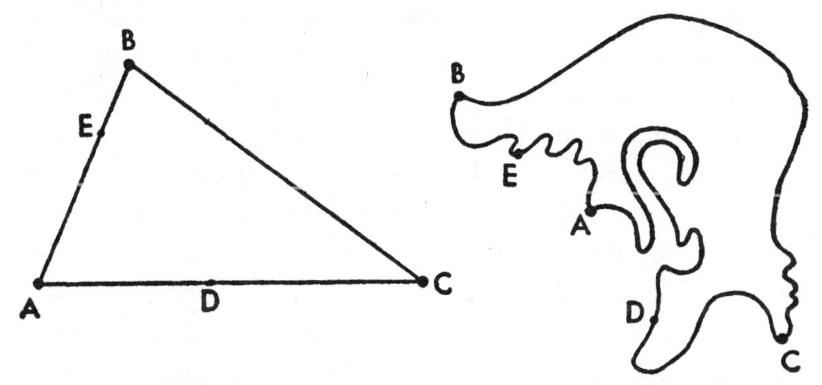

FIG. 91.—A plane triangle. FIG. 92.—Its surrealist twin.

In Fig. 91, the point D lies between points A and C, and E, between A and B. In Fig. 92 that order has been preserved. The *order* of the points is, therefore, invariant under the *transformation* which brought about this distortion. Figure 91 could have been transformed in some other way. If it had been cut from a sheet of rubber and the rubber triangle twisted, stretched and distorted in every possible way without tearing, the order of the points would still remain invariant.

The invariants of rigid bodies under ordinary motion are even more familiar, but they are so much a part of our lives that we never give them much thought. Yet our existence would be quite unthinkable without them. A rigid body suffers no change in size or shape when moved about. Its metric properties are invariant. In simple terms, ordinary motion has no distorting effect. The derby, bought in London, still fits when transported to New York. A measuring rod is the same in length after being moved from the top of a mountain to the bottom of the sea. A latch key fits a lock whether the door is swung open or shut. A steamer appears smaller on the horizon; but no one would maintain seriously that it shrinks as it steams away. And the philosopher's armchair fits him in every corner of the room, regardless of how he changes its position or his philosophy.

Such invariants we take for granted. To the mathematician, however, obvious things serve as valuable clues, and he rarely dismisses the obvious as unimportant. He carefully notes that the size and shape of *rigid* bodies are unaffected by motion, and reports in technical language that *the metric properties of rigid bodies are invariant under the transformation of motion.* He then considers those bodies which are *not* rigid, and which *do* change in size and shape when moved about, and seeks their geometric invariants. Topology embraces these invariants and integrates them into a mathematical system.

*

According to an ancient tale, a Persian Caliph, with a beautiful daughter, was so troubled by the number of her suitors that he was forced to set up qualifying rounds to determine the finals. The aspirants for his daughter's hand were presented with a problem (Fig. 93): To connect the corresponding numbers of the symmetric figures by lines which do not intersect.

That was simple. But the Caliph's daughter was not so easily won, for her father also insisted that the surviv-

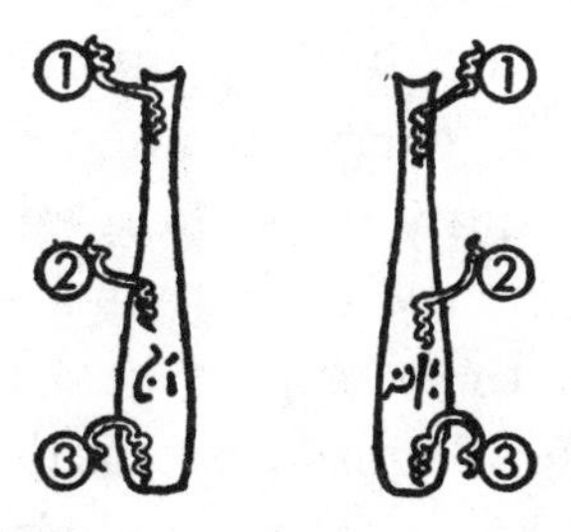

FIG. 93.—Connect 1 with 1, 2 with 2, and 3 with 3 by nonintersecting lines.

ing suitors join the corresponding numbers shown in Fig. 94.

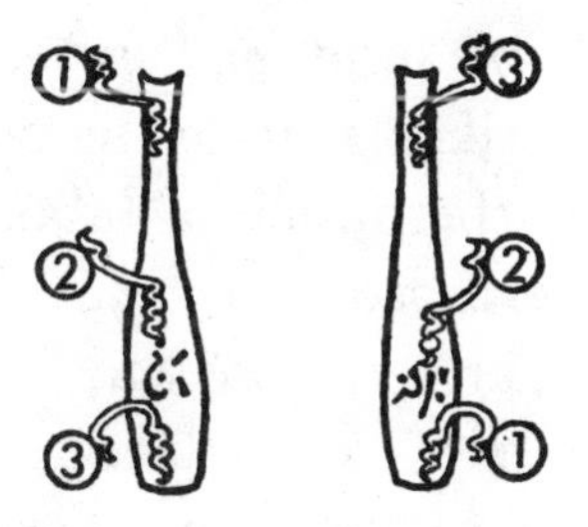

FIG. 94.—Try connecting the corresponding numbers by nonintersecting lines.

Unless the Caliph relented, we may assume that his daughter died an old maid, for this problem cannot be solved. Two lines may be drawn connecting any two corresponding numbers, but the third cannot be traced without crossing one of the other two. Again, we see why the mathematician never rejects the obvious. The problem of Fig. 93 is easy. That of Fig. 94 seems just as easy, but it is actually impossible. In what essential respects do the two differ?

As early as the nineteenth century, the physicist Kirchhoff recognized the importance of investigations in topology in order to aid in the solution of problems connected with the branching out and intertwining of wires or other conductors carrying an electric current. And, curiously enough, many important effects in physics have since been found exactly analogous to the spatial relationships displayed in the Caliph's problem.

The first real step in the systematic attack on all such problems was taken in the nineteenth century by the French mathematician Jordan. His theorem is as fundamental and important for topology as the Pythagorean theorem for metric geometry. It bears no resemblance to any previously stated mathematical theorem. It says simply that "*Every closed curve in the plane which does not cross itself divides the plane into one inside and one outside.*"

Doubtless this strikes you as being either idiotic or wonderful. Had mathematicians labored for centuries to bring forth such a mouse? But Jordan's theorem only *seems* idiotic, for when expressed in formal terms it looks so obvious as to be hardly worth repeating. In truth, it is a wonderful theorem, because it is so simple, so unassuming, and so important.

A curve which divides the plane into one inside and one outside is called *simple*. This is a simple curve:

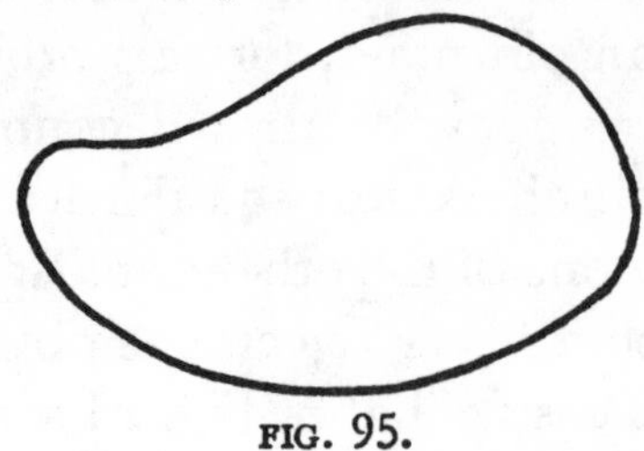

FIG. 95.

But these aren't:

FIG. 96(a). FIG. 96(b).

Nor is this:

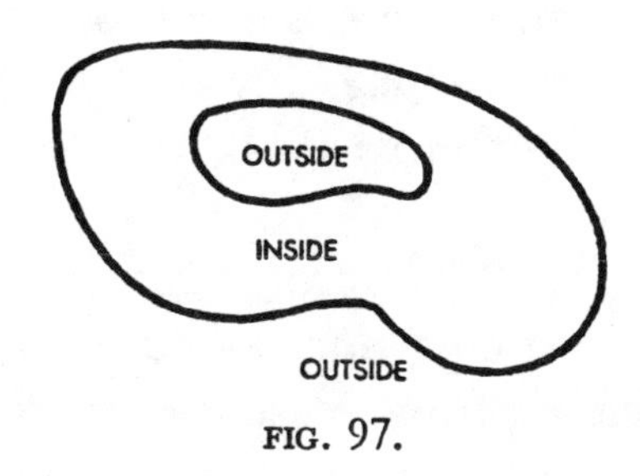

FIG. 97.

The three curves in Figs. 96a, 96b, and 97 do not fall within Jordan's definition of simple connectivity. The first has two insides and one outside, the second, several insides and one outside, and the area enclosed by the smaller curve in the third is also considered "outside," and not inside. It must be conceded that Jordan's theorem seems trivial when applied to easy figures. But it is not so easy to believe that the curve in Fig. 2, in spite of its tortuous appearance and labyrinthlike character, has only one inside. Strange as it may seem, such a curve may be regarded as a deformed circle. This might be demonstrated quite easily if it were made of a piece of string or a rubber band, for it could then be retransformed into a circle merely by smoothing out the twists and kinks. In metric geometry, a circle is defined as the locus of all points equidistant from a given point, which means that all the radii of a circle are of equal length. But in topology "equal length" has no significance. Thus a circle is thought of as a curve which has the fundamental property of dividing the entire plane into one inside and one outside. Any curve, however deformed, which has this property, may be regarded as the *topological equivalent of a circle.* It follows that *every simple curve in the plane is topologically equivalent to a circle.*

*

Jordan's theorem, when extended to three dimensions, states that any closed *surface*, any two-dimensional manifold[4] which does not cross itself, divides space into an inside and an outside.

Think of the room you are sitting in. The air in the room, all the furnishings, and you, are *inside.* The rest of the entire universe, from Vesuvius to the core of the earth, from Times Square to the Rings of Saturn and beyond, are *outside.* The gas in a balloon is *inside,* while everything else, in all possible directions, including the hopes and fears in the head of the balloonist, are *outside.* The circulatory system of the body is a two-dimensional manifold difficult to visualize. Nevertheless it is *simply connected.* It divides space into one inside and one outside. Inside flows the bloodstream, outside there are the countless cells of the body that twine and intertwine with the blood vessels, and beyond, the entire universe.

The restriction that the two-dimensional manifold shall not cross itself does not recall to mind any that do. Yet such manifolds are the center of attraction at the Institute for Advanced Study at Princeton where learned and famous mathematicians discourse strangely, almost like Alice's Walrus, on pretzels, knots, and doughnuts.

The pretzel is an object of interest, not only for its

gastronomical properties, but also for its topological ones. It is an example of a two-dimensional manifold which does *not* obey Jordan's theorem, for it crosses itself. But the pretzel is too difficult for our modest mathematical

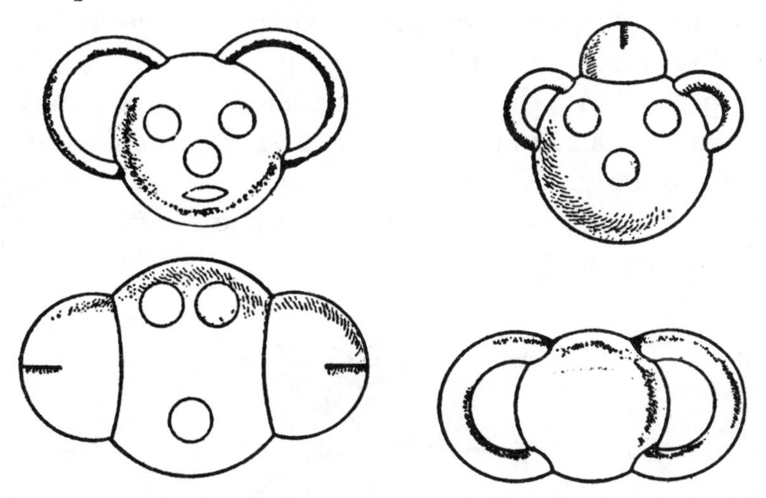

FIG. 97(a, b, c, d)—Not the creations of Walt Disney nor Picasso's impressions of the human form divine, but the objects of serious mathematical lucubrations at Princeton.

equipment. We must be content with manifolds which do obey Jordan's theorem. They cause enough trouble.

Figure 98 shows a *ring*—the portion of the plane bounded by two concentric circles. A ring is a figure

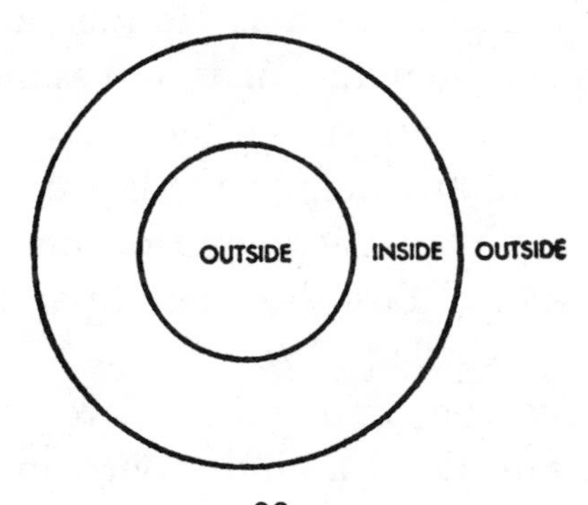

FIG. 98.

which is *not* simply connected since its boundary consists of *two* curves rather than one. How can we differentiate the inside from the outside?

Many of the difficulties we experience in explaining and analyzing spatial problems spring from the limitations of language revealed by such a question. One is apt to sympathize with the bibulous gentleman who was staggering around a cylindrical column on a Paris boulevard, weeping bitterly. "For heaven's sake," asked a curious passerby, "what's wrong?" "I'm walled in," wailed the toper, "walled in."

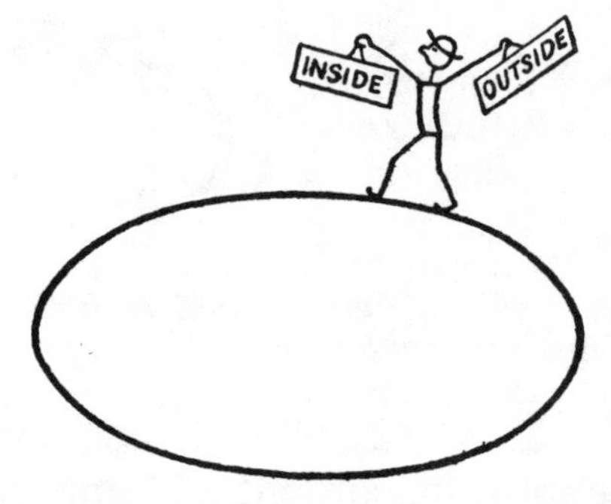

FIG. 99.—The man is walking counter-clockwise on the boundary of the curve. To the left of him is inside, to the right, outside.

Purely relative terms, such as "inside" and "outside," may confuse the mathematician as well as the melancholy boulevardier. The sole recourse is to agree upon a formal definition. A familiar analogy readily comes to mind: All parts of New York City lying on one side of Fifth Avenue are labelled "East," while all parts lying on the opposite side are labelled "West."

Intuitively, everyone knows the difference between the inside and the outside of a circle. But can this intuitive notion be translated into precise terms? Since no one has the slightest difficulty in distinguishing between left and right, and the notions of clockwise and counterclockwise also occasion little confusion, "inside," and "outside" may be redefined in terms of left, right, clockwise, and counterclockwise. Thus, for instance, starting on the circumference of a circle and proceeding in a counterclockwise direction, "inside" is defined as the region to the left, "outside," the region to the right.

The application of this definition to a nonsimply-connected manifold, such as the ring, requires a slight artifice. By cutting any nonsimply-connected manifold, it may be transformed into one which is simply connected.

FIG. 100.—Inside and outside of the cut ring.

Thus, while inside and outside appear to have little significance in relation to the ring (Fig. 98), the simple operation of cutting transforms the ring into a new manifold (Fig. 100) to which the definition is plainly applicable. The mathematician agrees that those regions which are "inside" after the ring is cut were "inside" before it was cut; and those regions which are "outside" after the cut were "outside" before. The doughnut presents the same problem in three dimensions as the ring in two. "Is the hole part of the inside or the outside of the doughnut?" If we relied entirely on the experience gained at the breakfast table, we might assert that the hole is inside. But the few facts thus far gathered would give rise to some doubts. It turns out that the hole *inside* the doughnut is *outside*. Of course, the first impression was not an optical illusion. The conclusion that the hole is outside is purely conceptual and must be regarded as the logical consequence of certain definitions.

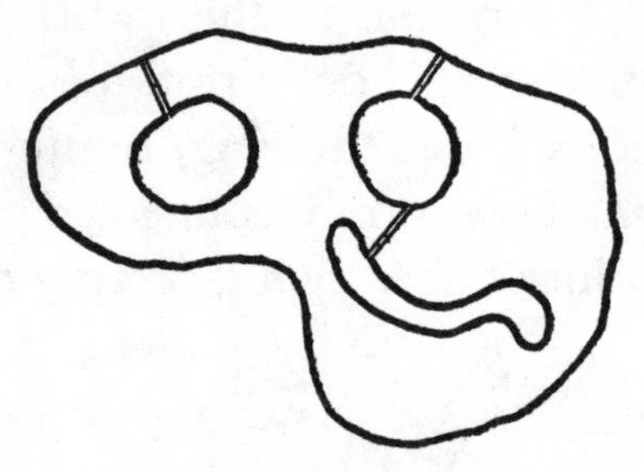

FIG. 100(a).—A triply-connected curve. It takes three cuts to make it simply connected.

As in *two* dimensions any simply-connected manifold is the equivalent of a circle, so in *three* dimensions any simply-connected surface is the equivalent of a *sphere*. By a gradual deformation, without tearing, any simply-connected three-dimensional object can be transformed into a sphere. A *doughnut* cannot be so transformed whence it follows that a doughnut is *not* simply connected. But an operation similar to the one performed on the ring—a simple cut—turns the doughnut into a sausage, which is

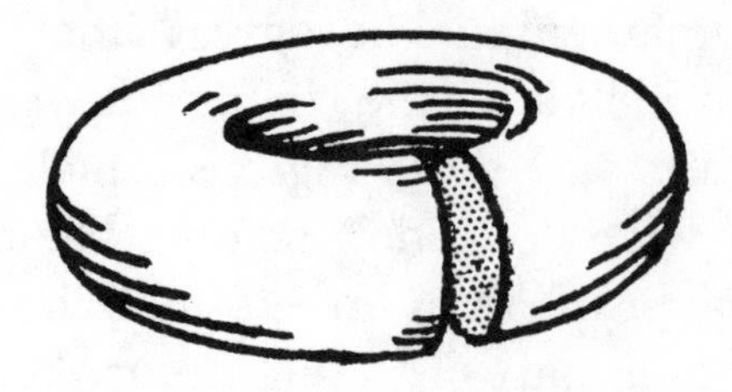

FIG. 101.—The doughnut becomes a sausage.

simply connected and the topological equivalent of a sphere.

The pretzel, together with the other objects shown (Fig. 102) are some of the more difficult manifolds studied in topology. None is simply connected, none can be transformed into a sphere. But by a number of cuts, similar to the ones performed on the ring and doughnut, these complex manifolds may be transformed into simply-connected configurations. Thus, with a sufficient number of cuts it

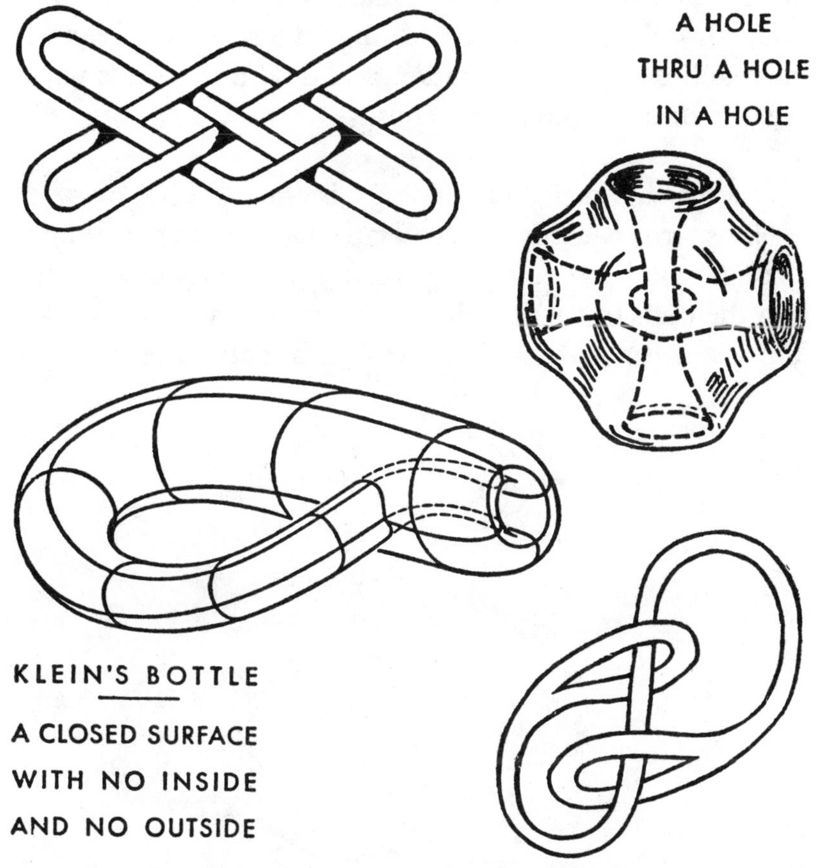

FIG. 102—Weird topological manifolds—exalted relatives of the pretzel.

is possible to change even the most tortuous pretzel into the equivalent of a sphere.

The number of cuts necessary to effect such a transformation is not a matter of chance, but perfectly determinate, and depends upon the connectivity of the manifold. A general rule may be formulated which will apply to both fantastic objects and easy ones. As in all mathematical inquiries, only such a rule will reveal the underlying principle; accordingly, topologists do not stop with the consideration of three-dimensional manifolds, however complicated and forbidding. They go far beyond the reaches of the imagination and devise theorems valid even for *n*-dimensional pretzels.

*

One of the curios of topology is the Möbius strip. A Möbius strip is easily constructed. Take a long rectangle (*ABCD*) made of paper (Fig. 103), give it a half-twist and join the ends so that *C* falls on *B*, and *D* on *A* (Fig. 104). This is a *one-sided* surface, and if a painter agreed to

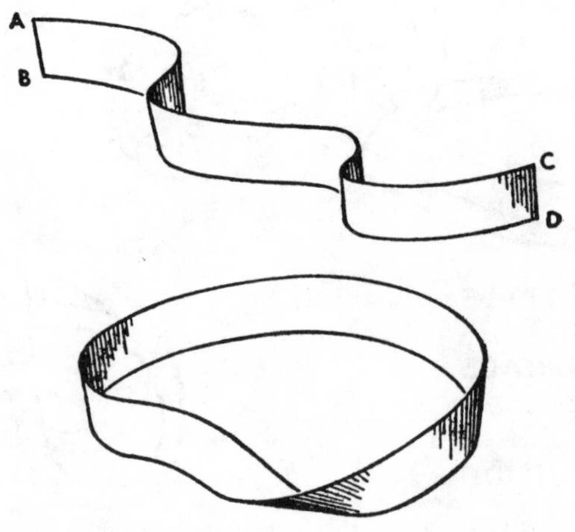

FIGS. 103, 104.—The Möbius strip—a one-sided, two-sided surface.

paint only one side of it, his union would interfere because in painting one side he would be painting both sides.[5] If the strip had not been twisted before gluing the ends together, a cylinder would have resulted—which is evidently a *two*-sided surface. However, the half-twist eliminated one of the sides. Incredible? You may convince yourself. Draw a straight line down the center of the strip, extending it until you return to the point at which you started. Now separate the ends of the strip and you will find that *both sides* are covered by the straight line even though in drawing it you did not cross any edges. Had you followed this same procedure with a cylinder, you would have had to cross *over the edge*, to get

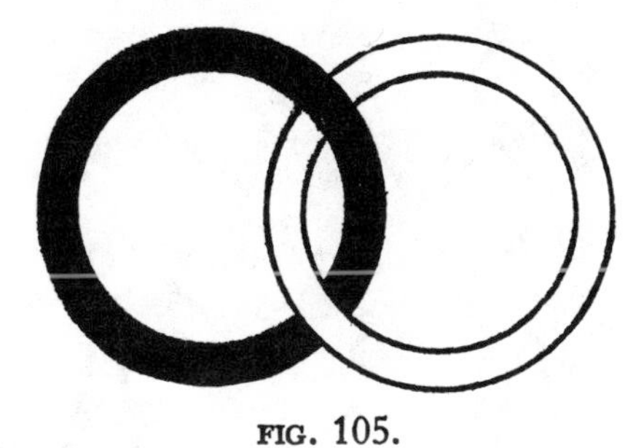

FIG. 105.

from one side to the other. Although every dictate of common sense indicates that the strip with the half twist has two bounding edges, we have proved it has only one. For any two points on the Möbius Strip may be connected by merely starting at one point and tracing a path to the other without lifting the pencil or carrying it over any boundary.

There is a good bit of amusement and interest in making such a strip for yourself. When you have studied the properties described, cut it in half with a pair of scissors along a line drawn down the center. The result will be astounding! And you can continue twisting and cutting a few more times for still further surprises.

Two interlocking iron rings are shown in Fig. 105. It is perfectly evident that they cannot be separated unless one of the rings is broken. But being perfectly evident, how shall we prove it? Before topology was invented, none of the existent tools of mathematics was suited for such a job. Only the creation of special tools made it possible to give an analytic proof of so evident a fact.

Here is a similar problem. Tie a piece of string to each of your wrists. Tie a second piece of string to each of the wrists of a partner in such a way that the second piece loops the first (Fig. 106).

FIG. 106.

Do you think you can separate yourself from your partner without tearing the string? Although this *looks* like the problem of separating the two rings, which we agreed is impossible, this feat can be accomplished. Try working it.

With a topologist and a pair of scissors handy (for accidents), you might try removing your vest *without removing your coat*. No fourth dimension is required. Merely remember the conditions of the problem. The coat may be unbuttoned, but at no point during the removal of the vest may your arms slip out of the coat sleeves.

FIG. 107.—The above portrays the trade-mark used by a well-known brewer. The three rings have this strange relation to one another. If any one ring is removed, the other two are found not to be joined. *Thus no two rings are joined, but all three are.* To put it more simply, no two rings are joined, but each holds the other two.

Topology is one of the youngest members of the mathematics family, but still it claims its problem child. While some mathematicians have been content to concentrate on the pretzels, knots, and doughnuts of analysis situs, a determined band of mathematical pediatricians have focused their attention exclusively on the Four-Color Problem. For a short while in the nineteenth century, it was thought that the child had been cured and its problem unraveled, but these were vain hopes, and the four-color puzzle continues to baffle the leading topologists.

At one time or another everyone has had experience in map coloring. Maps depicting the Holy Roman Empire, the cotton states of the South before the Civil War, or the rescrambling of Europe by the Treaty of Versailles, are painfully outlined every school day. Recently the business has become more hectic than ever. Stout crayons and a good eraser must always be at hand. Students discover early in their cartographical career that if a map is to be colored, countries having a common boundary, such as France and Belgium, must be colored differently so that they can be distinguished at a glance. The generalization of that idea led to the question "How many colors are necessary to color a map, with any number of countries, so that no two countries which adjoin on a frontier shall have the same color?" This problem had troubled cartographers for many years.

FIGS. 108, 109, 110.

Figure 108 illustrates an island in the sea. Each of two countries owns part of the island. Three colors are required for this map—one for the sea and one for each of the two countries.

For depicting the island in Fig. 109 four colors are required. The map with more regions, as in Fig. 110, also requires only four colors. The reason is not hard to find, for the country in the center, marked 1, may be the same color as the sea without causing any confusion.

Figures 111 and 112 respectively require three and four colors, even though they contain many more regions than any of the maps above.

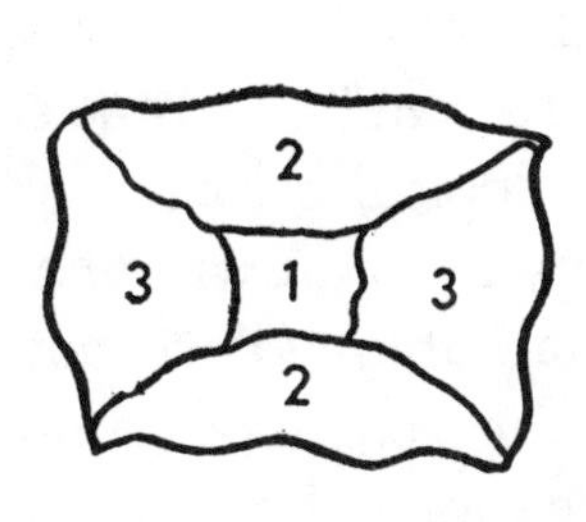

FIG. 111.—An island owned by five countries, requiring only three colors to map.

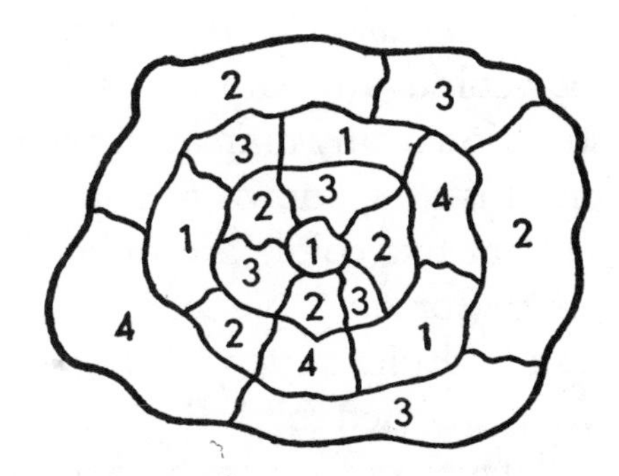

FIG. 112.—An island with nineteen counties. Only four colors are needed to map it.

It is quite natural to suppose that as maps grow more complicated, depict more countries, additional colors will be required to differentiate any two adjoining territories. Strangely enough, mathematicians have thus far found it impossible to construct a plane map for which four colors would *not* suffice. At the same time no one has been able to prove that four colors *would be sufficient* for any possible map.

The classical problem is concerned with the number of colors required to map any number of regions on a sphere. Though it has been shown that four colors are necessary, and five colors sufficient, the standard mathematical requirement, which is to find the one condition *both* necessary *and* sufficient, has not yet been satisfied.

Paradoxically, the problem has not been solved for a sphere or for a flat surface, although it has been solved for much more complicated surfaces, like the torus (doughnut), or the sphere with handles.

A. B. Kempe, English mathematician and barrister, author of the celebrated little book with the provocative title, *How to Draw a Straight Line*, offered a proof in 1879 that four colors are both necessary and sufficient for the construction of any map on a sphere. Unfortunately, Kempe's proof is now known to contain a fatal logical error.

That five colors are sufficient for any map drawn on a sphere, or on a plane, is in itself remarkable. The proof rests on Euler's even more remarkable theorem about simply-connected solids that states that the sum of the vertices and faces of any such solid is equal to the sum of the edges plus two:

$$V + F = E + 2$$

Euler's theorem is the simplest universal statement about solids. The underlying idea was familiar to Descartes, but very likely his proof was unknown to Euler.

We know that any three-dimensional solid which is simply connected is the topological equivalent of a sphere. From this fact and from Euler's theorem, there is one interesting consequence: Consider a hollow cube made of rubber. It is bounded by six faces, twelve edges, and eight vertices. Inflate this cube until it resembles a sphere. The faces of the cube are then regions of the sphere; the edges of the cube, the boundaries of these regions; and the vertices, points where three regions meet. The exercise of coloring the sphere is thus seen to be governed by Euler's theorem. For, if each region represents a country; each curved line, the boundary between two countries; and each vertex the juncture of three countries, the number of countries plus the number of points at which three countries meet is equal to the number of boundaries + two. In this way we see how Euler's theorem is extended to curved figures.

For a solid with a hole, such as a doughnut, the theorem fails. Indeed, it fails for any nonsimply-connected solid. In short, Euler's theorem is applicable in topology *only* when each of the faces of the figure is simply connected, and the entire surface is simply connected.

*

Of those who have made essential contributions to topology, L. J. Brouwer, the Dutchman, is one of the greatest. Particularly in the theory of point sets, Brouwer's to-

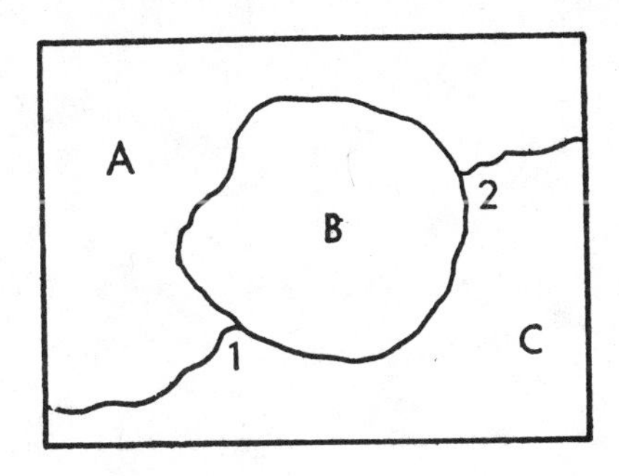

FIG. 113.—At the points 1 and 2, all three countries, A, B, C, meet.

pological theorems have proven of signal importance. But it is not his technical contributions which concern us here. In 1910 he published a problem, based on an idea of the Japanese mathematician Yoneyama, which illustrates beautifully the difficulties and subtleties of topology. The solution of this problem will perhaps leave you dissatisfied, but it cannot fail to challenge your imagination.

Figure 113 is a map of three countries. The points marked 1 and 2 are rather singular, for at both of these points all three countries meet. Manifestly such points are scarce on any map, no matter how complicated, for there are not many geographical instances of three coun-

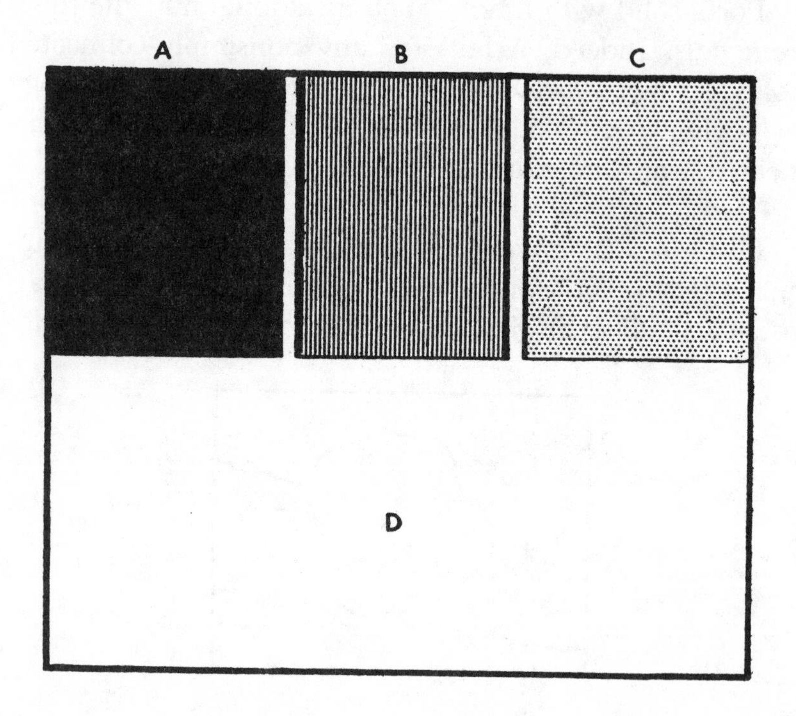

FIG. 114.—Countries A, B, C are separated by unoccupied corridors and D is unclaimed land.

tries meeting at a single point. But even if there were many such points, if it were a very queer map, their number would always be small compared to the totality of points along all the boundary lines. It is reasonably certain that a boundary point, chosen at random, on *any* map, will be the meeting place of at most two countries.

Now Brouwer concocted an example, at first sight

wholly unbelievable, of a map of three countries, on which *every single point along the boundary of each country is a meeting place of all three countries.*[6]

Consider the map in Fig. 114.

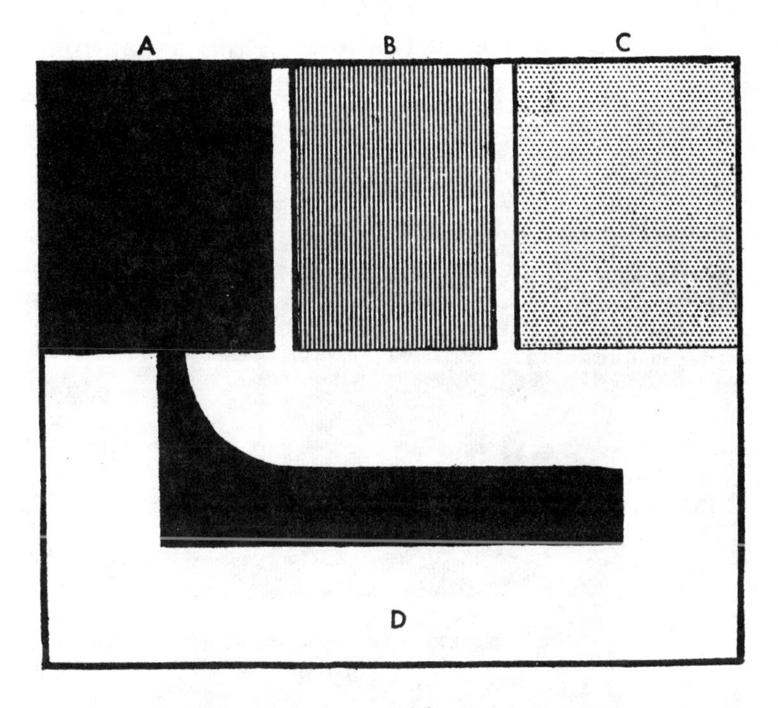

FIG. 115.

None of the nations borders on any of its neighbors, and the white unmarked portion of the map is intended to represent unclaimed territory. In keeping with the spirit of *Lebensraum*, Country *A* decides to extend its sphere of influence over the unclaimed land by grabbing a substantial portion. Accordingly, it sends out a corridor which does not touch the land of either of its neighbors, but leaves no point of the remaining, unclaimed land more than one mile from some point of the enlarged

Country *A*. It has now spread itself over the map as in Fig. 115.

Country *B*, instead of applying sanctions, decides to grab a share before it is too late. With becoming restraint, as well as with an eye to its neighbors' greater strength, *B* extends a corridor to within a half-mile of every point of the remaining unclaimed land. This corridor alters the map like this:

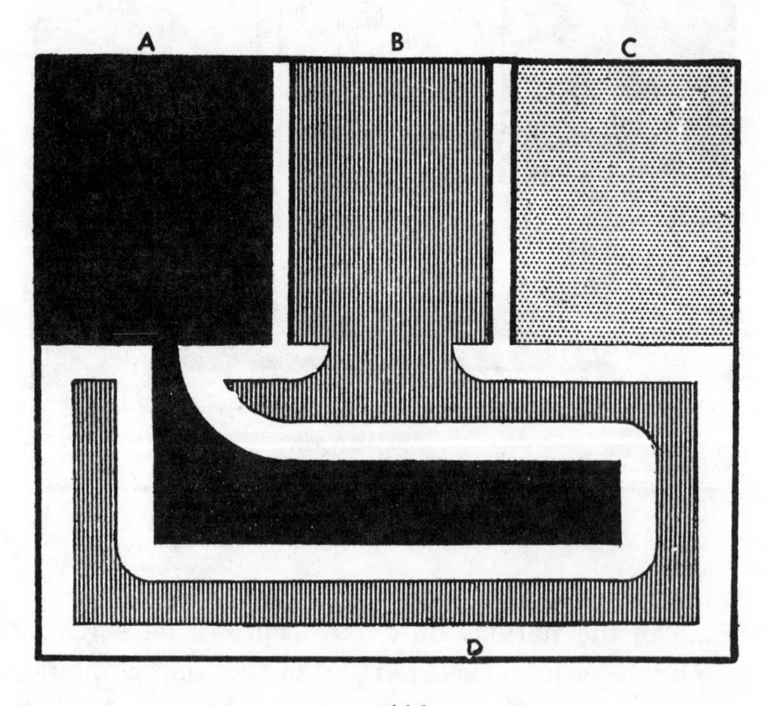

FIG. 116.

Of course Country *C* will not be left behind. It builds a corridor which approaches within a third of a mile of every point of the remaining unclaimed land but, just as the other two corridors, touches on no country but its own. The new map is shown in Fig. 117.

By now everyone should be quite content. On the contrary; this is only the beginning. Country *A* has the shortest corridor. Intolerable state of affairs which must be remedied *sofort*. It decides upon a new corridor to extend into the remaining territory which shall approach every point of that territory within a quarter of a mile (Fig. 118).

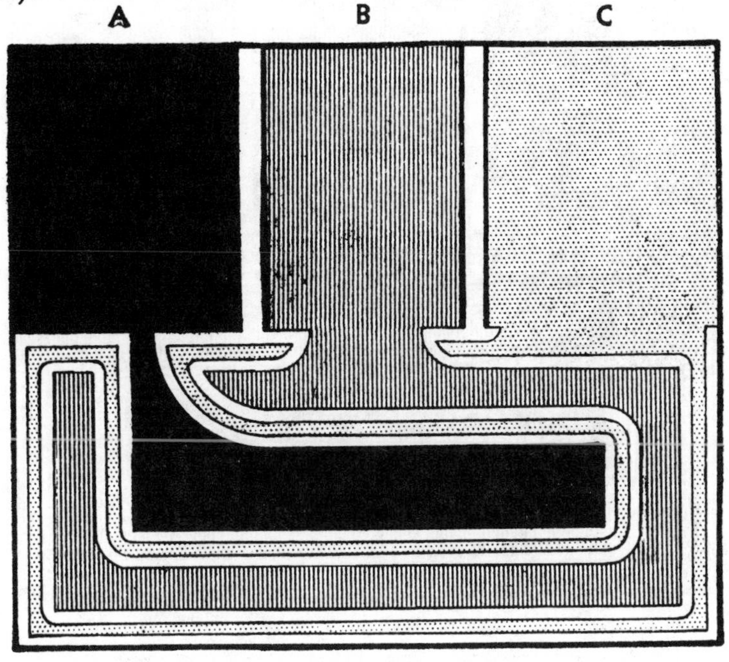

FIG. 117.

Country *B* follows with a corridor which approaches each unoccupied point within a fifth of a mile, Country *C*'s corridor comes within a sixth of a mile of each unoccupied point, and the merry-go-round goes round. More and more corridors! Never any contact between them, although they continue to come closer and closer, $\frac{1}{7}, \frac{1}{8}, \frac{1}{9}, \ldots \frac{1}{100}, \ldots \frac{1}{1000}, \ldots \frac{1}{1000000}, \ldots \frac{1}{\text{googol}}, \ldots$ of a mile.

We may assume, in order that this feverish program shall be completed in a finite length of time ("Two-Year Plan"), that the first corridor of Country A took a year to build, the first corridor of B a half-year, the first corridor of C a quarter-year, the second corridor of

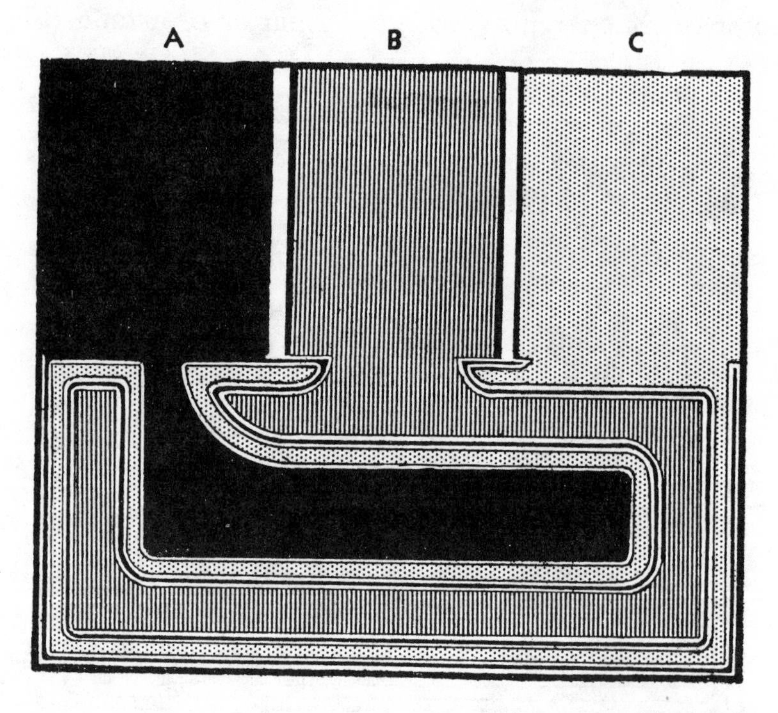

FIG. 118.

Country A an eighth of a year, and so on; each corridor took exactly half as long to build as its immediate predecessor. The total elapsed time then gives rise to the familiar series

$$1 + \frac{1}{2} + \frac{1}{4} + \frac{1}{8} + \frac{1}{16} + \frac{1}{32} + \ldots = 2.$$

Thus, at the end of two years, the once unclaimed territory has been entirely occupied, and not a speck of it remains unclaimed. Over each square inch there flies the flag of one of the three countries, either A, B, or C.

What of the new map which is to depict these boundaries? Actually it is impossible to draw but suppose we try to conceive what it would look like if it could be drawn. This conceptual map is put together out of pieces of sober mathematics and sheer fancy. For *every single boundary point on the map will be a meeting place, a boundary point, of not two, but of all three countries!*

*

In an apparently dynamic, incessantly changing world, one of perpetual novelty, the search for things which do not change constitutes one of the principal objectives of science. Philosophers since pre-Socratic times have been rummaging about for the unchanging essence of reality. Today, that is the job of the scientist.

In topology, as in other branches of mathematics, it takes the form of a search for invariants. Repeatedly, in the course of that search, the necessity arises for abandoning intuition, for transcending imagination. The invariants of 4, 5, 6, and n dimensions are purely conceptual. To fit them into our lives, to find use for them in the laboratory, to shape them for duty in the applied sciences seem impossible. There is nothing in experience to compare them with, not even a dream in which they could play a part.

Nevertheless, what is gathered by the mathematicians, slowly, painfully, bit by bit, in the weird world of beyond-the-make-believe, is in reality a part of the world of every-day, of tides, of cities, and of men, of atoms, of electrons, and of stars. All at once, what came from the land of n dimensions is found useful in the land of three. Or, perhaps, we discover that after all we live in a land of n dimensions. It is the reward for the courage and industry, for the fine, untrammeled, poetic, and imaginative sense common to the mathematician, the poet and the philosopher. It is the fulfillment of the vision of science.

FOOTNOTES

1. For two distinct journeys, the pencil must be lifted from the paper once; for three distinct journeys, twice; for n distinct journeys, $n - 1$ times.—P. 267.
2. Poincaré, *Science and Hypothesis.*—P. 273.
3. Invariant is a name invented by the English mathematician, Sylvester, who was called the mathematical Adam because of the many names that he introduced into mathematics. The terms "invariant," "discriminant," "Hessian," "Jacobian" are all his. In fact, he employed Hebrew characters in some of his mathematical papers, which, according to Cajori, caused the German mathematician Weierstrass to abandon him in horror.

   Invariants arise in other branches of mathematics. The theory of algebraic invariants, developed by Clebsch, Sylvester, and Cayley, lurks in the memory of everyone who studied quadratic equations. For example: The discriminant of the quadratic equation $ax^2 + bx + c = 0$ is the classical instance of an algebraic invariant. A quadratic equation under a linear transformation maintains unchanged a certain relation between its coefficients, expressed by the discriminant, $b^2 - 4ac$. The discriminant of the transformed equation remains equal to the discriminant of the original equation multiplied by a factor which depends only upon the coefficients in the transformation.—P. 273.
4. See Chap. 4, p. 119 and footnote 4, p. 154.—P. 278.
5. Osgood, *Advanced Calculus.*—P. 285.
6. We avail ourselves here of the version of the problem given by the distinguished Viennese mathematician, the late Hans Hahn, because it is more satisfying and clearer than Brouwer's own statement.—P. 293.

Change and Changeability—The Calculus

The ever-whirling wheele
Of Change, *the which all mortall things doth sway.*

—SPENSER

People used to think that when a thing changes, it must be in a state of change, and that when a thing moves, it is in a state of motion. This is now known to be a mistake.

—BERTRAND RUSSELL

"EVERYONE WHO understands the subject will agree that even the basis on which the scientific explanation of nature rests is intelligible only to those who have learned at least the elements of the differential and integral calculus . . . " These words of Felix Klein, the distinguished German mathematician, echo the conviction of everyone who has studied the physical sciences. It is impossible to appraise and interpret the interdependence of physical quantities in terms of algebra and geometry alone; it is impossible to proceed beyond the simplest observed phenomena merely with the aid of these mathematical tools. In the construction of physical theories, the calculus is more than the cement which binds the diverse elements of the structure together, it is the implement used by the builder in every phase of the construction.

Why is this branch of mathematics peculiarly suited for the precise formulation of natural phenomena? What powers can be attributed to the calculus that are not also shared by geometry and algebra?

Our most common impression of the world, whether erroneous or not, is its ever-changing aspect. Nature, as well as the artifacts we have invented to master it, seems to be in constant flux. Even the "absolutes"—space and time—contract and expand incessantly. Night and day repeatedly flow into one another, setting forth the vicissitudes of the seasons. Everywhere there is motion, flow, cycles of birth, death and regeneration. Everywhere the pattern moves.

For some strange reason, the subjects already considered, the many domains of mathematics already surveyed, have neglected this dynamism. With the exception of the exponential function, we have not spoken of the rate of change of a known or unknown quantity. Indeed, our equipment thus far could not have handled this concept. Fortunately, every problem was essentially static. Four-dimensional and non-Euclidean geometry treated of unchanging configurations; puzzles and paradoxes were solvable with the aid of ingenuity, logic and static arithmetic; topology sought out the invariant aspects of geometric forms independent of size and shape; and the concepts developed in the chapters on Pie, the Googol, and Probability were, with one or two exceptions, free of the ingredient of change. The conclusion is inevitable that the one indispensable means of attacking the vast majority of phenomena has been neglected—that our investigation has been confined to a peripheral aspect of the world scene.

*

The word "calculus" originally meant a small stone or pebble; it has acquired a new connotation. The calculus may be regarded as that branch of mathematical inquiry which treats of *change* and *rate of change*. The comfort with which one rides in an automobile is made possible, in part at least, by the calculus. While the planets would continue in their paths without the calculus, Newton needed it to prove that their orbits about the sun are ellipses. Shrinking from the celestial to the atomic, the solution of the very same equation used by Newton to describe the motion of the planets determines the trajectory of an alpha particle which bombards an atomic nucleus.* By means of the formula which relates the distance traversed by a moving body to the time elapsed, the velocity of the body, as well as its acceleration, at every instant of time is determined by the calculus.

Each of the above illustrations, whether simple or complex, involves change and rate of change. Without their exact mathematical enunciation none of the problems described would have meaning, much less be solvable. Thus, a mathematical theory has been created which takes cognizance of immanent and ubiquitous changes of pattern and which undertakes to examine and explain them. That theory is the calculus.

*

But had we not previously declared quite fervently that we live in a motionless world? And had we not shown at great length, by employing the paradoxes of Zeno, that motion is impossible, that the flying arrow is actually at rest? To what shall we ascribe this apparent reversal of position?

Moreover, if each new mathematical invention rests upon the old-established foundations, how is it possible to extract from the theories of static algebra and static geometry a new mathematics capable of solving problems involving dynamic entities?

* This holds only for alpha particles moving with relatively small velocities.

As to the first, there has been no reversal of viewpoint. We are still firmly entrenched in the belief that this is a world in which motion as well as change are special cases of the state of rest. There is no state of change, if change implies a state qualitatively different from rest; that which we distinguish as change is merely, as we once indicated, a succession of many different static images perceived in comparatively short intervals of time. An example may help to clarify the idea. Although in the cinema, a series of static pictures are projected upon the screen, one after the other, in rapid fashion, each picture differing only slightly from the one preceding it, there is not the slightest doubt in the mind of even the most intelligent moviegoer that motion is being portrayed on the screen. A completely convincing display of change is presented by a series of wholly static images. Let us pursue this with a more technical illustration. A steel rod, clamped in a horizontal position at one end, has a weight attached at the other. This system being at rest, it is said that the set of elements composing it are in equilibrium. If, when we next examine it, after some interval of time, we observe the same arrangement, the rod bent by the same amount, it is apparent that there has been no change. If, however, there is a new position of the rod, obviously a change has taken place. It is certain that the equilibrium could only have been disturbed and the position of the rod altered by a change in the attached weight. It is not hard to convince ourselves that additional weight would bend the rod further and that such additions, if made gradually, and just as quickly as motion pictures are projected on the screen, would give the impression that the rod is in motion. On the other hand, if we are aware of these additions of weight, we conclude that what we really have observed is not motion but merely a correlation of amount of bend with degree of weight and that for different weights there are different positions of the rod.

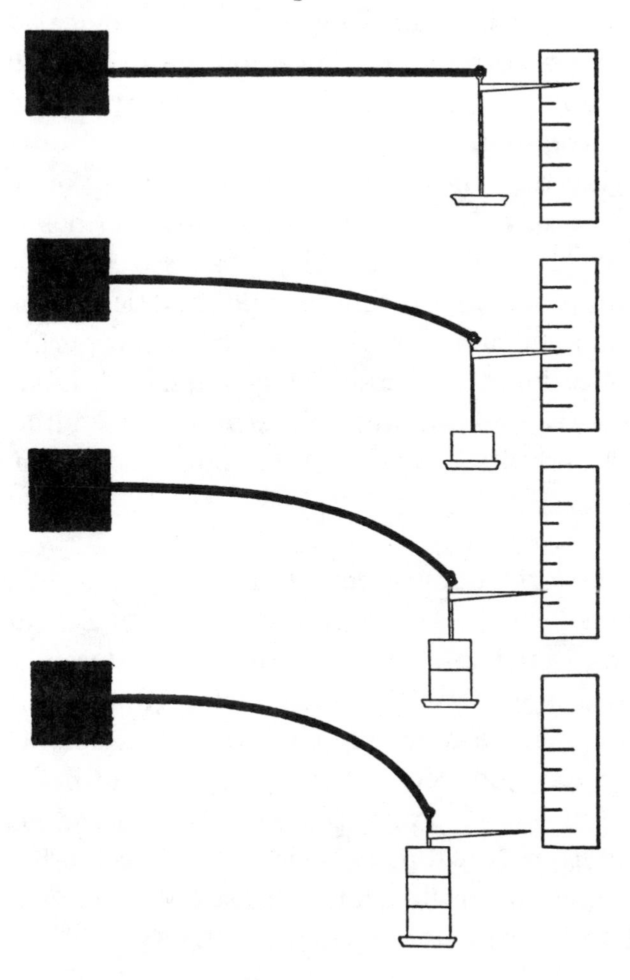

FIG. 119.—Each addition to the weight bends the rod a little further.

Intuitively convinced that there is continuity in the behavior of a moving body, since we do not actually see the flying arrow pass through every point on its flight, there is an overwhelming instinct to abstract the idea of motion as something essentially different from rest. But this abstraction results from physiological and psychological limitations; it is in no way justified by logical analysis. Motion is a correlation of position with time. Change is merely another name for *function*, another aspect of that same correlation.

For the rest, the calculus, as an offspring of geometry and algebra, belongs to a static family and has acquired no characteristics not already possessed by its parents. Mutations are not possible in mathematics. Thus, inevitably, the calculus has the same static properties as the multiplication table and the geometry of Euclid. The calculus is but another interpretation, although it must be admitted an ingenious one, of this unmoving world.

*

The historical development of the calculus did not follow such clear lines. The philosophic discussions as to the meaning of the subject came only after its usefulness had been indisputably established. Before that philosophers would not have deigned it worthy of attack. Unfortunately we cannot recount (though it would be amusing) the pitfalls which every philosopher and mathematical analyst from Newton to Weierstrass dug for his adversaries—and promptly fell into himself. We may, however, sketch the steps that preceded the theory as it is accepted today.

The calculus does not differ from other mathematical theories; it did not spring full-grown from the genius of any one man. Rather was it developed from a consideration of numerous questions essayed and successfully answered by the predecessors of Newton and Leibniz. "Every great epoch in the progress of science is preceded by a period of preparation and prevision . . . The conceptions brought into action at that great time had been long in preparation." [1]

The advent of analytic geometry furnished a powerful stimulus to the invention of the calculus, for the pictorial representation of a function revealed many interesting features. Kepler had noticed that, as a variable quantity approaches its maximum value, its rate of change be-

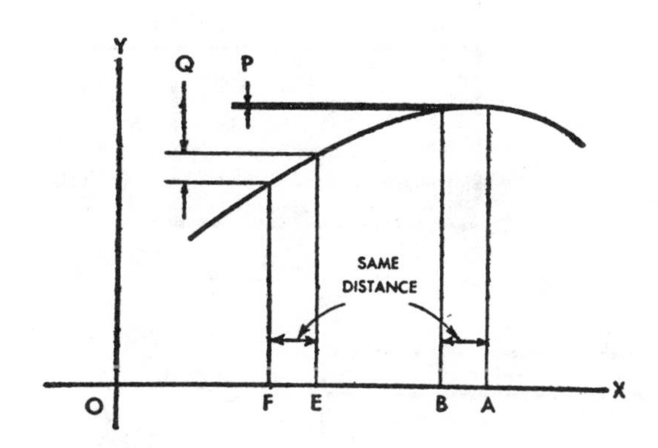

FIG. 120.—The rate of change of a variable quantity is smaller at a maximum point than elsewhere.

comes less than at any other value. It continues to choke off until, at the maximum value of the variable, the rate of change is zero.

In the above diagram, the values assumed by a variable quantity are measured by the distance from the straight line (the x axis) to the curve. The maximum value of the variable quantity (the greatest distance from the x axis to the curve) is attained at the point labeled A; when moving slightly either to the right or to the left of A, for instance to the point B, the change in the value of the variable quantity is very small, and is measured by P. If we move to the right or left of some other point E the same distance that we moved from A to B, so that the distance EF is equal to the distance AB, the change in the value of the variable quantity in the neighborhood of E is measured by Q. But obviously, this second width, Q, is greater than the first width, P. In this, which is Kepler's contribution, we have a geometric illustration of the

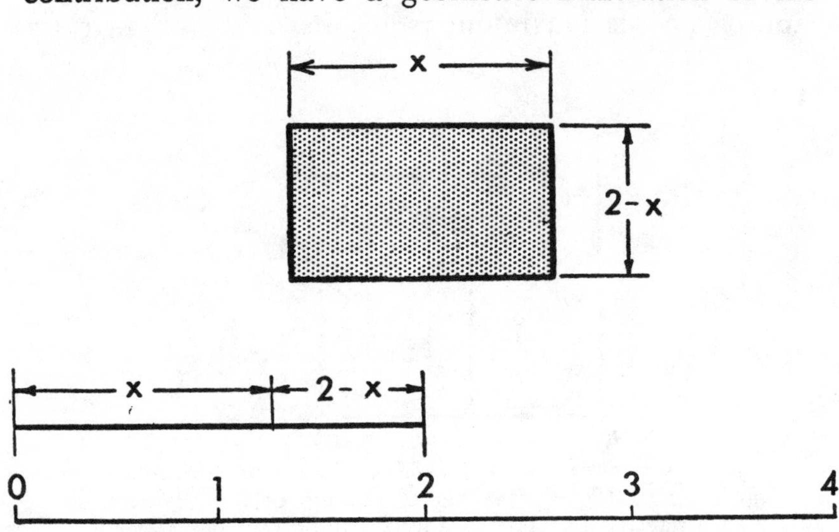

FIG. 121.—Using the scale, the perimeter of the rectangle is clearly 4 units.

principle of maxima and minima: the rate of change of a variable quantity is smaller in the neighborhood of its maximum (and minimum) value than elsewhere. In fact, at the maximum and minimum values, the rate is zero.

Pierre de Fermat, who shares with Descartes the distinction of discovering analytic geometry, was one of the first mathematicians to devise a general method applicable to the solution of problems involving maxima and minima. His method, used as early as 1629, is substantially that applied today to problems of this type. Let it be required to draw a rectangle such that the sum of the sides is four inches and such that the area* shall be a maximum. If we denote one side of the maximum rec-

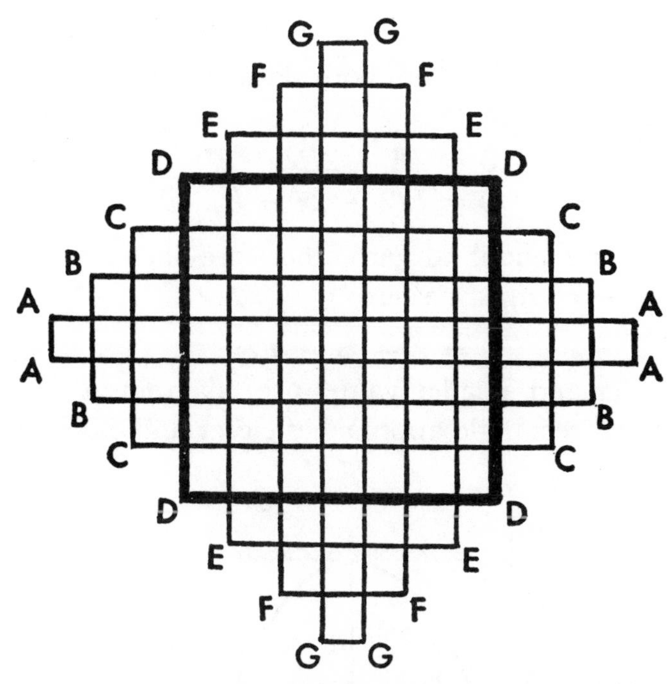

FIG. 122.—The perimeter of each of the seven rectangles viz. *AAAA*, *BBBB*, *CCCC*, etc., is the same. But obviously the rectangle of maximum area is the square *DDDD*.

tangle by x, the adjacent side, as may be seen from Fig. 121, will be $2 - x$; and the area of the rectangle will be $x(2 - x)$. If the side x is increased by a small amount E, the side $2 - x$ will have to be diminished by E in order to maintain a constant perimeter. The new area will then be $(x + E)(2 - x - E)$. Since the original area

* The area of a rectangle is the product of two adjacent sides.

was a maximum, this slight alteration in the relation of the sides can have produced only a slight change in the area. Thus, considering the two areas *approximately* equal, we have

$$x(2 - x) = (x + E)(2 - x - E)$$

whence $\quad 2x - x^2 = 2x - x^2 - Ex + 2E - Ex - E^2.$

Subtracting $2x - x^2$ from both sides of this equation and factoring:

$$0 = 2E - 2Ex - E^2$$
$$0 = E(2 - 2x - E).$$

But E is *not* equal to zero, therefore the other factor $(2 - 2x - E)$ must be zero:

$$0 = 2 - 2x - E.$$

As smaller and smaller values are taken for E, (i.e., as the altered rectangle approaches closer and closer to the

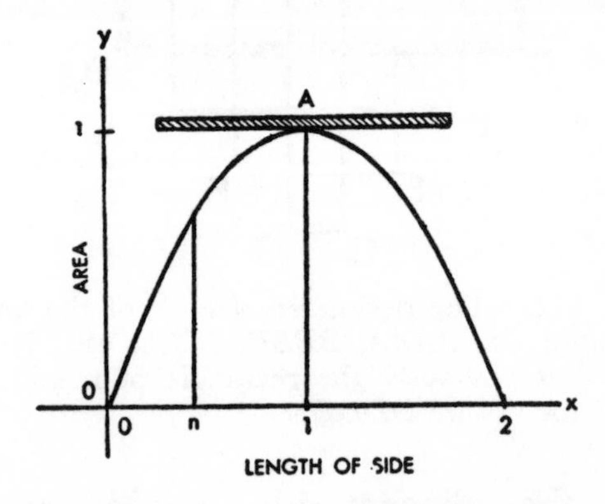

FIG. 123.—The curve is a parabola representing the areas of all rectangles whose perimeter is 4 units long. Erect a perpendicular at any point n along the x axis to the curve. The length of this perpendicular will be the area of the rectangle, one side of which equals n. The maximum area corresponds to the point A on the graph, i.e., the perpendicular erected at $x = 1$. Thus the rectangle of maximum area, with a perimeter $= 4$, has a side $= 1$ and is therefore a square.

original maximum rectangle) the expression on the right-hand side of the equation approaches closer and closer to the expression obtained by setting E equal to zero, namely, $2 - 2x$. Solving this resulting equation:

$$0 = 2 - 2x$$

we find that: $x = 1$; or, in terms of the original problem, the rectangle with the maximum area is a square.

It is well to note that the area of the rectangle is a function of the lengths of the sides, and this function can be portrayed by a curve. (Fig. 123.)

The highest point of this curve is at $x = 1$. This is the *maximum* of the function. To use a crude analogy, since this point is neither "uphill" nor "downhill," a small steel ball would be in equilibrium, or a ruler could be balanced at this point. If we think of a straight line being "balanced" at this point, such a line would be called the *tangent to the curve*.[2] The interesting fact is that the tangent to a curve at its maxima and minima points will *always* be horizontal (Fig. 124). To this idea, so important in the calculus, we shall return later.

Sir Isaac Newton and Baron Gottfried Wilhelm von

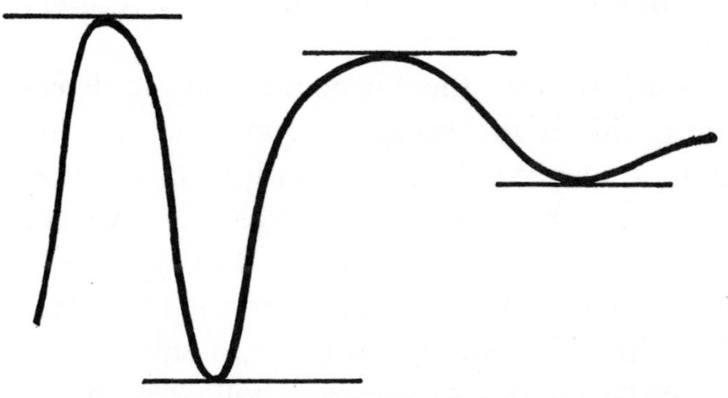

FIG. 124.—The horizontal lines are tangent to the relative maxima and minima of the curve.

Leibniz share the credit in the history of mathematics as independent discoverers of the differential and integral calculus. Their conflicting claims gave rise to a controversy which raged in Europe for more than a century. This monumental invention made simultaneously by these men now commends itself to our attention.

*

A tiny flame, lit by Archimedes and his predecessors, burst forth into new brilliance in the intellectually hospitable climate of the seventeenth century to cast its light over the entire future of science. The fertile concept of *limit* revealed its full powers for the first time in the development of the differential calculus.

We are already acquainted with the *limit of a variable quantity*. The sequence of numbers 0.9, 0.99, 0.999, 0.9999, . . . converges to the limiting value 1. The series $1 + \frac{1}{2} + \frac{1}{4} + \frac{1}{8} + \frac{1}{16} + \ldots$ converges to the limiting value 2. Nor are geometric examples unfamiliar. If a regular polygon is inscribed in a circle, the difference between the perimeter of the polygon and the circumference of the circle can be made as small as one wishes merely by taking a polygon with a sufficient number of sides. The limiting figure is the circle, the limiting area, the area of the circle.

In these instances, there is no difficulty in determining the limit; this is the exception, however, not the rule. Usually, a formidable mathematical procedure is required to determine the limit of a variable quantity. Consider this: Draw a circle with a radius equal to one. In it inscribe an equilateral triangle. In the triangle inscribe another circle; in the second circle, a square. Continue with a circle in this square, and follow with a regular five-sided figure in the circle. Repeat this procedure, each time increasing the number of sides of the regular polygon by one.

At first glance, one might suppose that the radii of the shrinking circles approach zero as their limiting value.

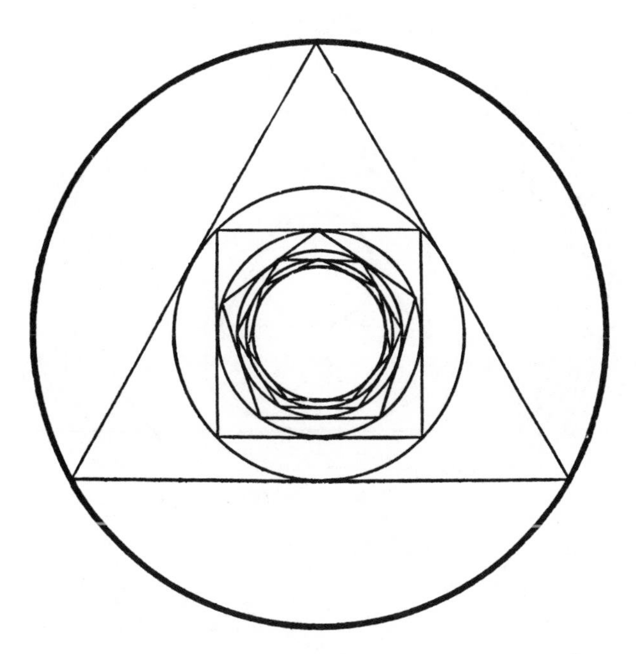

FIG. 125.—The diminishing radii approach a limit approximately $\frac{1}{12}$ that of the radius of the first circle.

But this is not so; the radii converge to a definite limiting value different from zero. As an explanatory clue, it need only be remembered that the shrinking process itself approaches a limit as the circles and inscribed polygons become approximately equal. The limiting value of the radii is given by the infinite product: [2]

$$\text{Radius} = \cos\frac{\pi}{3} \times \cos\frac{\pi}{4} \times \cos\frac{\pi}{5} \times \ . \ . \ . \ \times \cos\frac{\pi}{(n-2)}$$

Closely related to this problem is the one of circumscribing the regular polygons and the circles instead of inscribing them.

FIG. 126.—The increasing radii approach a limit approximately 12 times that of the original circle.

Here it would seem that the radii should grow beyond bound, become infinite. This, too, is deceptive, for the radii of the resulting circles approach a limiting value given by the infinite product:

$$\text{Radius} = \frac{1}{\cos\frac{\pi}{3} \times \cos\frac{\pi}{4} \times \cos\frac{\pi}{5} \times \ldots \times \cos\frac{\pi}{(n-2)}}$$

Interestingly enough, the two limiting radii are so related that one is the reciprocal of the other.

So much for the *limit of a variable quantity*. Let us now turn to the *limit of a function*, recalling briefly the meaning of function.* It is often found that two variable quantities are so related that to each value of one there corresponds

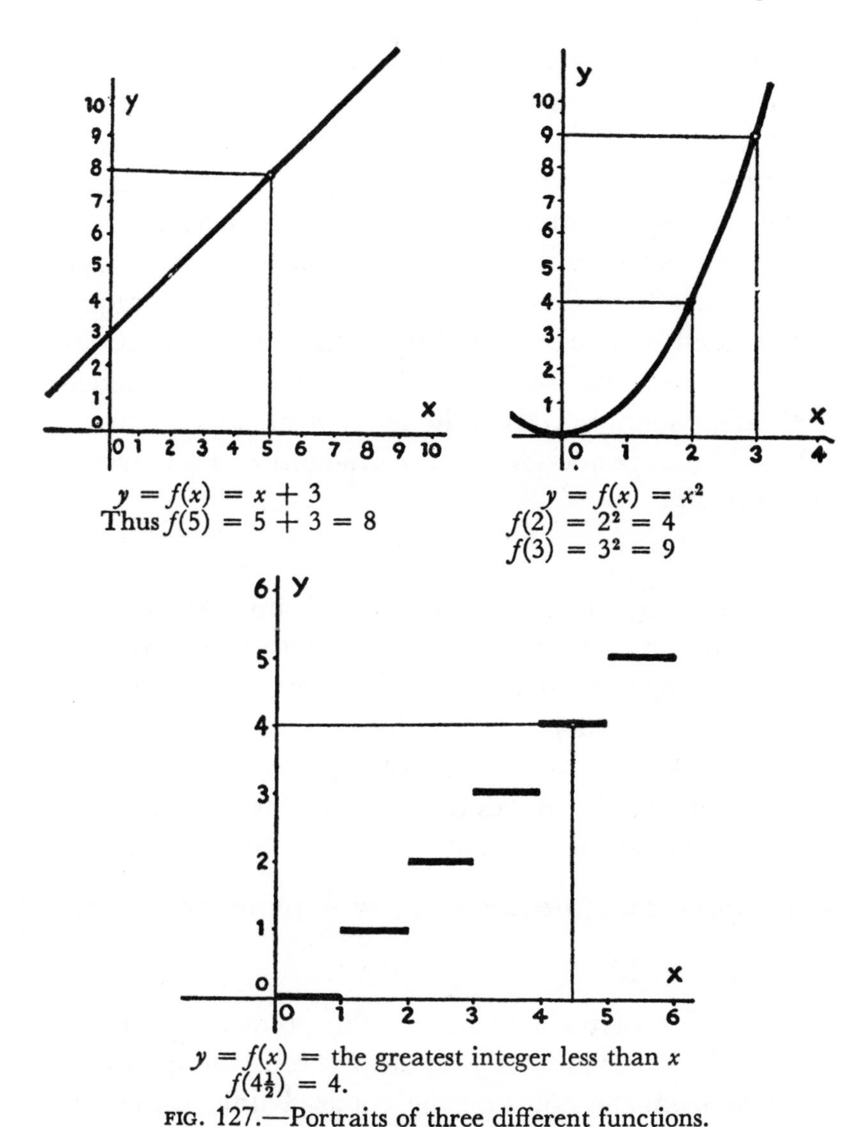

FIG. 127.—Portraits of three different functions.

* Though we have done this before, the notion of function is so important, so all-pervading in mathematics, that it is worth going over again.

a value of the other. Under this condition, the two variable quantities are said to be functions of one another, or functionally related. Thus, the force of attraction (or repulsion) between two magnets is a function of the distance between them. The greater the distance between the magnets, the less the force; the less the distance, the greater the force. If the distance is permitted to assume arbitrary values, it may be considered as an *independent variable*. The force then becomes the *dependent variable*, dependent upon the distance (and the functional relation) and is uniquely determined by assigning values to the independent variable. In functional relations, the letter x usually denotes the independent variable, the letter y the dependent variable. The dependency "y is a function of x" is written symbolically:

$$y = f(x).$$

The graphic representation of a point has been discussed in the section on *analytic geometry*. The equation $y = f(x)$ determines a value of y for every value of x. Each pair of values which satisfies this equation is considered as the Cartesian co-ordinates of a point in a plane; the curve depicting the function is composed of all such points.

In discussing the concept "limit of a function," let us study specifically the function $y = \frac{1}{x}$, represented graphically in Fig. 128.

The value of the function at the point $x = \frac{1}{2}$ is $y = f(\frac{1}{2}) = 2$. This value is graphically represented by the distance from the point on the x axis, $\frac{1}{2}$ unit to the right of the origin, to the curve. Likewise, the value of the function at each point along the curve is represented by its distance from the x axis.

For the function $y = \frac{1}{x}$, take two neighboring points,

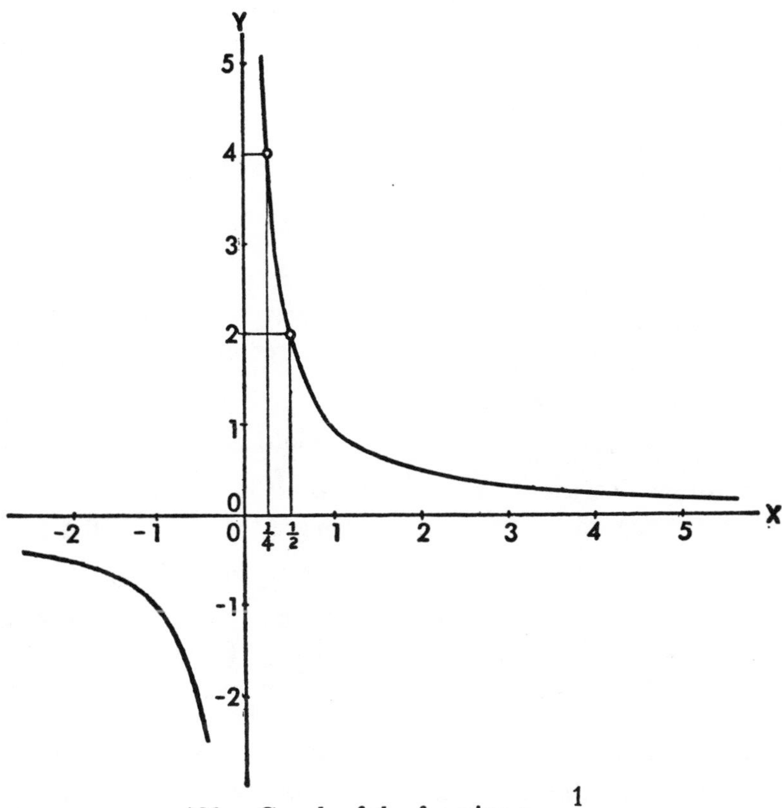

FIG. 128.—Graph of the function $y = \frac{1}{x}$.

$x = \frac{1}{4}$, and $x = \frac{1}{2}$. As the independent variable moves along the x axis from the point $x = \frac{1}{4}$ to $x = \frac{1}{2}$, the dependent variable is "forced" along the curve from the point $y = f(\frac{1}{4}) = 4$ to $y = f(\frac{1}{2}) = 2$. In other words, as the independent variable x approaches as its limit the value $\frac{1}{2}$, the dependent variable, the function, approaches as its limit the value 2. Generally, as an independent variable x *approaches* a value A, its dependent variable y (the function of x) *approaches* a value B. Thus, the limit of

$f(x)$, as x approaches A, is B. This is what is meant by "limit of a function."

Recalling the example of the steel rod flexed under a weight, we may construct a parallel dictionary of terms.

MATHEMATICS	PHYSICS
Independent variable, x.	Amount of weight.
Dependent variable, y.	Amount of bend of steel rod.
Function is the relation between x and y.	Function is the relation between the weight and the degree of bend.
Increase or decrease of x (i.e., change).	Addition or diminution of weight (i.e., change).
Increase or decrease of y (i.e., change).	Increase or decrease in the degree of bend of the steel rod (i.e., change).
Limiting value of y (the function of x) equals a number.	Limiting value of degree of bend (function of the weight) equals a position.

With the concepts limit, function, and limit of a function in mind, there remains to define an idea embracing all three—"rate of change."

Consider the determination of the speed [3] of a moving body at a given instant of time. A bomb is dropped from a stationary airship at an altitude of 400 feet. Five seconds will elapse before it hits the ground. Its average speed is thus $\frac{400 \text{ feet}}{5 \text{ seconds}} = 80$ feet per second. Hence, the average rate of change of distance with respect to time is 80. We are aware, however, from the most elementary knowledge of physics that a body gathers speed as it falls.

Throughout the fall the bomb was not moving at a constant rate of 80 feet per second; the speed with which it fell varied from point to point, increasing at each successive instant (disregarding air resistance). Suppose, for the sake of simplicity, we limit our interest to the speed of

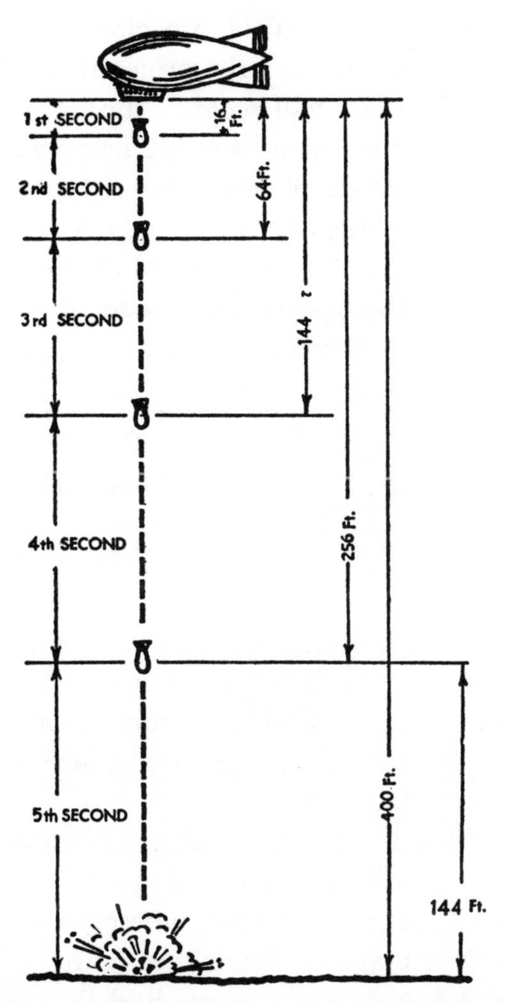

FIG. 129.—The diagram shows the distance covered by a falling projectile at the end of 1, 2, 3, 4, and 5 seconds.

the bomb at the exact moment of striking the ground. Evidently, its speed during the last second before striking will give a fair approximation to its speed at the instant of striking. The distance covered during this last second being 144 feet, the rate of change of distance with respect to time is 144. If we now take smaller and smaller intervals of time, we may expect to obtain closer and closer approximations to the speed of the projectile at the moment of impact. In the last half second, the distance covered was 76 feet, so that the speed was 152 feet per second. The table lists the intervals of time, the distance covered in those intervals, and the average speed over each interval. It is readily seen that as the interval of time approaches zero, we obtain the approximation to the speed of the body at the instant it hits the ground.

Interval of time in seconds.	1	$\frac{1}{2}$	$\frac{1}{4}$	$\frac{1}{8}$	$\frac{1}{16}$	$\frac{1}{32}$	$\frac{1}{64}$	$\frac{1}{160}$	$\frac{1}{1600}$	$\frac{1}{16000}$
Distance covered in feet	144	76	39	$19\frac{3}{4}$	$9\frac{15}{16}$	$4\frac{63}{64}$	$2\frac{127}{256}$	$\frac{1599}{1600}$	$\frac{15999}{160000}$	$\frac{159999}{16000000}$
Average speed in feet per second	144	152	156	158	159	$159\frac{1}{2}$	$159\frac{3}{4}$	$159\frac{9}{10}$	$159\frac{99}{100}$	$159\frac{999}{1000}$

These approximations approach a limiting value, 160 feet per second, which is defined as the *instantaneous speed* of the bomb upon striking the ground, or what is the same thing, its rate of change of distance with respect to time at that instant.

We may discuss the same example from an algebraic standpoint. The distance covered by a falling body is given by the function $y = 16x^2$, where y is the distance, and x, the time elapsed. From this formula, merely by substituting 5 (seconds) for x, we find that y is equal to 400 (feet). How shall we make use of this formula to find the speed at the end of five seconds? Let us fix our attention upon a short interval of time just before the falling object strikes the ground and the correspondingly short interval of distance traversed in that period of time. We shall call this small interval of time Δx*, and the

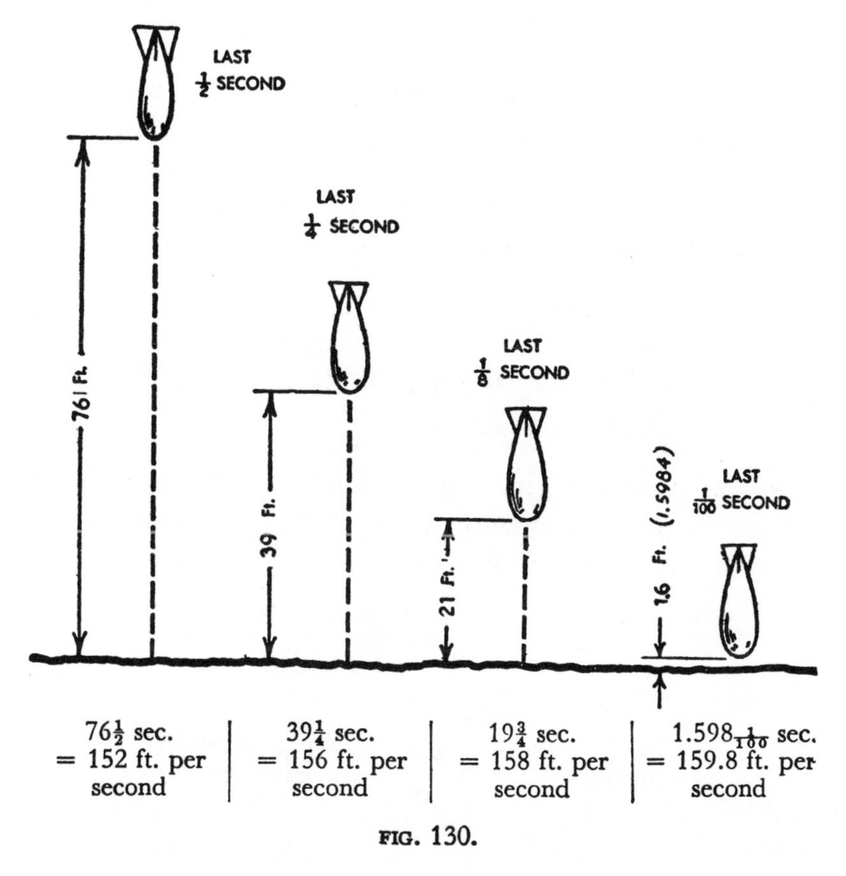

FIG. 130.

distance traversed in that period Δy. Knowing the value of Δx, having chosen it arbitrarily, the problem is to find the value of Δy. At the beginning of the space interval, Δy, the exact elapsed time since the falling body left the

* Read "delta *x*," not "delta times *x*;" for Δ is merely a symbol, a direction for performing a certain operation, to wit, taking a small portion of *x*.

airship was $(5 - \Delta x)$ seconds. The distance covered in the time $(5 - \Delta x)$ seconds is $(400 - \Delta y)$ feet. Our functional relation indicates that

$$\text{Distance} = 16(\text{Elapsed Time})^2.$$

Thus, for the entire fall

$$400 = 16(5)^2,$$

and for the incompleted journey

$$(400 - \Delta y) = 16(5 - \Delta x)^2.$$

This may be simplified to

$$\begin{aligned} &400 - 16(5 - \Delta x)^2 = \Delta y \\ &400 - 16(25 - 10\Delta x + \Delta x^2) = \Delta y \\ &400 - 400 + 160\Delta x - 16\Delta x^2 = \Delta y \\ &160\Delta x - 16\Delta x^2 = \Delta y. \end{aligned}$$

The last equation gives the distance Δy in terms of Δx units. To find the *average* speed during the entire time interval Δx, we must form the fraction

$$\text{Average Speed} = \frac{\text{Distance Interval}}{\text{Time Interval}}$$

or

$$\text{Average Speed} = \frac{\Delta y}{\Delta x} = \frac{160\Delta x - 16\Delta x^2}{\Delta x}.$$

Thus,

$$\frac{\Delta y}{\Delta x} = 160 - 16\Delta x.$$

Now as the interval of time Δx is made smaller, that is, as we take closer and closer approximations to the speed at the instant the body strikes the ground (5 seconds having elapsed) the limit of the ratio $\Delta y / \Delta x (= 160 - 16\,\Delta x)$ is 160. In other words, as Δx approaches zero in value, the function of Δx (the expression $160 - 16\,\Delta x$) approaches 160. Thus, the *instantaneous speed* at the end of five seconds is 160 feet per second. We indicate that the ratio $\Delta y/\Delta x$ approaches a limit by writing its limiting value as dy/dx. In technical terms

$$\lim_{\Delta x \to 0} \frac{\Delta y}{\Delta x} = \frac{dy}{dx}$$

which may be read, "The limit of $\Delta y/\Delta x$ as Δx approaches zero is dy/dx."

*

Let us pause for a moment to get our bearings. What have we accomplished? It may seem trivial that with all the elaborate machinery at our disposal we have succeeded only in ascertaining the instantaneous speed of a falling body as it strikes the earth. Yet if our accomplishment is trivial, then motion is trivial as well, for we have whether we realized it or not, trapped Zeno's arrow in its flight and established the changelessness of our universe. With the aid of the concepts of limit and function, we have made meaningful the notion of change and rate of change. *Change is a functional table.* As an item (independent variable) on one side of the table varies, its corresponding item (dependent variable) on the other side shows a correlative variation. The quotient of change, i.e., the limiting ratio of the two variations, is denoted by *rate of change.* All the vagaries, the mysteries, and uncertainties indissolubly linked with the idea of motion, are thus swept away or, more appropriately, transformed into a few precise and definable aspects of the idea of function. The limit of a function is exemplified quite simply by the ratio $\Delta y/\Delta x$ as Δx approaches zero. It is easy to see that $\Delta y/\Delta x$ is a function of Δx, in other words, that this ratio is a function of the independent variable Δx. As we assign arbitrary values to Δx, its dependent variable, Δy, assumes a corresponding set of values, and as we have seen, that ratio approaches a limit. It follows that we have not only revealed the meaning of the limit of function but have already made practical use of this concept.

It is now possible to define the fundamental process of the differential calculus, computing the limit of a function, or what is the same thing, determining its *derivative.* For, in effect, the rate of change of a function is itself a function of that function, and in getting at the limit of the rate of change, the derivative, we are getting at the heart of the machinery of our primitive function.

Assume we wish to determine the rate of change of a function $y = f(x)$ at an arbitrary point x_0. The average change in the function $f(x)$ over an interval extending from x_0 to $x_0 + \Delta x$ is the difference in the value of the function $y = f(x)$ at the two end points, x_0 and $x_0 + \Delta x$, divided by the length between these two end points, $(x_0 + \Delta x) - x_0$. Thus,

$$y_0 = f(x_0)$$

and

$$y_0 + \Delta y = f(x_0 + \Delta x).$$

Whence a change in a function, from the purely algebraic standpoint, is given by

$$\Delta y = f(x_0 + \Delta x) - f(x_0),$$

and the *average rate of change of a function*, obtained by dividing the change, Δy, by the length of the interval over which that change is taken, Δx, is

$$\frac{\Delta y}{\Delta x} = \frac{f(x_0 + \Delta x) - f(x_0)}{\Delta x}.$$

In order to obtain better approximations to the instantaneous rate of change at the point x_0, it is only necessary to take smaller intervals, that is, to let Δx approach zero. As Δx approaches zero, the expression $\frac{f(x_0 + \Delta x) - f(x_0)}{\Delta x}$ approximates as closely as may be desired to the instantaneous rate of change at x_0. Thus, in the *limit* as Δx approaches zero, the quotient $\frac{f(x_0 + \Delta x) - f(x_0)}{\Delta x}$ approaches a limiting value, denoted by dy/dx. It is this which is called *the derivative of the function* $f(x)$ *at the point* x_0. But since x_0 is an arbitrary point, *the derivative may be said to represent the instantaneous rate of change of a function as the independent variable ranges through an entire set of values.*

For the sake of clarity, a geometric interpretation of the derivative may be helpful. Chronologically, the geometric interpretation preceded the analytic. One of the outstanding problems of the seventeenth century was that of drawing the tangent to a curve at an arbitrary point. It was solved by Newton's predecessor and teacher at Cambridge, Isaac Barrow. On the basis of the geometrical researches of Barrow, Newton developed the concept of the rate of change along analytic lines. The close connection between algebra and geometry, epitomized by the fact that every equation has a graph and every graph an equation, thus bore fruit once more. In the Cartesian plane, let the graph of the function $y = f(x)$ be the curve in Fig. 131.

Consider the points P_1 and P_2 on this curve; their x co-ordinates are denoted by x_0 and $x_0 + \Delta x$, where Δx is the distance between the projection of the two points on the x axis. The y co-ordinates of the points P_1 and P_2

are then determined from the equation of the curve and are $f(x_0)$ and $f(x_0 + \Delta x)$ respectively. The slope[2] of the line joining P_1 and P_2 (the tangent of the angle Θ) is precisely the quotient

$$\frac{f(x_0 + \Delta x) - f(x_0)}{\Delta x}$$

As we let Δx approach zero, the point P_2 is carried along the curve so that it approaches the point P_1, and the slope of the line (the above quotient) approaches as its

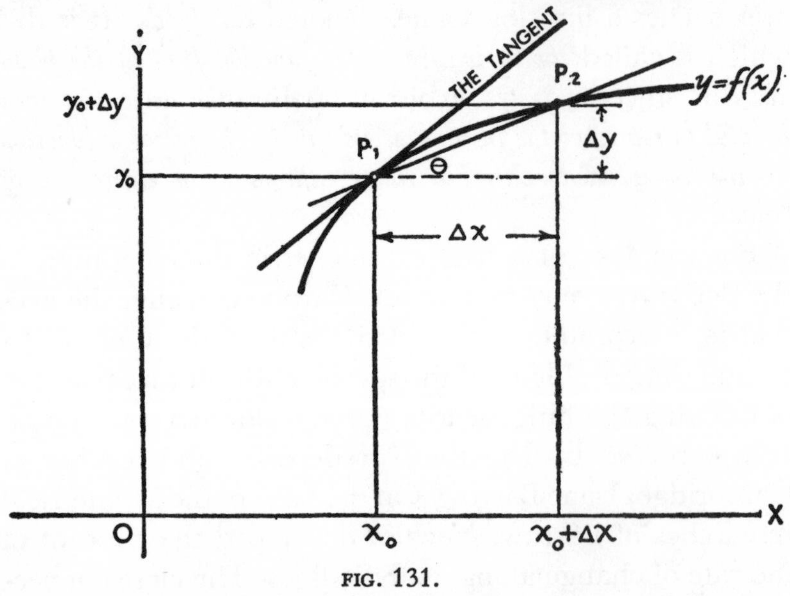

FIG. 131.

limiting value the slope of the tangent to the curve at the point P_1. But the slope of the tangent at that point is numerically equal to $\frac{dy}{dx}\left(\text{since } \lim_{\Delta x \to 0} \frac{\Delta y}{\Delta x} = \frac{dy}{dx}\right)$. In other words, the slope of the tangent at every point along a curve is identical with the derivative at that point. Or, to put it differently, the slope of the tangent to a curve gives the direction the curve is taking (i.e., whether it is rising or falling), and thus its rate of change. Thus, the geometric equivalent of the derivative is the slope of the tangent.

We may now recall our statement that the values for which a function attains its maximum or minimum correspond to the points on the curve at which the tangent is horizontal. The slope of a horizontal line is, of course, zero. Since the derivative is identical with the tangent, we may conclude that the maximum and minimum values of a function are those for which the derivative of the function is equal to zero. Many interesting problems can be solved in this way.

The previously discussed problem of determining the rectangle with greatest area and given perimeter falls into this category. One side of the rectangle was denoted by x, the adjacent side by $2 - x$, and the area, y, by $x(2 - x)$. Since the area is a function of x, its derivative will be equal to zero when the function attains its maximum value. Finding the rectangle with maximum area by means of the calculus entails these steps: (1) Differentiate the function, i.e., find its derivative; (2) Set the derivative equal to 0; (3) Solve the resulting equation for x.

Step I:

$$y = x(2 - x)$$

$$y + \Delta y = (x + \Delta x)(2 - x - \Delta x)$$

$$(y + \Delta y) - y = (x + \Delta x)(2 - x - \Delta x) - x(2 - x)$$

$$\Delta y = 2x - x^2 - x\Delta x + 2\Delta x - x\Delta x - \Delta x^2 - 2x + x^2$$

$$\Delta y = 2\Delta x - 2x\Delta x - \Delta x^2$$

$$\frac{\Delta y}{\Delta x} = 2 - 2x - \Delta x$$

$$\underset{\Delta x \to 0}{\text{Limit}} \frac{\Delta y}{\Delta x} = \frac{dy}{dx}$$

$$\text{and } \frac{dy}{dx} = 2 - 2x$$

Step II:

$$\frac{dy}{dx} = 2 - 2x = 0$$

Step III:

$$2 - 2x = 0$$
$$2 = 2x$$
$$1 = x$$

This checks with the result obtained before without the aid of the calculus: the rectangle of maximum area, with a perimeter of 4, is a square each of whose sides equals 1.

More elaborate examples, drawn from the fields of chemistry, economics, physics, etc., require a greater sophistication with respect to mathematical technique, but not with respect to the ideas involved.

*

By considering the derivative at every point of the interval over which it is defined, we have seen that the derivative is in turn a function of the independent variable. Differentiation need not stop here, for the derived function may also have a derivative, the second derivative of the original function. The notation for the second derivative of $y = f(x)$ is $\frac{d^2y}{dx^2}$. The nth derivative of a function is obtained by differentiating the function n times. Its symbol is $\frac{d^ny}{dx^n}$. What do these higher derivatives mean?

Usually it is possible to give to the second derivative a physical and geometrical interpretation. If the function $y = f(x)$ represents the distance covered by a falling body in the time x, the first derivative represents the rate of change of distance, with respect to time. The second derivative is the rate of change of the rate of change of distance with respect to time, and is commonly known as the *acceleration* of the body. For a falling body, the distance $y = 16x^2$ must be differentiated once to obtain the speed and once again to obtain the acceleration. The mathematical details of both differentiations are:

(I)

$$
\begin{aligned}
y &= 16x^2 \\
y + \Delta y &= 16(x + \Delta x)^2 \\
(y + \Delta y) - y &= 16(x + \Delta x)^2 - 16x^2 \\
&= 16(x^2 + 2x\Delta x + \Delta x^2) - 16x^2 \\
&= 16x^2 + 32x\Delta x + 16\Delta x^2 - 16x^2 \\
\Delta y &= 32x\Delta x + 16\Delta x^2 \\
\frac{\Delta y}{\Delta x} &= 32x + 16\Delta x \\
\underset{\Delta x \to 0}{\text{Limit}} \frac{\Delta y}{\Delta x} &= \frac{dy}{dx} \\
\frac{dy}{dx} &= 32x.
\end{aligned}
$$

(II)

$$
\begin{aligned}
\left(\frac{dy}{dx}\right) &= 32x \\
\left(\frac{dy}{dx}\right) + \Delta\left(\frac{dy}{dx}\right) &= 32\ (x+\Delta x) \\
\left(\frac{dy}{dx}\right) + \Delta\left(\frac{dy}{dx}\right) - \left(\frac{dy}{dx}\right) &= 32\ (x+\Delta x) - 32x \\
\Delta\left(\frac{dy}{dx}\right) &= 32\Delta x \\
\frac{\Delta\left(\frac{dy}{dx}\right)}{\Delta x} &= 32 \\
\underset{\Delta x \to 0}{\text{Limit}} \frac{\Delta\left(\frac{dy}{dx}\right)}{\Delta x} &= \frac{d^2y}{dx^2} \\
\frac{d^2y}{dx^2} &= 32
\end{aligned}
$$

The second derivative is a constant, the number 32. This constant is called the gravitational constant of a falling body due to the earth's gravitational pull. It expresses the remarkable fact that any body, regardless of its mass, dropped from a height 16 feet above the earth (and neglecting air resistance), will strike it in one second, moving at a speed of 32 feet per second at the instant of impact.

So far as the geometric interpretation of the second derivative goes: For curves drawn in the plane, at every point the *curvature* is directly proportional to the second derivative. To determine the curvature of a given arc, draw the circle which best fits that arc.

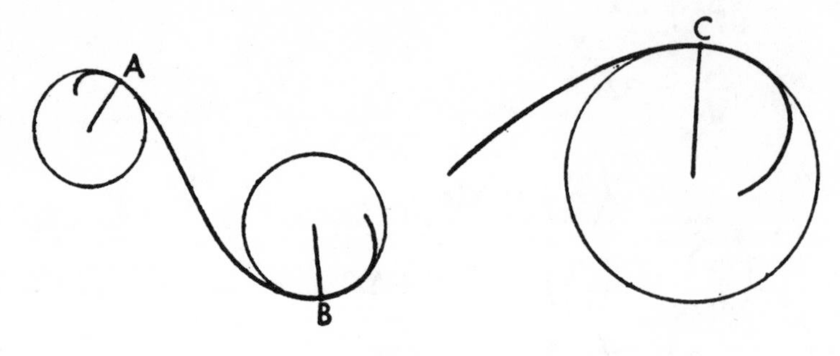

FIG. 132.

The radius of that circle is the *radius of curvature*, and its reciprocal the *curvature*.

Let us see how this is applied, for example, to the straight line. The curvature of a straight line is zero. Any function, the graph of which is a straight line, has an equation of the form $y = mx + b$, where m and b are constants.

Differentiating gives $dy/dx = m$. When m is differentiated, its rate of change or derivative equals zero, since m is a constant. Thus, the first derivative tells us that the slope of a straight line is a constant; the second derivative, that its curvature is zero.

Simple physical or geometric interpretations of third, fourth, and higher derivatives do not exist. Higher derivatives do occur, however, in many problems arising in physics. Automobile engineers are interested in third

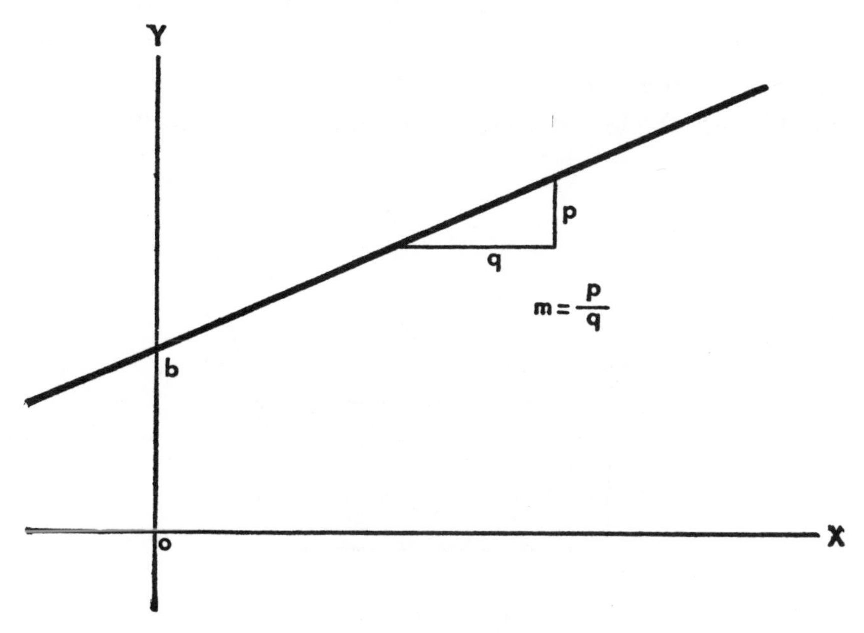

FIG. 133.—The graph of the equation $y = mx + b$.

derivatives because they yield information about the riding quality of a car. Structural engineers, concerned with the elasticity of beams, the strength of columns, and any phase of construction where there is shear and stress, find first, second, third, and fourth derivatives indispensable; and there exist innumerable other examples in the fields of the physical sciences and statistical applications to the social sciences.

*

The questions answered by the integral calculus stem from a much earlier period than those of the differential calculus. By this it is not meant that the mathematical devices used in the one preceded those used in the other, for the concepts of limit, function, and limit of a function, as *they appear in the calculus*, were developed at about the same time for both its branches. But the type of problem which the integral calculus seeks to solve is easier to propose and, therefore, it is not surprising to find among the writings of the Greek mathematicians problems which we

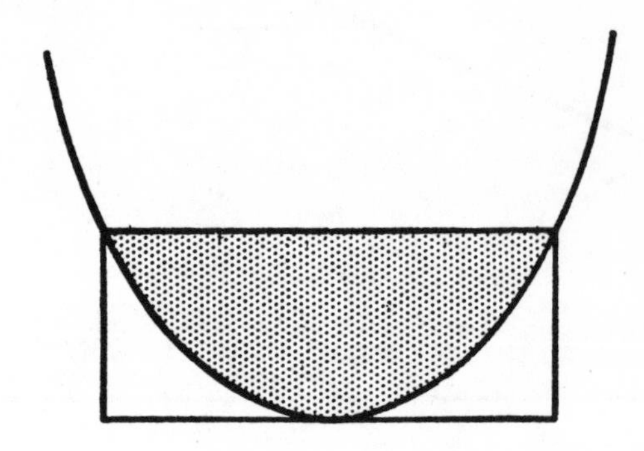

FIG. 134.—Squaring the parabola. The shaded area = $\frac{2}{3}$ the area of the rectangle.

now identifiy as belonging under the head of integration.

Much more astounding is the close relationship which exists between the two divisions of the calculus, the differential and the integral. It is one thing to determine the rate of change of a function or the slope of the tangent to a curve; to compute the area under a given curve appears to be an inquiry of a wholly different order. As wonderful as the link between these apparently unconnected inquiries may seem, it is secondary to the satisfaction that the mathematician experiences from the complementary character of two such powerful tools.

"Squaring the circle" had challenged Greek mathematicians. Another aspect of this problem, perhaps not as well known, but of equal importance, is the rectification of the circle. It is concerned with the determination of the length of the circumference of a circle in terms of the length of the radius. While they never mastered the circle, they partially tamed the parabola. In this, as in other things, they drew upon their fertile ingenuity. They succeeded, with profoundly beautiful methods, in squaring * the parabola, but not in rectifying it.[4]

A discussion of their methods would disclose more of the genius of Archimedes than of the general theory of the integral calculus. Undoubtedly, Archimedes' plan foreshadowed the technique of the calculus, but in the comparatively barren centuries that followed the seed he had planted found little nourishment. Not until Kepler appeared was there an attempt to deal systematically with the determination of the areas and volumes of curvilinear figures. Melancholy to relate, his incentive was less a thirst for learning than the commercial requirements of the thirst-slaking industry. "Kepler was originally led to make the calculations . . . by a desire to improve upon the crude methods then in use for estimating the contents of wine casks and other vessels. While buying wine he noticed that the vintners determined the contents of the cask by passing a measuring rod through the bunghole as far as the opposite staves without taking account of the curvature of the latter. By rotating the longitudinal section of the cask about its axis, a body equal in volume to the cask would be formed. Kepler's plan was to divide up such solids of rotation into an in-

* Squaring the parabola, as we saw earlier, means computing the area bounded by a parabolic segment and a straight line.

finite number of elementary parts, and to sum these; and in his *Stereometria* he applies this method to some ninety special cases. Kepler regarded infinitely small arcs as straight lines, infinitely narrow planes as lines, and infinitely thin bodies as planes. His conception of infinitely small magnitudes was one which the ancients had in general avoided, but which a little later formed the basis of Cavalieri's method." [5]

Perhaps it should be stressed at this point that in our discussion of the calculus every reference to the infinite, whether small or large, has been sedulously avoided. Because Weierstrass disposed of the infinitesimal djinn, the calculus rests securely on the understandable and nonmetaphysical foundations of limit, function, and limit of a function. Nothing prevents the extension of these concepts to the integral calculus. Indeed, the banishment of the infinitely small means more to the integral calculus than to the differential. It was precisely this refinement of thought which raised the calculus to a very exact science.

The work of Cavalieri marked progress in showing greater generality and a more abstract method of treatment than that of Kepler. One of the leading theorems still bears his name. If two solids have the property that when they are cut by planes, their areas are exposed in constant proportion throughout, their volumes will be in the same proportion.

The initial question, then, of determining the areas under curves was well on the way to solution in so far as crude machinery made that possible. Yet the design of the machinery made it unsuitable for the computation of the length of a curved line. A different device was required.

All simple problems in mathematics share a common feature—one cannot anticipate what difficulties they may conceal. Certainly nothing appears easier than measuring the length of a line. Take a piece of paper and mark two points on it. If you connect these two points with a straight line, all that is required to ascertain its length is a ruler. Nor need we be led astray by the yawning regress of philosophical discourse: What means shall be employed to measure the length of the ruler; what means shall be employed to measure the measuring rod which shall measure the ruler, etc., etc. It is agreed that we can measure the length of a straight line. Suppose, however, that we connect the two points by a curve; finding its length presents an altogether different story. One way of proceeding might be to take a piece of string, fit it along the curve, then remove the string and measure its length with a ruler. But this puts us back where we started, for it appears that the only lines that can be measured are straight. To measure the length of a curved line, it becomes necessary, in effect, to uncurve it.

By now, another means might suggest itself for measuring curves. Frequent recourse has been had, particularly in this chapter, to methods of approximation. Thus, we might divide the arc into a number of smaller parts and connect the end points of the small arcs by straight lines. The small straight lines will differ in sum from the sum of the small arcs less than a single straight line would differ from the length of the entire curve.

In other words, the sum of the lengths of the small straight lines will approximate the length of the curve. By choosing a large enough number of lines (and making them individually small), we should succeed in making the sum of their lengths differ from the length of the curve by as little as we please. The more numerous the little lines, the more accurate the approximation.

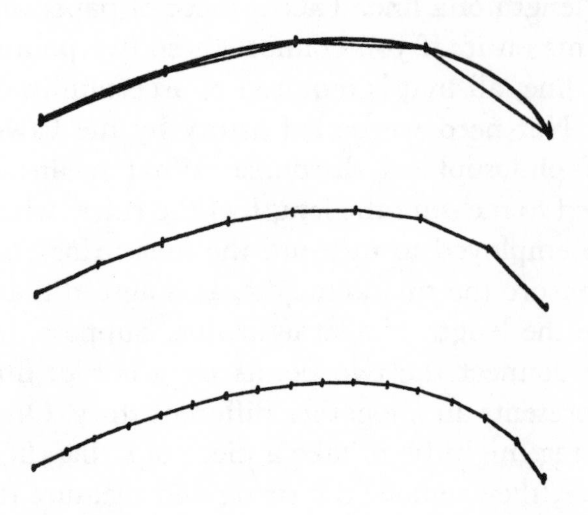

FIG. 135.—Approximating to the length of a curve by straight lines.

If we conceive of the number of lines as increasing beyond bound, it may be said that their sum approaches a limit—the length of the curve. Let us try to formulate this in terms of limits and limits of functions.

Suppose $y = f(x)$ is the equation of the curve which connects the two points A and B in a Cartesian plane. Let the x axis under the curve be subdivided into n equal parts. The x co-ordinate of the initial point A is a_0; the x co-ordinate of the next point is a_1; of the third point, a_2; and so on, so that the x co-ordinate of the last point is a_n, or B. The difference between two adjacent values of x may be denoted by Δx; the difference between two adjacent values of y (obtained by erecting perpendiculars from adjacent values on the x axis) is Δy. In Fig. 136, every pair of selected points on the curve bounds a hypotenuse of a shaded right-angle triangle, the base of which is Δx and the altitude Δy. Thus, the hypotenuse of each of these triangles will be an approximation to the length of that portion of the curve which bounds it. It follows that the sum of the hypotenuses of all the little triangles approximates to the length of the curve. By the use of the Pythagorean theorem, the value of each hypotenuse is easily obtained. Increasing the number of subdivisions

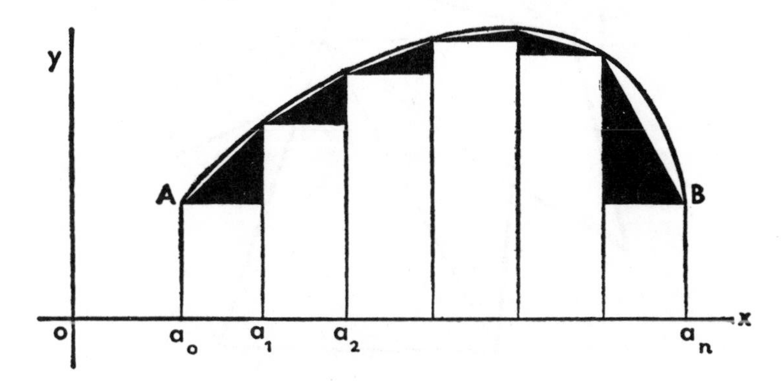

FIG. 136.—Approximating to the length of a curve by the hypotenuses of right-angle triangles. The base of each triangle is Δx, its altitude Δy.

will make the approximations more accurate. Thus, as Δx approaches 0, as the intervals along the x axis are made smaller, the sum of the hypotenuses of the right-angle triangles approaches a limit, which is the length of the curve. It should be noted that the length of each small hypotenuse is a function of its corresponding Δx.

*

We may now turn to the determination of the area under a curve, for it is in this problem that the ideas of the integral calculus are first vividly set forth.

Estimating the area of a figure bounded by straight lines, no matter how irregular, is comparatively easy. One need only introduce auxiliary lines so that the original figure is broken up into a number of triangles. By summing the areas of these triangles, the area of the original figure is measured.

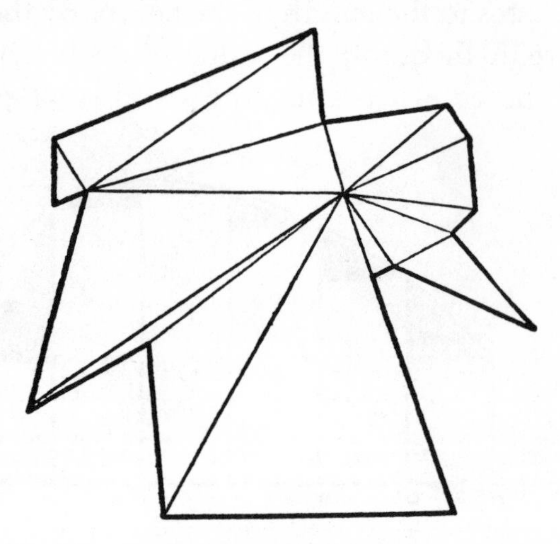

FIG. 137.—The area of this irregular polygon is determined by forming the triangles indicated and computing the area of each.

When the boundary of a figure is not straight but curved, this procedure is inadequate, and one must again resort to approximation. If we divide the curved sides of the figure into a great many parts, connecting their end points by straight lines, exactly as we did above, the resulting figure, a polygon bounded by straight edges, has an area which may be determined by elementary means. By increasing the number of sides of the polygon, its area may be made to differ from that of the original figure by as little as we wish and thus yield an approximation as close as we desire.

But a more effective means of dividing a curvilinear figure is by the use of rectangles. Precisely this device was invented by Archimedes. Figure 138 illustrates a circle divided into rectangular strips. According to the method of constructing these strips, it should be noted that not one, but two approximations can be obtained. The first gives the area of the rectangles inscribed in the circle, the

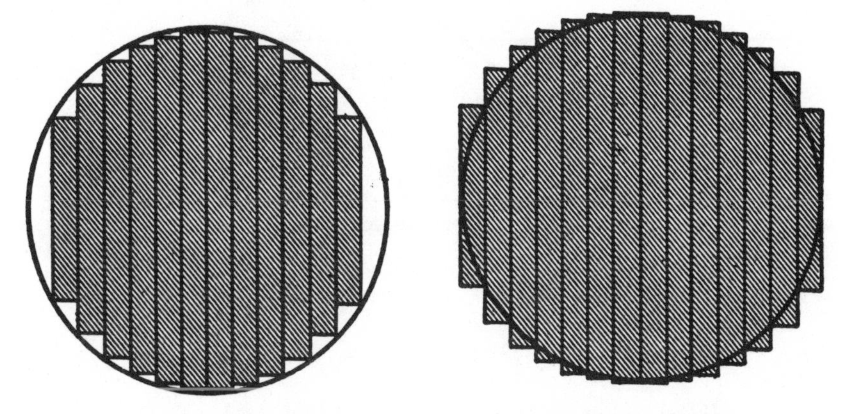

FIG. 138.—Approximating to the area of a circle by the use of rectangles.

second the area of the rectangles circumscribing the circle. The discrepancy between the two rectangulated areas becomes smaller and smaller as the number of rectangles is increased, in other words, as they are diminished in width. Their common limit, as the inner area increases and the outer area decreases, is the area of the circle.

Instead of confining ourselves to this special example, if we discuss the general problem of finding the area under the segment of an arbitrary curve, the method just described can perhaps be made even clearer. We wish to find the area of the shaded section in Fig. 139 below. It is bounded on top by the curve $y = f(x)$, below by the portion of the x axis from $x = A$ to $x = B$, and on the right and left by two straight lines parallel to the y axis. Divide the x axis into n equal subintervals, as in Fig. 136. Erect at each of the dividing points a perpendicular from the x axis to the curve. At each point where a

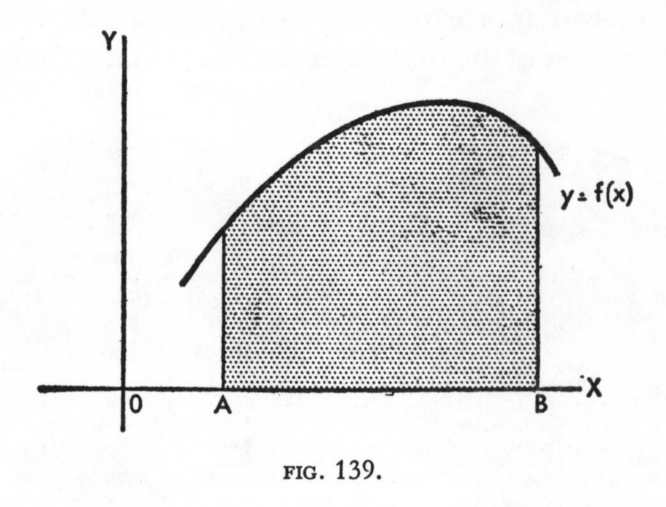

FIG. 139.

perpendicular intersects the curve, draw a horizontal line to the adjacent vertical lines. For every little subinterval on the x axis there will be two rectangles, one under the curve, the other protruding above it and containing part of the area outside. Consider a typical subinterval (see Fig. 140).

The area of the smaller rectangle *ABCD* is the base *AB* times the altitude *AD*, where the altitude is the value of the function at the initial point of the subinterval, *A;* the area of the larger rectangle *ABEF* is the product of the same base *AB*, by the altitude *BE*, the value of the function at the terminal point of the subinterval, *B*. The area under the curve lies between the areas of these two rectangles. An excellent approximation to the desired area is obtained by taking the *average* value of the two rectangles.

Repeating this process for each subinterval and forming the sum of the average rectangles gives the approximation to the entire area under the curve. Again enlisting the aid of the concept of limit of a function, it may be seen that as the number of subintervals on the x axis is increased, the

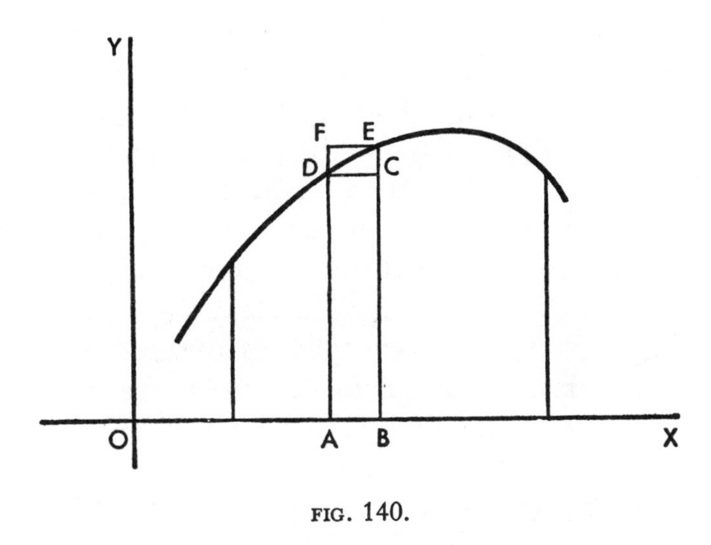

FIG. 140.

sum of the corresponding areas necessarily approaches the area of the shaded figure (Fig. 139). In the limit, this sum of the many tiny elements of area is called the *definite integral of the function* $y = f(x)$ *between the values of* $x = A$ *and of* $x = B$, and in the shorthand of Leibniz is:

$$\int_A^B f(x)dx.$$

Briefly recapitulating: Each of the subintervals along the x axis is Δx, which is the base of every one of the tiny rectangulated areas. The altitude of the average rectangle is represented by a perpendicular line drawn from a typical interior point of the interval Δx to the curve. Its value is, of course, $f(x)$. The area of each such average rectangle is $f(x) \cdot \Delta x$, and the sum of these areas is the sum of all such products. In technical symbolism the limiting area is written $\int_A^B f(x)dx$, where dx replaces Δx, since $\Delta x \to 0$.

*

Our interpretation of the definite integral is that it is an area. To assign such a meaning is always possible, but there exist integrals of certain functions which have additional physical significance. Mainly this is because the definite integral is a number, a *sum*, as well as an area. Whenever, in science, a function is summed to the limit, the definite integral plays a role. One of the achievements of the integral calculus has been the determination of the *moment of inertia* of all solids. Again, it is to the definite integral that structural engineers must render thanks for the Golden Gate Bridge, for it rests on this even more than on concrete and steel. Restraining the force on our gigantic dams, with their curved and uneven faces, represents another problem in the integration of a function. By determining the water pressure at an arbitrary point and summing it over the whole face of the dam, the total force is uniquely determined. The centroid, that is, the center of gravity of any plane or solid figure, is easily reckoned by means of the integral calculus when applied to the particular function defining that figure. Such examples might be multiplied indefinitely.

*

Beyond the concept of the definite integral, with its many uses and richly studded field of application, there is the notion of the *indefinite integral*, of even greater intrinsic value to the mathematician. Its chief theoretical interest is that it enables us to exhibit the astounding relationship between the derivative and the integral.

Consider the function $y = f(x)$. Instead of limiting the interval as before, from $x = A$ to $x = B$, imagine it to extend from $x = A$ to $x = x_0$, where x_0 may assume *any* value. For different values of x_0, the definite integral will also take on different values. True, we will no longer have under consideration a limited area, but we will have all the requisites for preparing a functional table. On one side there will be listed successive values of x_0; on the other, corresponding values of the definite integral. This correspondence between values of x_0 and values of the definite integral is itself a function called "the indefinite integral" of the function $y = f(x)$. Here is the crux of the matter: The definite integral of the function $y = f(x)$ is a number determined by an interval of definite length and a portion of the curve $y = f(x)$ defined over that interval. When the interval is extended from a fixed point through a succession of others, to each of these there corresponds a value of the definite integral. This correspondence, this function, is the *indefinite integral of the original function* $y = f(x)$ and is symbolized by

$$\int f(x)dx.$$

From this you may perhaps guess what the two seemingly diverse branches of the calculus have in common. For the relation between differentiation and integration is reminiscent of elementary arithmetic. It is the same relation that exists between addition and subtraction, multiplication and division, involution and evolution. The one operation is the *inverse* of the other. Starting with the function $y = f(x)$, upon differentiating we obtain $\frac{dy}{dx}$. What do we get upon *integrating* the function $\frac{dy}{dx}$? The motif of the calculus is hereby revealed, for we obtain the original function, $y = f(x)$. The indefinite integral of the function $y = f(x)$ is another function of x which we shall denote by $y = F(x)$. Of course, the derivative of $y = F(x)$ is $f(x)$. Every function may thus be regarded as the derivative of its integral and as the integral of its derivative.

*

Earlier we alluded to the exponential function, $y = e^x$, and its usefulness in describing the phenomenon of growth. It is the only function, the rate of change of which is equal to the function itself. Differentiating $y = e^x$ yields $\frac{dy}{dx} = e^x$. Integrating yields the same result. It follows that the life history of any organism—amoeba, man, or redwood—of any phenomenon which exhibits properties of organic growth—is aptly described by the integral of e^x. This stirring conception is not difficult to visualize. Proportionality of rate of growth to state of growth may be embodied in the exponential function. If this is integrated, the total growth over any given period is given by the definite integral, and the general character of growth succinctly set forth by the indefinite integral.

In conclusion, let us re-examine the problem of a falling body. We started with the distance that the body fell in a period of time and derived its speed at every instant by differentiation. Acceleration at every instant was obtained, in turn, by differentiating the first derivative, finding the rate of change of the speed with respect to the time. Galileo and Newton did the same thing—backwards. They shrewdly guessed that the acceleration of a falling body was a constant, the gravitational constant. Upon integrating the function expressing this hypothesis, they made the classical discovery of the laws of motion:

(1) the speed of a falling body is gt, where g is the gravitational acceleration, 32, and t the time elapsed since the body was dropped.

(2) the distance covered by a falling body is $\frac{1}{2}gt^2$.

This and the other laws of motion governing every particle in the universe are derivable simply and elegantly by means of the calculus. But this is not all, for the calculus not only helped release some of nature's most intimate secrets; it gave the mathematician more new worlds to conquer than Alexander ever sighed for.

APPENDIX

PATHOLOGICAL CURVES

The curves treated by the calculus are normal and healthy; they possess no idiosyncrasies. But mathematicians would not be happy merely with simple, lusty configurations. Beyond these their curiosity extends to psychopathic patients, each of whom has an individual case history resembling no other; these are the pathological curves of mathematics. We shall try to examine a few in our clinic.

Before we can do so, it will be necessary to introduce the idea of a curve as the limit of a sequence of polygons. Let an equilateral triangle be inscribed in a circle. This triangle may be considered as a curve—C_1. Let C_2 be the regular hexagon obtained by bisecting the three arcs in Fig. 141, and by joining, in order, the six vertices. (Fig. 142).

C_3 is the regular dodecagon formed by bisecting the six arcs of Fig. 143, and joining the twelve vertices in order. Repeat this process each time by bisecting the arcs, doubling the number of sides. The curve approached as

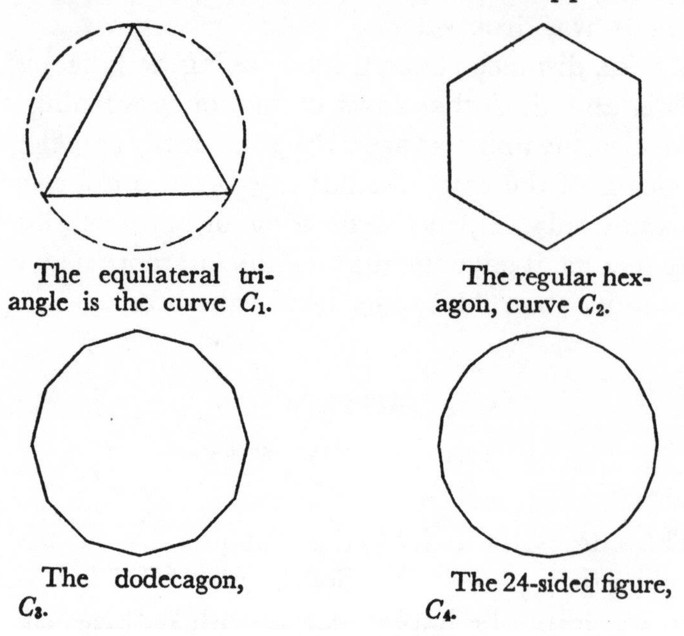

The equilateral triangle is the curve C_1.

The regular hexagon, curve C_2.

The dodecagon, C_3.

The 24-sided figure, C_4.

FIGS. 141, 142, 143, 144.—The circle as the limit curve of a sequence of curves.

limit is the circle. Thus, the circle is described as the limit curve of a sequence of curves or polygons.

(1) *The Snowflake Curve.* Start with an equilateral triangle, with a side one unit in length. This triangle is curve C_1. (Fig. 145.)

Trisect each side of the triangle and on each of the middle thirds erect an equilateral triangle pointing outward. Erase the parts common to the new and the old triangles. This simple polygonal curve is called C_2.

Trisect each side of C_2, and again upon each middle third erect an equilateral triangle pointing outward.

Erase the part of the curves common to the new and old figures. This simple curve is C_3.

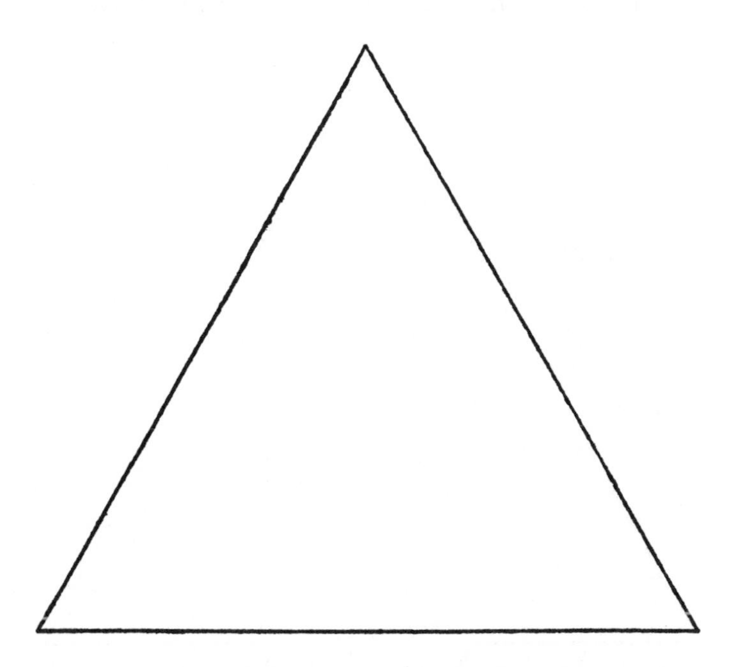

FIG. 145.—The first stage of the Snowflake Curve—C_1.

Repeat this process, as shown in Figs. 148–150.

What is the *limit curve* of this sequence of curves? Why is it called the Snowflake Curve, and why is it described as pathological?

It derives its name from the shape it assumes in the successive stages of its development. Its pathological character is borne out by this incredible feature: Although one may conceive that the limit curve can be drawn on a piece of paper, it is hard to imagine that this is possible, because, though the area is finite, the length of its perimeter is infinite! But it is clear that at each stage of the construction the perimeter increases, and since the sequence of numbers representing the length of the perimeter at each stage does not converge, i.e., does not choke off, the perimeter must grow beyond all bounds. We are

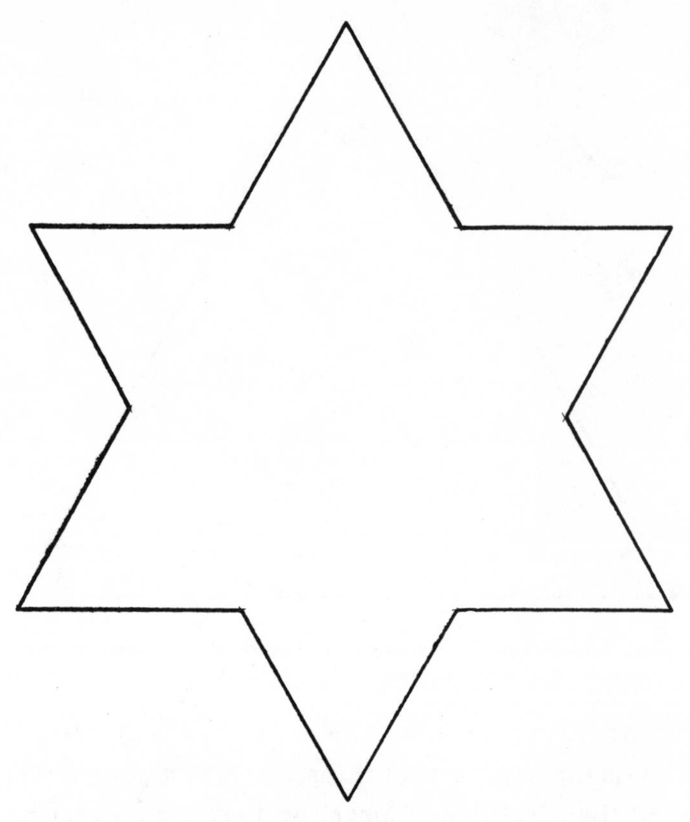

FIG. 146.—The second stage of the Snowflake Curve—C_2.

thus confronted by the amazing fact that a curve of infinite length may be drawn on a small sheet of paper—for example, on a postage stamp.

The proof is simple: The perimeter of the original triangle was 3. The perimeter of curve C_2 is $3 + 1$; of C_3, $3 + 1 + \frac{4}{3}$; of C_4, $3 + 1 + \frac{4}{3} + \frac{4^2}{3^2}$. The perimeter of C_n is $3 + 1 + \frac{4}{3} + \frac{4^2}{3^2} + \ldots + \frac{4^{n-2}}{3^{n-2}}$. Thus, as n

grows, so grows the sequence, for we are dealing with an infinite series which does not converge.

The fact that the curve remains on the paper proves that the area of the snowflake is finite. Explicitly, the area of the final curve is $1\frac{3}{5}$ times that of the original triangle. And if this is not weird enough, consider that

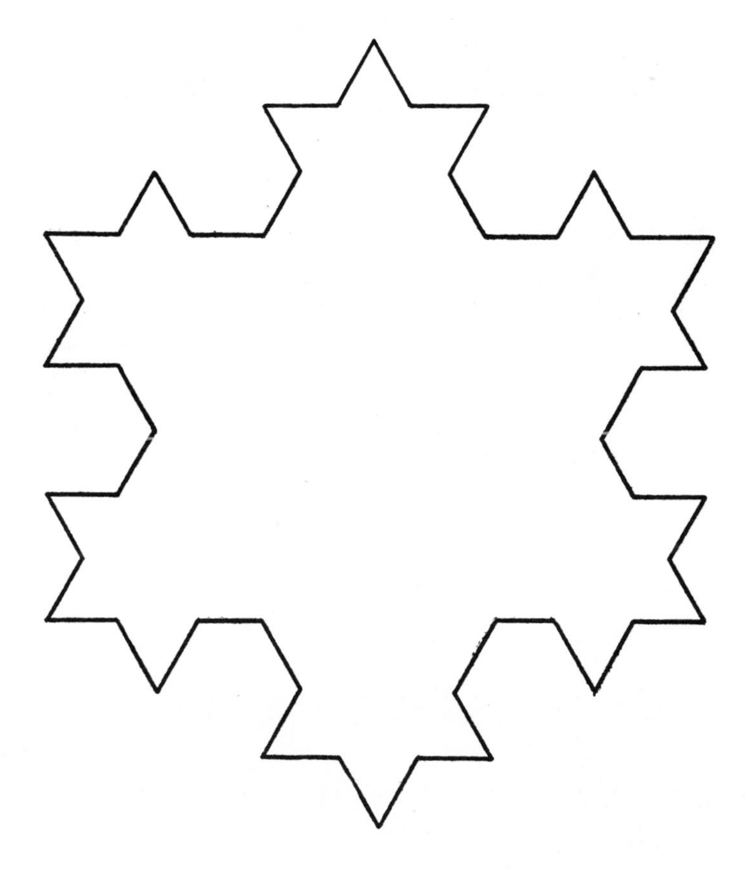

FIG. 147.—The third stage—C_3.

it is not possible to tell at any point on the limit curve the direction in which it is going, that is, the tangent line does not exist.[6]

(2) The *Anti-Snowflake Curve* is obtained by drawing the triangles inward, not outward, and has many of the same properties as its brother. Its perimeter is infinite,

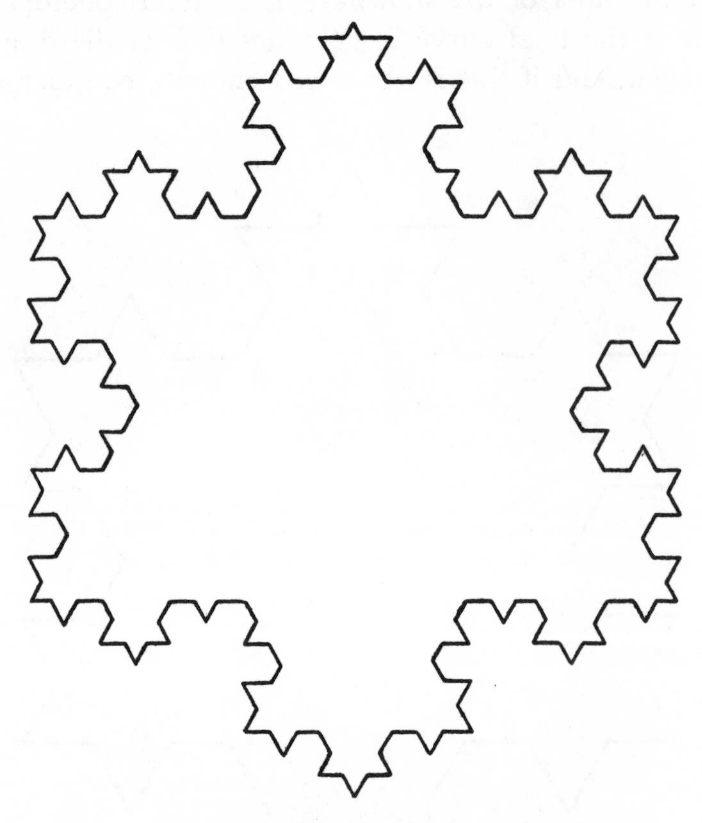

FIG. 148.—The fourth stage—C_4.

while its area is finite, and no tangent can be drawn to it at any point. (Figs. 151–154.)

(3) Another pathological curve is the *In-And-Out Curve.* Draw a circle (with radius = 1) and choose six points on it so as to divide the circumference into six equal parts. Take three alternate arcs and turn them inward. The original circle is C_1, the new figure C_2. (Figs. 155–156.)

The perimeter of C_2 is the same as the perimeter of C_1, because its length is not altered by turning three arcs inward.

Next, trisect each arc, and turn the middle one outward if it is now turned inward, inward if it is now

FIG. 149.—The fifth stage—C_5.

turned outward. This new curve is C_3. Its perimeter is also equal to that of the original circle. Moreover, the area of C_3 is the same as that of C_2 because we alternately added and subtracted the same size segments. (Fig. 157.)

Repeat this process. The limit curve has a perimeter equal to the perimeter of the circle. Its area is equal to that of C_2, which, in turn, is equal to the area of a regular hexagon. Like the Snowflake and Anti-Snowflake, this curve, too, has its pathological features.

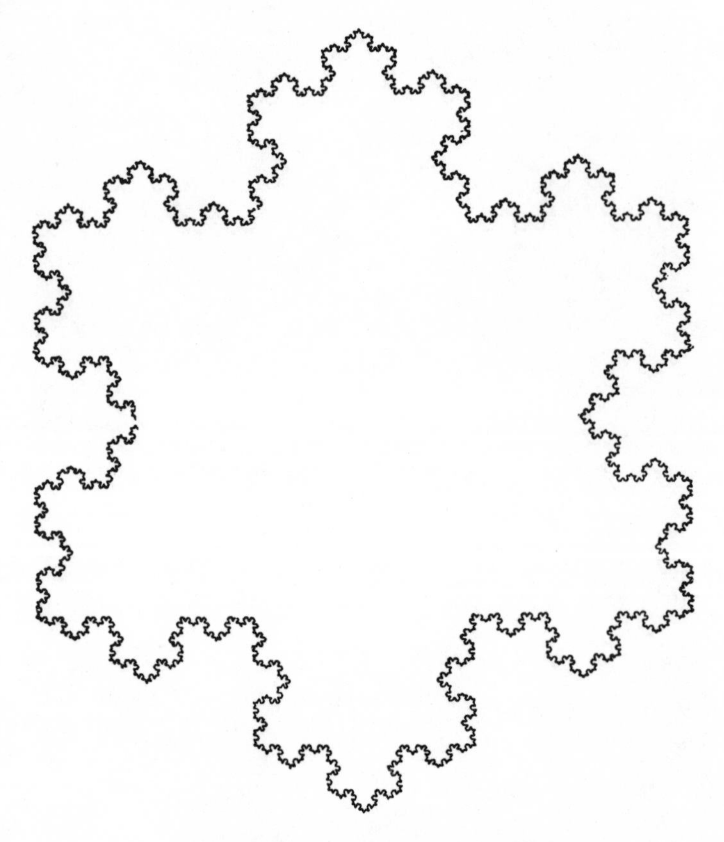

FIG. 150.—The sixth stage—C_6.

While the curvature of a circle is computed without difficulty, the In-And-Out Curve presents a different aspect. Consider an arbitrary point upon it. In which direction, toward the center of the circle or away from

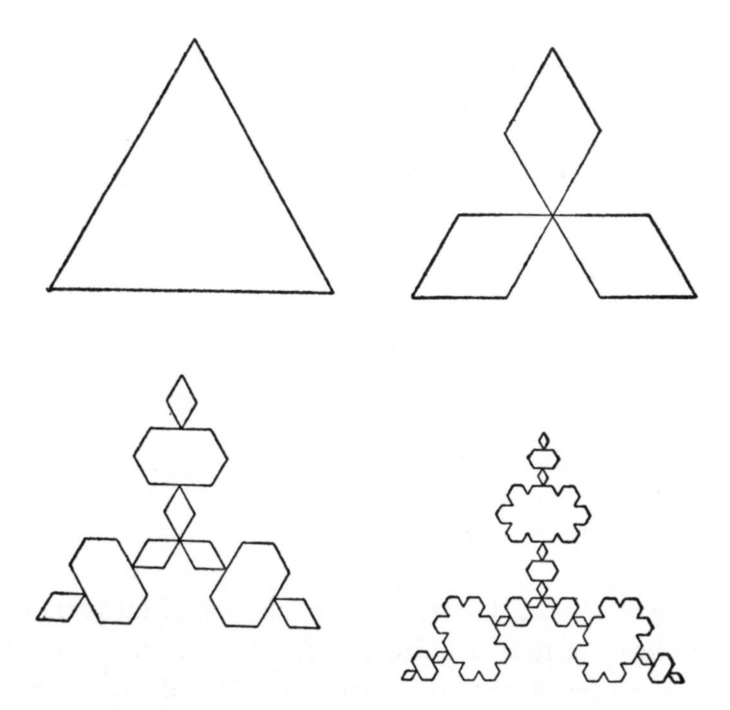

FIGS. 151, 152, 153, 154.—The first four stages of the Anti-Snowflake Curve.

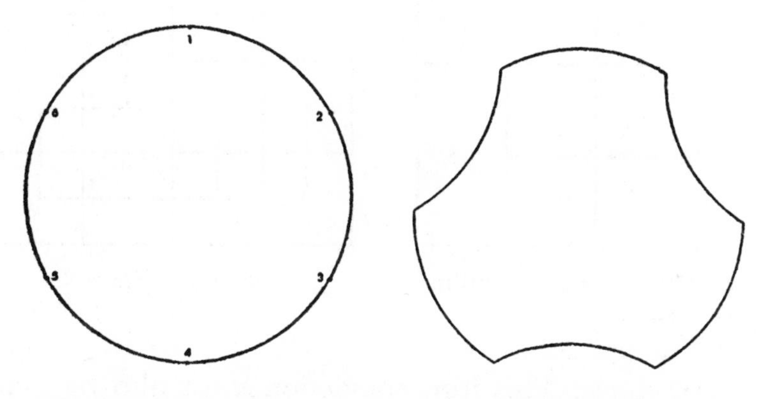

FIG. 155.—The In-And-Out Curve —stage C_1.

FIG. 156.—Stage C_2.

the center, shall we measure its curvature? We find there is no definite curvature. The second derivative does not exist.

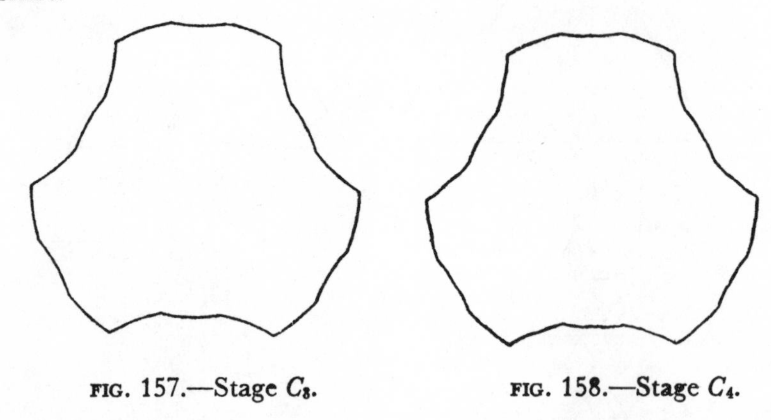

FIG. 157.—Stage C_3. FIG. 158.—Stage C_4.

(4) *Space-Filling Curves:* One of the cardinal principles of geometry is that a point has no dimensions, and that a curve is one-dimensional and can, therefore, never fill

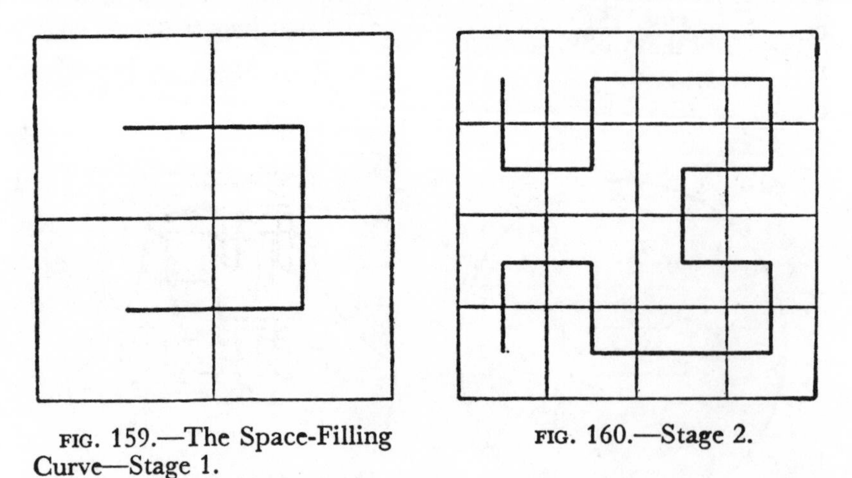

FIG. 159.—The Space-Filling Curve—Stage 1. FIG. 160.—Stage 2.

a given space. This iron conviction must also be shattered. For behold the pathological specimen supreme, the Space-Filling Curve, which will not only occupy the interior of a square, but gobble up the space in an entire cubical box.

The successive stages of such a curve are illustrated in Figs. 159–164. Select *any* point in the square or cube.

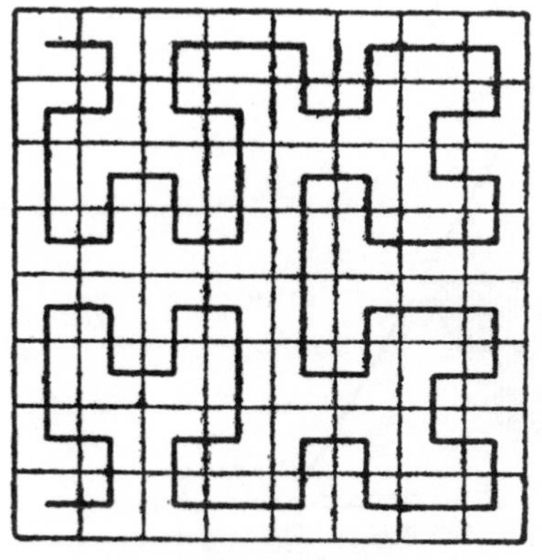

FIG. 161.—Stage 3.

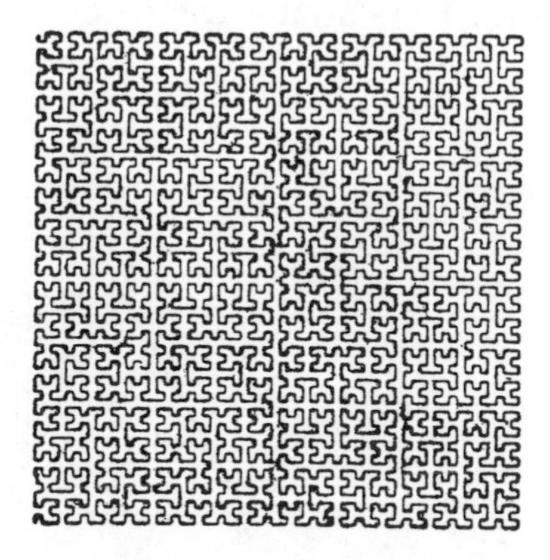

FIG. 162.—An advanced stage.

It can be shown that eventually, when the curve has been completed, it will pass through that point. Since this reasoning extends to *every point*, it follows logically that the curve must fill the entire square or cube.

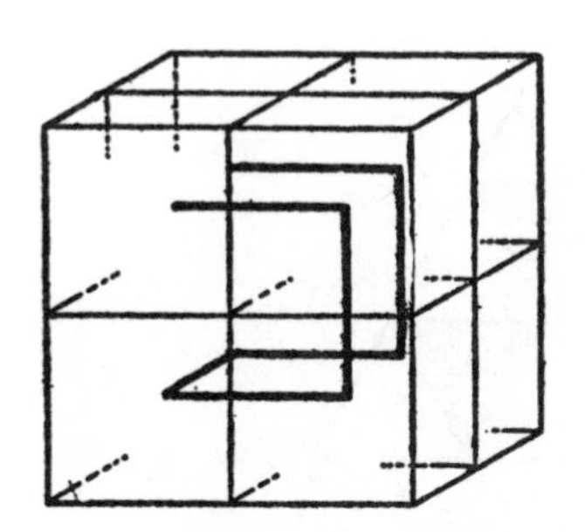

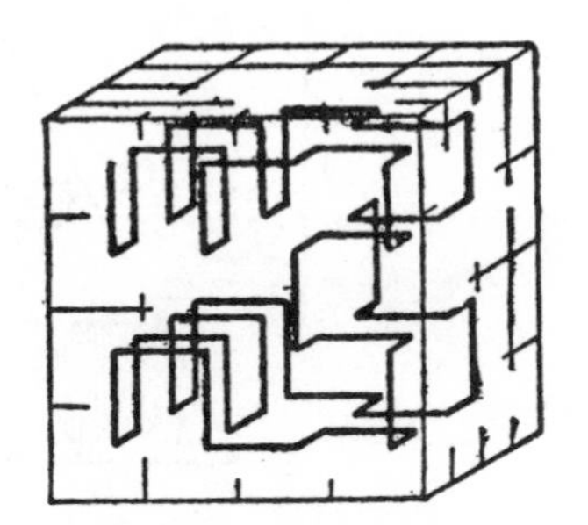

FIGS. 163, 164.—The first two stages of a curve which fills an entire cubical box.

(5) *The Crisscross Curve:*

This curve has the property that it crosses itself at

every one of its points. We are certain that you don't believe us—and never will—but here are the directions for making it:

1st Step: Inscribe a triangle within a triangle as in Fig. 165. Shade the interior triangle.

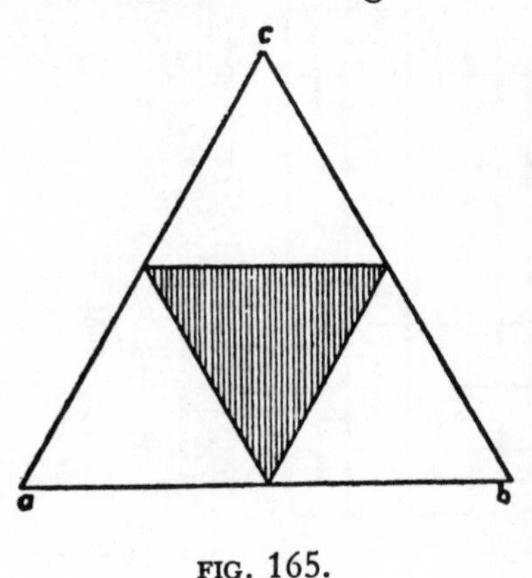

FIG. 165.

2nd Step: Continue the process for each of the three remaining triangles as in Fig. 166.

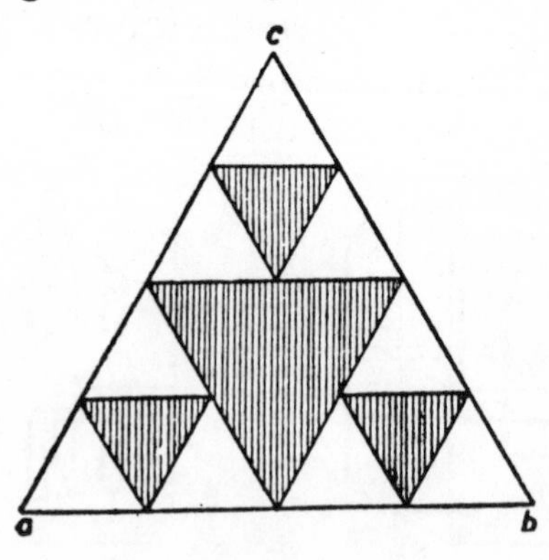

FIG. 166.

3rd to *n*th Step: Repeat the process indefinitely (Fig. 167 is the 5th stage). Then join the points of the original triangle remaining unshaded and distort the original triangle so that the three points A, B, and C are brought together.

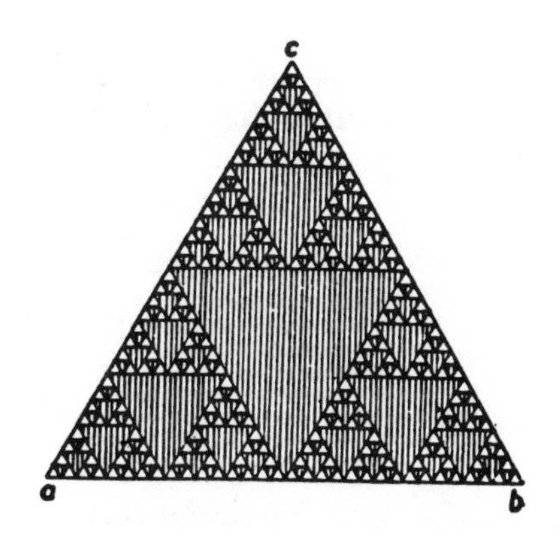

FIG. 167.

There you have the Crisscross Curve.

FOOTNOTES

1. Cajori, *History of Fluxions*.—P. 305.
2. Protocol on trigonometry for those who have forgotten:

In the right-angle triangle below the following are the trigonometric ratios (functions of an angle):

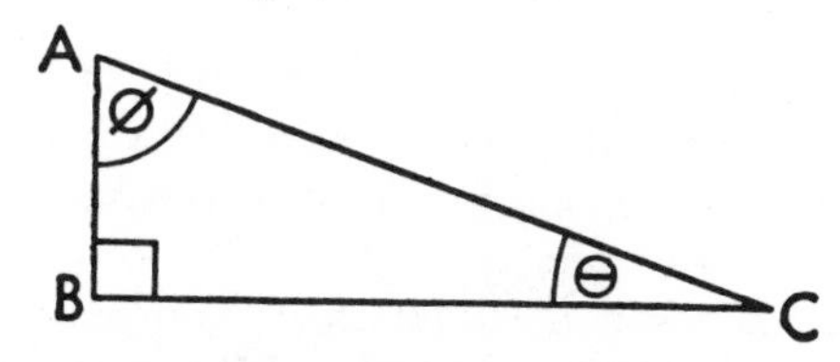

$$\text{Sine } \theta = \frac{\text{Side } AB}{\text{Side } AC} = \text{Cosine } \phi$$

$$\text{Cosine } \theta = \frac{\text{Side } BC}{\text{Side } AC} = \text{Sine } \phi$$

$$\text{Tangent } \theta = \frac{\text{Sine } \theta}{\text{Cos } \theta} = \frac{\frac{AB}{AC}}{\frac{BC}{AC}} = \frac{AB}{BC} = \text{Cotangent } \phi.$$

In other words, the trigonometric functions are the ratios of the sides of a right-angle triangle to one another, and depend in turn upon the angles.

The concept of *tangent* has immediate application in analytic geometry and the calculus. In the diagram below, the *slope* of the line AB is the ratio P/Q which is none other than the *tangent* of the angle Θ.

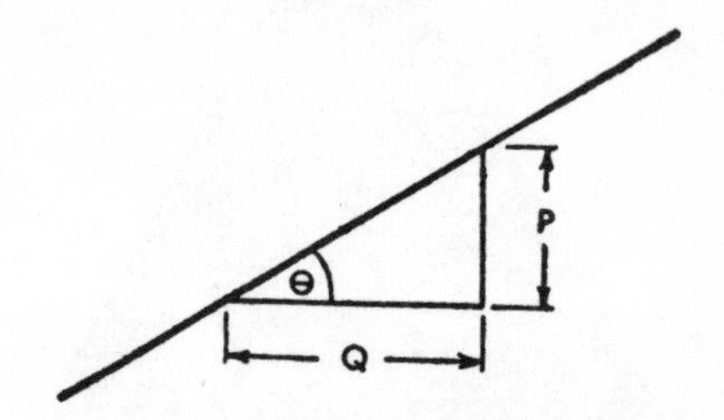

FIG. 169—The slope of a straight line is the ratio $\frac{P}{Q}$.

But the word tangent has another meaning quite different from the one above. This new meaning is essential to the calculus.

In the Cartesian plane draw the curve ABC. Consider two points P_1 and P_2 on this curve joined by the straight line passing through them. (See Fig. 131, page 324.) As P_2 moves along the curve to P_1, the line joining these two points approaches a limiting value called *the tangent to the curve ABC at the point* P_1. The slope of this tangent line at the point P_1 is the derivative of the function, the graph of which is the curve ABC.—Pp. 309, 311, 324.

3. With apologies to the physicist for the use of the term speed, instead of the technically correct term "velocity."—P. 316.
4. The length of a parabolic segment is expressible only in terms of logarithms and consequently could not be computed by means of the elementary methods known to the ancients.—P. 331.
5. Wolf, *History of Philosophy, Science and Technology in the Sixteenth and Seventeenth Centuries*.—P. 332.
6. Thus, we have, in essence, a continuous function without a derivative.—P. 347.

Mathematics and the Imagination

There is no conclusion. What has concluded that we might conclude in regard to it? There are no fortunes to be told, and no advice to be given. Farewell.

—WILLIAM JAMES

"Cheshire-Puss," she began, rather timidly. . . "Would you tell me please, which way I ought to go from here?"

"That depends a good deal on where you want to get to," said the Cat.

"I don't much care where—" said Alice.

"Then it doesn't matter which way you go," said the Cat.

—LEWIS CARROLL

WHAT IS mathematics? A large and varied body of thought which has grown from the earliest times purports to answer this question. But upon examination, the opinions which range from those of Pythagoras to the theories of the most recent schools of mathematical philosophy reveal the sad fact that it is easier to be clever than clear. Particularly of late has there been a tendency to present aphorisms in place of straightforward replies, aphorisms which, unfortunately, shed little light. In the method of approaching the problem lies the main obstacle to a satisfactory answer. If one were to ask, "What is biology?" it would be comparatively simple to start with an etymological definition and then to group together the great body of imformation comprised in the biological sciences. Next would be a conclusion on how all the items are synthesized into an integrated science. Even a crude explanation, such as "Biology discusses horses, bats, daffodils, and whales," would give a fair idea of what it is about. On the other hand, the study of mathematics—arithmetic, algebra, geometry, the calculus—does not imply anything more about its nature than that it is concerned with numbers and that it is a useful technique. So far as the concept of numbers is concerned, no definition has yet been given which in itself would simplify the task of defining mathematics.

Here, then, in mathematics we have a universal language, valid, useful, intelligible everywhere in place and in time—in banks and insurance companies, on the parchments of the architects who raised the Temple of Solomon, and on the blueprints of the engineers who, with their calculus of chaos, master the winds. Here is a discipline of a hundred branches, fabulously rich, literally without limit in its sphere of application, laden with honors for an unbroken record of magnificent accomplishment. Here is a creation of the mind, both mystic and pragmatic in appeal. Austere and imperious as logic, it is still sufficiently sensitive and flexible to meet each new need. Yet this vast edifice rests on the simplest and most primitive foundations, is wrought by imagination and logic out of a handful of childish rules. Even though no definition thus far has encompassed either its scope or its nature, can it be that the question "What is mathematics?" must go unanswered?

*

That mathematics enjoys a prestige unequaled by any other flight of purposive thinking is not surprising. It has made possible so many advances in the sciences, it is at once so indispensable in practical affairs and so easily the masterpiece of pure abstraction that the recognition of its pre-eminence among man's intellectual achievements is no more than its due.

In spite of this pre-eminence, the first significant appraisal of mathematics was occasioned only recently by the advent of non-Euclidean and four-dimensional geometry. That is not to say that the advances made by the calculus, the theory of probability, the arithmetic of the infinite, topology, and the other subjects we have discussed, are to be minimized. Each one has widened mathematics and deepened its meaning as well as our comprehension of the physical universe. Yet none has contributed to mathematical introspection, to the knowledge of the relation of the parts of mathematics to one another and to the whole as much as the non-Euclidean heresies.

As a result of the valiantly critical spirit which engendered the heresies, we have overcome the notion that mathematical truths have an existence independent and apart from our own minds. It is even strange to us that such a notion could ever have existed. Yet this is what Pythagoras would have thought—and Descartes, along with hundreds of other great mathematicians before the nineteenth century. Today mathematics is unbound; it has cast off its chains. Whatever its essence, we recognize it to be as free as the mind, as prehensile as the imagination. Non-Euclidean geometry is proof that mathematics, unlike the music of the spheres, is man's own handiwork, subject only to the limitations imposed by the laws of thought.

*

The philosophy which goes under the name of logical positivism has prepared a formidable program: first, to eliminate metaphysics from philosophy; and second, to exhibit the mutual relations between the laws of thought (i.e., logic) and mathematics. There are some who believe that logical positivism affords an advance, beyond that made by non-Euclidean geometry, in evaluating the nature of mathematics. Quite modestly the hope has been expressed that here at least is a doctrine that faces squarely those essential and inherent difficulties blocking the road to the summit.

In purging mathematical philosophy of metaphysics, there has been (in our judgment) a real gain. No longer is mathematics to be looked upon as a key to the truth with a capital *T*. It may now be regarded as a woefully incomplete, though enormously useful, Baedeker in a mostly uncharted land. Some of the landmarks are fixed; some of the vast network of roads is made understandable; there are guideposts for the bewildered traveler.

On the other hand, one cannot suppress the feeling that this new appraisal of mathematics is so incomplete, so devoid of color, as to be almost trivial and inconsequential. In regarding it as merely a handful of primitive, undefined propositions, coupled with a methodology for manufacturing new ones, something of the spirit, of the flavor and color of mathematics seems to have been lost. While those who oppose logical positivism grant that it serves some purpose, they assail the stultification of discourse, the narrowing of horizons, which it inevitably entails. We share the feeling that mathematics is more than a factory of tautologies, rather that it is a vehicle to carry on the highest aspirations of the creative intellect.

Briefly, here is what the positivists say: Logic is concerned with the formal rules for manipulating the symbols of language. Mathematics is concerned only with equations, i.e., literally statements of equivalence of number.

All the essential, internal relations of meaning are the province of mathematical science. An omniscient being would thus require neither logic nor mathematics since the relations between all entities would, to him, be self-evident. Though he might still find other sciences useful, for example, biology to provide him with a catalogue of living things, or a telephone directory to help him find his friends, his need for logic and mathematics would have vanished. For once all meaning and all relationships were fully disclosed, these disciplines would be superfluous.

Is there not reason to think that in such an interpretation, though we have scourged ourselves pitilessly and driven out the confusing spirit of metaphysics, we may also have drained the vitality of mathematics? May we not well have lost "the spirit in the word"?

*

As we have already indicated, the creation of non-Euclidean geometry signalized the realization that mathematics in no sense depends upon our environment. Although many similarities exist between the behavior of the fecund little symbols which we place on paper or juggle in our heads and those phenomena which take place in the physical world, mathematics is to be recognized as an autonomous discipline, restricted only by the formal rules of thinking. The development of mathematics is a counterpart of the everlasting struggle for greater comprehensiveness and greater freedom: from the particular to the general; from configurations bounded by straight lines to pathological curves; from the properties of this or that specific figure to the properties of *all* figures; from one dimension to *n* dimensions; from the finite to the infinite. In this march the imagination has played a notable role. For imagination has the pragmatic value that it leaps ahead of the slow-moving caravan of well-ordered thought and often scouts out reality long before its ponderous master. Therein lies its essential contribution to one of the strangest collaborations of thought, staid mathematics and volatile imagination.

Mathematics is an activity governed by the same rules imposed upon the symphonies of Beethoven, the paintings of Da Vinci, and the poetry of Homer. Just as scales, as the laws of perspective, as the rules of metre seem to lack fire, the formal rules of mathematics may appear to be without lustre. Yet ultimately, mathematics reaches pinnacles as high as those attained by the imagination in its most daring reconnoiters. And this conceals, perhaps, the ultimate paradox of science. For in their prosaic plodding both logic and mathematics often outstrip their advance guard and show that the world of pure reason is stranger than the world of pure fancy.

Bibliography

This descriptive bibliography contains a number of works allied in interest to the contents of this book. It is not (nor is it intended to be) full, authoritative, or exhaustive. It is merely an expression of personal tastes which may be helpful to the reader whose curiosity has been stimulated.

ABBOTT, EDWIN A. *Flatland—A Romance of Many Dimensions*. Boston: Little Brown, 1929.

The celebrated little book about a world of two dimensions.

AHRENS, W. *Mathematische Unterhaltungen und Spiele*. Leipzig: B. G. Teubner, Vol. I, 1910; Vol. II, 1918.

For those who can read German, this work will be a rare treat. Although parts of it are quite technical, it is an exhaustive, definitive study of puzzles with a wealth of related stories, anecdotes, and historical sketches.

BALL, W. W. ROUSE. *Mathematical Recreations and Essays*. 11th Edition. New York: Macmillan, 1939.

This is a standard work, packed with information, but arranged more as a handbook than for continuous reading. It is indispensable for anyone who likes puzzles.

BELL, E. T. *Men of Mathematics*. New York: Simon and Schuster, 1937.

The various portions of this book devoted to non-Euclidean geometry make stimulating and engaging reading. On this as well as other subjects of mathematics, Bell is clear, his style racy and entertaining. There are biographical sketches of the leading mathematicians from Zeno to Poincaré, sometimes distinguished, often delightful, always readable. The exposition of difficult phases of mathematics is particularly good.

———. *The Queen of the Sciences*. New York: G. E. Stechert, 1938.

This is a reprint of a charming little volume, published originally for the World's Fair at Chicago, 1933. Simple, worth reading.

BLACK, MAX. *The Nature of Mathematics*. New York: Harcourt Brace, 1935.

An account of the mathematical philosophies of the three principal contemporary schools, concise and clear, but by no means easy reading.

BLISS, G. A. "Mathematical Interpretations of Geometrical and Physical Phenomena," *American Mathematical Monthly*, Vol. 40 (October, 1933).

A readable account of the application of mathematics to physical phenomena.

BRIDGMAN, PERCY W. *The Logic of Modern Physics*. New York: Macmillan, 1927.

Not a mathematics book, but a crystal-clear exposition of modern physical theories, the laws of cause and effect, and probability, by a distinguished physicist.

BALL, W. W. ROUSE. *History of Mathematics*. New York: Macmillan, 1925.

CAJORI, FLORIAN. *History of Mathematics*. New York: Macmillan, 1919.

Both of these are standard histories of mathematics. The work by Cajori is quite detailed, though nothing in comparison to Moritz Cantor's monumental German history of mathematics in four large volumes. Ball's work, while in some respects out of date and discursive, is sufficiently authoritative, and being less detailed, can be read more pleasantly with far less of a mathematical background.

CAJORI, FLORIAN. "History of Zeno's Arguments on Motion," *American Mathematical Monthly*, Vol. XXII. (1915) Pp. 1–6, 292–297.

———. *A History of Elementary Mathematics*. New York: Macmillan, 1917.

A standard work, interesting, not too technical, informative.

CANTOR, GEORG. *Contributions to the Founding of the Theory of Transfinite Numbers*. Edited by P. B. Jordain. Chicago: Open Court, 1915.

A translation of Cantor's work on the infinite, particularly valuable for the historical introduction.

CARSLAW, HORATIO S. *The Elements of Non-Euclidean Plane Geometry and Trigonometry*. London: Longmans Green & Co., 1916.

A good elementary text on non-Euclidean geometry. Technical.

COHEN AND NAGEL. *Introduction to Logic and Scientific Method*. New York: Harcourt Brace, 1934.

An admirable book in every way. A textbook in a class by itself. Refreshing and lucid chapters on the entire problem of scientific method and a special chapter on probable inference, well reasoned, well written.

COOLEY, GANS, KLINE, and WAHLERT. *Introduction to Mathematics*. New York: Houghton Mifflin Co., 1937.

A clear, unpretentious introduction to many advanced branches of mathematics, including infinite classes. It is also suitable for several other topics discussed in this book. A text.

COURANT, RICHARD. *Differential and Integral Calculus*. Vol. I. London: Blackie & Son, 1934.

An excellent text with unusually good examples of the applications of the calculus.

DANTZIG, TOBIAS. *Number, the Language of Science*. New York: Macmillan, 1933.

An interesting nontechnical account of the development of the number concept up to and including the transfinites.

De Morgan, Augustus. *Budget of Paradoxes*. Chicago: Open Court, 1940.

A collection of letters, stories, anecdotes, paradoxes, and puzzles, all relating to mathematics.

Dresden, Arnold. *An Invitation to Mathematics*. New York: Henry Holt & Co., 1936.

Unusually praiseworthy text on higher mathematics, covering analytic geometry, number theory, the calculus, projective geometry, the mathematics of the infinite. A text with a well-executed idea of making mathematics attractive, it presupposes a knowledge of only elementary algebra and geometry.

Eddington, Sir Arthur S. *Space, Time, and Gravitation*. Cambridge University Press, 1920.

Brilliant discussion of the problems of relativity including chapters on "kinds of space" and "What is geometry?" Nontechnical, but by no means simple.

———. *New Pathways of Science*. New York: Macmillan, 1935.

Contains interesting discussions on probability by a famous astronomer and equally famous writer on popular science. Beyond that, the rest is also worth reading, as is anything by Eddington.

Enriques, Federigo. *Historic Development of Logic*. New York: Henry Holt, 1929.

The best work of its kind. The "principles and structure of science in the conception of mathematical thinkers." Largely nontechnical but occasionally difficult.

Galileo. *Two New Sciences*. New York: Macmillan, 1914.

Not only one of the greatest books of all times, but a perfectly fascinating one. It contains, among much else, Galileo's views on infinity. For anyone interested in the history of science or mathematics, or, for that matter, for anyone who enjoys reading. Besides, this edition is a fine piece of bookmaking.

Granville, William A., Smith, P. F., and Longley, W. R., *Elements of the Differential and Integral Calculus*. Boston: Ginn & Co., 1934.

A good elementary textbook. Well printed, clear.

Hardy, G. H. *A Course of Pure Mathematics*. 6th Edition. Cambridge: Cambridge University Press, 1933.

A standard work, this elementary introduction to higher mathematics is a tough morsel but will repay whatever you invest in it in terms of effort.

Hogben, Lancelot T. *Mathematics for the Million*. New York: W. W. Norton, 1937.

The "Gone With The Wind" of mathematics. Well-written, interesting, this is by no means an easy book. The Marxian interpretation of the history of science and mathematics in Hogben's pungent style pervades the whole.

JEANS, SIR JAMES H. *The Mysterious Universe.* New York: Macmillan, 1930.

Lucid exposition of modern science. Sir James has an inimitable gift for making hard things easy, for making gigantic numbers easier to remember than telephone numbers, for making tenuous scientific theories more transparent than invisible glass.

KEYNES, JOHN MAYNARD. *A Treatise on Probability.* London: Macmillan, 1921.

It would be hard to find a more intelligent and comprehensive work on the philosophy and mathematics of probability, although many portions are only for the trained mathematician. Parts of it, nevertheless, are easily understandable, and the style, like Bertrand Russell's, is sparkling.

KEYSER, CASSIUS, J. *Mathematical Philosophy.* New York: Dutton, 1922.

Well-written, not too technical, and clear. In addition, Professor Keyser has written many other brilliant essays on mathematics and the humanizing of mathematics. He would deserve the title of the Grand Old Man of Mathematics—were it not for the fact that it would never occur to anyone that Keyser at 78 is anything but youthful.

KLEIN, FELIX. *Elementary Mathematics from a Higher Standpoint.* New York: Macmillan, 1932.

A mathematical classic, although not easy to follow for the non-mathematician. Contains a wealth of material, including a discussion of transfinite mathematics.

LEVY, HYMAN. *Modern Science.* Knopf, 1939.

This volume devotes considerable portions to the applications of mathematics to science. Professor Levy possesses unusual gifts of simplification. Highly recommended.

LIEBER, LILLIAN R. and HUGH G. *Non-Euclidean Geometry.* New York: Academy Press, 1931.

A delightful little book with charming illustrations, making the elements of non-Euclidean geometry readily understandable.

MANNING, H. P. *Non-Euclidean Geometry.* Boston: Ginn & Co., 1901.

A brief, rather simple text.

———. *The Fourth Dimension Simply Explained.* New York: Munn & Co., 1910.

A collection of essays on the fourth dimension, submitted in a prize contest run by *The Scientific American.* Many of the essays are amusing, much ingenuity being displayed in finding analogues of four-dimensional figures in a three-dimensional world.

MARK, THIRRING, NÖBELING, HAHN, and MENGER. *Krise und Neuaufbau in den exakten Wissenschaften.* Vienna: Franz Deuticke, 1937.

A collection of German essays by well-known physicists and

mathematicians on the revolutionary aspects of modern science. All the contributors to this splendid work were members of the celebrated "Vienna Circle." The lecture by Hans Hahn is an interesting presentation of some of the more startling paradoxes of modern mathematics.

MERZ, J. T. *History of European Thought in the 19th Century*. Edinburgh: Blackwood & Sons, 4th Edition, 1923.

In this monumental and always eminently readable work, there is a very full treatment of the development of mathematics in the nineteenth century—particularly statistics and the theory of probability.

PEIRCE, CHARLES S. *Chance, Love, and Logic*. New York: Harcourt, Brace, Inc., 1923.

A collection of philosophical essays, particularly on the subject of probability, by one of America's most distinguished philosophers—the founder of pragmatism.

POINCARÉ, HENRI. *The Foundations of Science*. Garrison, New York: The Science Press, 1913.

Anything Poincaré wrote is worth reading. The pellucid quality of his style makes all of his nontechnical writings readily comprehensible to the layman.

RUSSELL, BERTRAND. *Introduction to Mathematical Philosophy*. London: Allen and Unwin, 1919.

A standard work, but sometimes tough sledding. No discussion of this subject is easy, but Russell is always readable.

———. *Mysticism and Logic*. New York: Longmans Green, 1919.

The essay, "Mathematics and Metaphysicians" in this collection is in Russell's best style: brilliant, impudent, and jaunty. The problems of infinite classes and Zeno's paradoxes are particularly well treated. The essay "A Free Man's Worship" is one of the finest and noblest expressions of faith in science and reason in the English language.

SMITH, DAVID E. *History of Mathematics*, Vols. I and II. Boston: Ginn & Co., Vol. I, 1923; Vol. II, 1925.

A good history, profusely illustrated and well-suited for the nonprofessional.

STEINHAUS, H. *Mathematical Snapshots*. New York: G. E. Stechert, 1938.

Intended to reveal some of the more unusual aspects of mathematics, including mathematical paradoxes. Requires no mathematical training whatsoever.

SULLIVAN, J. W. N. *Aspects of Science*, 2nd Series. London: W. Collins Sons, 1926.

Stimulating essays for the layman on a variety of subjects, including mathematics. Recommended, by the same author, *Limitations of Science*, Chatto and Windus, London, 1933.

SWANN, W. F. G. *The Architecture of the Universe*. New York: The Macmillan Co., 1934.

A physicist tells how probability is used in studying the laws of gases. Well written, nontechnical.

WHITEHEAD, ALFRED N. *An Introduction to Mathematics*. New York: Henry Holt, 1911.

This might well serve as a model for all popular books on mathematics. Nothing quite as good appeared before or since its publication. Designed for the layman, it is simple but not condescending, witty but free from "epigramitis," perfectly clear, informative, full of verve, good humor and understanding. A first-rate job in every respect.

YOUNG, J. W. *Fundamental Concepts of Algebra and Geometry*. New York: Macmillan, 1911.

A most enjoyable collection of lectures. Suitable for the neophyte.

Index

Abbott, Edwin Abbott, 128, 363
Abel, Niels Henrik, 17
Advanced Calculus, 298
Absolute rest, 24
Absolute zero, 24
Acceleration, 301, 326, 342–343
Achilles and the tortoise paradox, 37, 38, 57–58, 62
Ahrens, Wilhelm Ernst Martin Georg, 189, 191, 363
Air-raid casualties, statistics of, 262–263
Alcuin, 159
d'Alembert, Jean le Rond, 246
Aleph-Null, 45, 54
Alephs, 45, 46–47, 62, 63
Alexander of Macedon, 343
Algebraic equations with integer coefficients, 6, 49, 64, 71, 79, 84, 109, 111
Algebraic invariant, 298
Algebraic numbers, 49, 50, 110
Alte Probleme—Neue Lösungen, 219
American Journal of Mathematics, The, 178, 191
American Mathematical Monthly, 363, 364
Amusements in Mathematics, 189
Analysis situs, 266, 271–274, 287
Analytical geometry, 95–99, 103, 120–123, 305, 306, 314
Analytical Theory of Heat, The, 254
Analytic theory of probability, 240
Annals of Mathematics, 191
Anthropology, statistics in, 260–261
Anti-snowflake curve, 348–351
Apollo, 71
Apollonius, 12, 13
Applied mathematics, 114, 116, 150, 151
Approximation, methods of, 318, 333–334, 336–337
Arabs, 17, 185
Arago, Dominique François Jean, 190
Archimedean number (*see* π), 75
Archimedes, 33–34, 69, 74, 310, 331
Architecture of the Universe, The, 367
Area under a curve, 335–339
Argand, Jean Robert, 100
Aristotle, 62, 117, 213, 230–231
Arithmetic, 28, 43, 46, 68, 81, 92, 300, 341
Arithmetic fallacies, 208–211
Arithmetic tricks, 163–164
Arithmetical progression, 82
Arrow in flight paradox, 37, 38-39, 58–60, 62, 301, 321
Aspects of Science, 367
Augustine, Saint, 112, 150
Axiom, reducibility, 63, 218, 221

Babylonians, 190
Bachet, Claude-Gaspard, Sieur de Meziriac, 157
Ball, Walter William Rouse, 189, 190, 191, 219, 221, 363
Banach, Stephen, 205–207, 219
Banach and Tarski's theorem, 205–207, 219
Barrow, Isaac, 323
Beethoven, Ludwig van, 362
Bell, Eric Temple, 142, 363
Beltrami, Eugenio, 142
Bergson, Henri, 65, 108

Berkeley, George, Bishop, 40
Bernouilli, James, 240
Binary notation, 165–173
Binomial theorem, 249–251, 264
Bismarck, Otto Edward Leopold, Prince von, 177
Black, Max, 363
Bliss, Gilbert Ames, 363–364
Bolsheviks, 8
Boltzmann, Ludwig, 95, 258
Bolyai, Johann, 118, 135
Bolzano, Bernhard, 39, 40–42
Bombelli, Raphael, 91
Booth, John Wilkes, 236
"Boss puzzle" (*see* "15 Puzzle")
Bouton, 191
Boyle, Robert, 257–258
Bridgman, Percy Williams, 84, 364
Briggs, Henry, 81, 84
Brouncker, William, Viscount, 78
Brouwer, Luitzen Egbertus Jan, 37, 221, 291, 292, 298
Brouwer's problem, 287–291
Brownian movement, 24–25
Budget of Paradoxes, 79, 261, 365
Buffon, George Louis Leclerc, Count de, 110, 246–247
Bürgi, Jobst, 110

Cabalists, 269
Cajori, Florian, 81, 298, 355, 364
Calculus (299–343):
  differential, 14, 75, 299, 310, 322–329, 330, 341
  integral, 75, 299, 310, 329–342
  of probability, 228, 229, 230, 238–251, 252, 253
Cantor, Georg, 39, 41, 42–57, 62, 63, 194, 364
Cantor's diagonal array, 52–53
Cantor's paradox, 43, 44
Cantor's theorem, 43–44, 55
Card tricks, 163
Cardan, Girolamo, 91, 168
Cardinality, 44, 45, 46, 48, 49, 54, 55
Cardinal numbers, 64
Cardinal of the continuum, 53, 55
Carlyle, Thomas, 39
Carnera, Primo, 23
Carroll, Lewis, 61
Carslaw, Horatio Scott, 364
Casting out nines, 164–165
Catenary, 107
Catherine of Aragon, 236
Catherine II of Russia, 265
Cauchy, Augustin Louis, Baron, 14
Cavalieri, Bonaventura, 332
Cavalieri's theorem, 332
Cayley, Arthur, 118, 156, 186, 298
Centroid, 340
Ceulen, van, Ludolph, 75
Chaldeans, 28
Chance, 223–264
  calculus of, 239–251
Chance, Love, and Logic, 264, 367
Change, 300, 301, 302, 304, 321
Charlemagne, 157, 159
Chaucer, Geoffrey, 3
Chess, 32–33, 68, 115, 173, 222
Chinese, 90
Chinese ring puzzle, 169
Circle, 10–14, 17, 41–42, 66–67, 73, 78, 80, 277, 278, 279, 280, 282, 310–312, 337, 343–344
  squaring of the, 12, 65, 66–69, 71–79, 109, 331
Classes, 28–31, 42–58, 61, 62, 63, 120, 132, 151–152, 154, 216–217
Class of all classes not members of themselves, 216–217
Clausius, Rudolf Julius Emmanuel, 258
Clebsch, Rudolf Friedrich Alfred, 298
Clock, 14
Coat and vest problem, 286–287
Coefficients of an equation, 109–110
Cohen, Morris Raphael, 137, 150–154, 224, 264, 364
Colerus, Egmont, 155
Collected Scientific Papers (of P. G. Tait), 191
Combinations, 242–244

Combinatorial analysis, 242, 243
Complex numbers, 95, 101–102, 103, 210
Complex plane, 101–102
Compound interest, 86–87, 111
Compound probability, 248–251
Comte, Auguste, 67
Condorcet, Marie Jean Antoine Nicolas, Marquis de, 253–254
Confessions (of St. Augustine), 150
Congruence, 126, 202, 204, 205, 206, 219
Conic sections, 17, 107
Connectivity:
  nonsimple, 279–283, 284
  simple, 276–277, 281, 282, 283, 290
Continued fractions, 75, 78, 85–86
Continuum, 53, 55, 56, 58
Continuum, cardinal of the, 53, 55
Contributions to the Founding of the Theory of Transfinite Numbers, 364
Convergence of a series, 64, 69, 70, 109
Cooley, Hollis Raymond, 364
Co-ordinates, 96–99, 102, 120, 121, 122, 123, 314
Cosine, 355–356
Countable classes, 45, 47, 49
Counting, 27, 28–30, 34
Courant, Richard, 364
Course of Pure Mathematics, A, 365
Craig, John, 261
Creative Evolution, 108
Cretans, 62–63, 221
Crime of Sylvester Bonnard, The, 189
Crisscross curve, 353–355
Cryptograms, 233–235
Cubic equation, 17
Cubit, 173, 191
Curtate cycloid, 200
Curvature, 28, 147–148, 328
Curves:
  anti-snowflake, 348–351
  catenary, 107
  circle, 10–14, 17, 41–42, 66–67, 73, 78, 80, 277, 278, 279, 280, 282, 310–312, 337, 343–344
  clock, 14
  crisscross, 353–355
  curtate cycloid, 200
  cycle, 11–13
  cycloid, 196–200
  ellipse, 17
  of error, 257–260
  hyperbola, 17
  nonsimply-connected, 7, 281
  parabola, 17, 106–107, 308, 330, 331
  patho-circle, 13
  pathological, 343–355
  prolate cycloid, 198–199
  simply-connected, 7, 276–282
  snowflake, 344–350
  space-filling, 352–353
  tractrix, 141, 142, 143
  turbine, 9–10
Cycle, 11–13
Cycloid, 196–200

Dampier, William, Sir, 259, 260
Dantzig, Tobias, 111, 264, 364–365
Darwin, Charles Galton, 264
Dase, Johann Martin Zacharias, 77
Da Vinci, Leonardo, 362
Davis, Watson, 25
Decimal, nonterminating, 51–53, 64, 167
Decimal notation, 164, 165, 166, 167
Dedekind, Richard, 41
Deductive method, 232
Definite integral, 339–341
Delphic oracle, 71, 157
De Moivre, Abraham, 103
De Morgan, Augustus, 79, 228, 261, 365
Denumerable classes, 45, 49, 50
Denumerably infinite classes, 45, 49

Dependent variable, 314, 315, 316, 321
Derivative, 14, 111, 322–329, 341–342, 356
Descartes, René, 65, 95, 96, 290, 306, 359
Diagonal proof, Cantor's, 52–53
Dice, 237, 239, 244, 245
Dichotomy paradox, 37–38
Die Paradoxien des Unendlichen, 40
Differential calculus, 14, 75, 299, 310, 322–329, 330, 341
Differential and Integral Calculus, 364
Differentiation, 5, 322–323, 325–326, 341, 342
Dimension, 119–121
Diophantus, 187
Discriminant, 298
Divergence of a series, 109
Doughnut, 281, 282, 287
Drawing cards from a pack, 237, 243–244
Dresden, Arnold, 191, 365
Doyle, Arthur Conan, Sir, 263
Dudeney, Henry Ernest, 189
Dürer, Albrecht, 185
Duodecimal notation, 190
Duplication of the cube, 12, 66, 68, 71–72, 109
Dyadic notation, 165–173

e, 49, 50, 66, 80, 84–86, 87–88, 89, 110, 111
Eddington, Arthur Stanley, Sir, 23, 32, 131, 154, 365
Egyptians, 8, 17, 66, 74, 167
Einführung in das mathematische Denken, 219
Einstein, Albert, 22, 23, 29
Elementary Mathematics from a Higher Standpoint, 366
Elements, 9–10, 151–152, 154
Elements (of Euclid), 4, 108, 113
Elements of the Differential and Integral Calculus, 365
Elements of Non-Euclidean Plane Geometry and Trigonometry, 364
Ellipse, 17
Ellipsoid, 147
Encyclopédie, 246
Enriques, Federigo, 365
Epimenides paradox, 62–63, 221
Equations:
  algebraic, 6, 298
    with integer coefficients, 6, 49, 64, 71, 79, 84, 109, 111
  cubic, 17
  quadratic, 17
  quartic, 17
  quintic, 17–18
Equiprobability, 241–242, 248, 251
Error, probability curve of, 257–260
Essai philosophique sur la probabilité, 264
Euclid, 4, 62, 69, 108, 113–114, 115, 134, 135, 137–140, 143, 147–150, 187, 192, 271, 304
Euclidean geometry, 116, 134, 137, 138, 139, 140, 143, 149, 150
Euclidean manifold, 123–124
Euler, Leonhard, 85–86, 92, 93, 103, 156, 185, 265–268, 269, 290–291
Euler's theorem, 266–268, 290–291
Evolution, 6, 93
Existence, 61–62, 63
Exponential function, 88–89, 111, 300, 342
Exponents, 82, 110

Factorial, 111
Fallacies:
  arithmetic, 208–211
  geometric, 211–213
  mathematical, 207–213
Falling bodies, 316–321, 326–328, 342–343
Fermat, Pierre de, 68, 156, 187, 188, 239–240, 306
Fermat's last theorem, 68, 187–188
"15 Puzzle," 177–180

Finite classes, 43
Finite numbers, 19, 22, 23, 32–33, 47, 63
Flatland—A Romance of Many Dimensions, 128, 363
Flaubert, Carolyn, 158
Flaubert, Gustave, 158
Formalists, 62, 222
Forsyth, Andrew Russell, 154
Foundation of Physics, 154
Foundations of Science, The, 367
Four-color problem, 287–291
Four-dimensional geometry, 93, 113, 115, 116, 118, 119, 124, 300, 359
Four-dimensional manifold, 123–124, 154
Fourier, Jean Baptiste Joseph, 254–255
Fourth dimension, 116, 117, 118, 119, 124, 125, 127–128, 130, 131
Fourth Dimension Simply Explained, The, 366
Fractions:
  continued, 75, 78, 85–86
  rational, 48
France, Anatole, 189
Franklin, Benjamin, 186
Frege, Gottlob, 218
Function, 5, 14, 304, 309, 312–316, 318–327, 332
  exponential, 88–89, 111, 300, 342
  monogenic, 14
  one-valued, 64
  polygenic, 14
  trigonometric, 355–356
Functions of a real variable, theory of, 201
Fundamental Concepts of Algebra and Geometry, 154, 368

Galileo, 41, 198, 252, 365
Galois, Évariste, 71
Gans, David, 364
Gases, kinetic theory of, 257–260
Gauss, Karl Friedrich, 68, 77, 92, 101, 108, 134, 135, 258
Geodesic, 146, 147, 181
Geometric fallacies, 211–213
Geometric mean, 99–100
Geometric progression, 82
Geometry:
  analytical, 95–99, 103, 120–123, 305, 306, 314
  Euclidean, 116, 134, 137, 138, 139, 140, 143, 149, 150
  four-dimensional, 93, 113, 115, 116, 118, 119, 124, 300, 359
  Lobachevskian, 136–139, 140, 142–143, 146, 150
  non-Euclidean, 113, 132, 134–150, 155, 300, 359, 360
  nonquantitative, 272–297
  Riemannian, 139–140, 142, 143, 144, 145, 146, 149, 150
  rubber-sheet, 265–298
Geometry of Four Dimensions, 154
Gestapo, 216
"Going to Jerusalem," 29–30
Goldbach, C., 187
Goldbach's theorem, 187
Gold Bug, The, 233
Googol, 20–25, 27, 32, 33, 47, 81, 221
Googolplex, 23–25, 31, 32, 44, 81
Granville, William Anthony, 365
Graph (topological), 266
Graphic representation of points, 96–99, 314
Grassmann, Hermann, 118
Gravitational constant, 327–328, 342–343
Great circle, 143, 146, 147, 181
Greeks, 8, 16, 34, 40, 66, 71, 90, 95, 108, 213
Grimaldi, Francis Maria, 166
Groups, 4, 5, 8
Growth, phenomenon of, 88, 342
Guessing numbers, 163, 164

Hahn, Hans, 298, 366
Haldane, John Burdon Sanderson, 262–263

Hall, Henry Sinclair, 190
Hamilton, William Rowan, Sir, 156
Hardy, Godfrey Harold, 32, 365
Hausdorff, Felix, 204–205, 207
Hausdorff's theorem, 204–205, 207, 219
Henry IV of France, 75
Henry VIII of England, 236
Heraclitus, 27
Hermite, Charles, 111
Hessian, 298
Higher Algebra, 190
Higher derivatives, 326–329
Hilbert, David, 222
Hindus, 17, 90, 268
Historic Development of Logic, 365
History of Elementary Mathematics, A, 364
History of European Thought in the 19th Century, 367
History of Japanese Mathematics, A, 191
History of Mathematics, 364, 367
History of Philosophy, Science and Technology in the Sixteenth and Seventeenth Centuries, 356
History of Science and its Relations with Philosophy and Religion, A, 259
Hitler, Adolf, 228
Hogben, Lancelot Thomas, 365–366
Homer, 362
Hottentots, 19
Houdini, Harry, 127
How to Draw a Straight Line, 290
Hyperbola, 17
Hypercube (*see* Tesseract)
Hyperplane, 124
Hyperradical, 16
Hypervolume, 126

i, 65, 66, 89, 91, 93–95, 99, 100, 101–103, 110, 210
Illusions, optical, 211, 220–221
Imaginary numbers, 65, 66, 90–95, 99, 100, 103, 117, 210
Impossibility, mathematical, 67–68
Indefinite integral, 340–342
Independent events, 244–247
Independent variable, 314, 315, 316, 321, 326
Induction, mathematical, 35–36, 232
Infinite classes, 31, 43–44, 45–46, 48–50, 53, 55–58, 61, 62, 63, 132
Infinite numbers, 19, 20, 22, 23, 33–36
Infinite series, 38, 64, 65, 66, 69–70, 75, 76, 77, 78, 79, 80, 87, 109, 346–347
Infinitesimal, 40, 43, 332
Infinity, 19, 35, 36, 40, 41, 43, 56, 61, 63, 132, 222, 332
Integers, 34–35, 43–45, 46–49, 50, 51, 52, 54, 63, 92, 108, 110
Integral calculus, 75, 299, 310, 329–342
Integrals:
definite, 339–341
indefinite, 340–342
Integration, 5, 339–340, 341, 342
Interlocking rings, 285, 286, 287
Introduction to Logic and Scientific Method, An, 150–154, 264, 364
Introduction to Mathematical Philosophy, 367
Introduction to Mathematics, 364
Introduction to Mathematics, An, 368
Intuitionists, 62, 221–222
Invariants, 273–274, 297, 298, 300
Invitation to Mathematics, An, 190, 365
Involution, 54
Irrational numbers, 49
Italians, 17

Jacobian, 298
Japanese, 175–176

Jeans, James, Sir, 258, 366
Jeu du Taquin (*see* "15 Puzzle")
Johnson, Samuel, 28
Johnson, William Woolsey, 191
Jordan, Camille, 7, 276–279
Jordan's theorem, 7, 276, 278, 279
Josephus, 173, 174
Josephus problem, 173–176

Kant, Immanuel, 6, 118
Kasner, Edward, 23
Kempe, Alfred Bray, Sir, 290
Kepler, Johann, 156, 252, 331–332
Keynes, John Maynard, 238, 263, 366
Keyser, Cassius Jackson, 13, 366
Kinetic theory of gases, 257–260
Kirchhoff, Gustav Robert, 276
Klein, Felix, 299, 366
Kline, Morris, 364
Knight, Samuel Ratcliffe, 190
Kries, Johann von, 263
Krise und Neuaufbau in den exakten Wissenschaften, 154, 367
Kronecker, Leopold, 35

La Fontaine, Jean de, 7, 217
Lagrange, Joseph Louis, Count, 156, 253
Laguerre, Jean Henri Georges, 11, 12, 13
Laplace, Pierre Simon, Marquis de, 221, 252–253, 264
Lazzerini, 247
Lebensraum, 293
Lee, Samuel, 261
Legouvé, Ernest, 183–184
Leibniz, Gottfried Wilhelm, Baron von, 39, 40, 75–77, 91, 156, 165, 166, 240, 304, 309
Length of a curved line, 331, 333–335, 356
Les problèmes plaisants et délectables, 157
Levy, Hyman, 366
Lieber, Hugh Gray, 366
Lieber, Lillian R., 366
Lietzmann, Walther, 189, 219
Limit, 64, 69, 70, 235, 310, 332
  of a function, 312–316, 321–327, 332, 341
  of a variable quantity, 310–312
Lincoln, Abraham, 236
Lindemann, Ferdinand, 71, 72, 79, 111
Lindsay, Robert Bruce, 154
Listing, Johann Benedict, 272
Lloyd, Sam, 177, 178
Lobachevsky, Nikolai Ivanovich, 118, 135–143, 146–147, 150, 181, 271
Lobachevskian geometry, 136–139, 140, 142–143, 146, 150
Locke, John, 225, 252
Logarithms, 50, 80, 81–84, 110, 210–211
Logic, 300, 360–361
Logic of Modern Physics, The, 364
Logical paradoxes, 194, 213–219
Logical positivism, 359–361
Logistic school, 62, 63, 222
Longley, William Raymond, 365
Lucas, Edouard, 189
Ludolphian number (*see* π), 75
Lustiges und Merkwürdiges von Zahlen und Formen, 189, 219

Magic squares, 185–186
MacFarlane, Alexander, 184
Machin, John, 75–76, 78, 118
Manifold, 119–124, 133, 149, 154
  Euclidean, 123–124
  four-dimensional, 123–124, 154
  nonsimply-connected, 281
  simply-connected, 282
  three-dimensional, 120–124, 154, 282–284
  two-dimensional, 278, 279, 282
Manning, Henry Parker, 366
Map-coloring problem, 186, 287–291
Margenau, Henry, 154
Mark, Hermann, 366
Matching, 28
Materialism, 254–256

Mathematical Excursions, 189
Mathematical fallacies, 207–213
Mathematical impossibility, 67–68
Mathematical induction, 35–36, 232
Mathematical Philosophy, 366
Mathematical possibility, 242
Mathematical recreations, 156–192
Mathescope, 25–26
Mathematical Recreations and Essays, 189, 363
Mathematical Snapshots, 367
Mathematics for the Million, 365
Mathematische Unterhaltungen und Spiele, 189, 363
Maxima and minima, 306–309, 325–326
Maxwell, James Clerk, 258
Measuring, 28, 66, 133, 134, 202–203
Mécanique Céleste, 221, 253
Men of Mathematics, 363
Mendel, Gregor Johann, 40
Menger, Karl, 219, 366
Méré, Antoine Gombault, Chevalier de, 239–240
Merrill, Helen Abbot, 189
Mersenne, Marin, 187
Merz, John Theodore, 367
Metamathematics, 222
Metaphysics, 359–361
Methods of approximation, 318, 333–334, 336–338
Mikami, Yoshio, 191
Miracles, 24, 261–262
Mirifici Logarithmorum Canonis Descriptio, 80, 110
Möbius, August Ferdinand, 118, 186
Möbius strip, 118, 186, 284–286
Modern Science, 366
Mohammedans, 261, 268
Moment of inertia, 340
Monkey and rope puzzle, 192
Monogenic functions, 14
More, Henry, 118
Motion, 37, 58–60, 194–195, 274, 300, 301–304, 321, 342–343
Mutually exclusive events, 243–244
Mysterious Universe, The, 366
Mysticism and Logic, 37, 367

Nagel, Ernest, 150–154, 224, 264
Napier, John, 80–84, 110
Napoleon, 253
Nature, 262
Nature of Mathematics, The, 363
Needle problem, 110, 246–247
Negative curvature, 147–148
Negative numbers, 65, 90–91, 92, 111, 117, 210
Newcomb, Simon, 78
New Pathways of Science, 365
Newton, Isaac, Sir, 75, 80, 252, 261, 304, 309, 323
Nim, 172, 191
Nines, casting out, 164–165
Nöbeling, Georg, 154, 366
Nondenumerable classes, 49, 50, 53
Non-Euclidean geometry, 113, 132, 134–150, 155, 300, 359, 360
Non-Euclidean Geometry, 366
Nonquantitative geometry, 272–297
Nonsimple curves, 7, 281
Nonterminating decimals, 51–53, 64, 167
Notation:
  binary, 165–173
  decimal, 164, 165, 166, 167
  duodecimal, 190
  dyadic, 165–173
  positional, 80, 110
  sexagesimal, 190
Number, the Language of Science, 111, 264, 364–365
Numbers:
  cardinal, 64
  complex, 95, 101–102, 103, 210
  imaginary, 65, 66, 90–95, 99, 100, 103, 117, 210
  infinite, 19, 20, 22, 23, 33–36
  irrational, 49
  negative, 65, 90–91, 92, 111, 117, 210

Numbers (*Continued*):
ordinal, 64
prime, 64, 108, 187–188, 192
rational, 49, 50
real, 49, 50, 51, 53, 54, 55
transcendental, 6, 49, 50, 53, 64, 78, 79, 84, 111, 117
transfinite, 45, 47, 49, 53, 54, 55, 63, 194, 359

Omar Khayyám, 90, 111
On the Hypotheses Which Underlie the Foundations of Geometry, 139
One-sided surface, 118, 186, 284–286
One-to-one correspondence, 29, 31, 34, 44, 45, 48, 49, 50, 56, 57, 58, 59, 60, 206
One-valued functions, 64
Optical illusions, 211, 220–221
Ordinal numbers, 64
Osgood, William Fogg, 298

Parabola, 17, 106–107, 308, 330, 331, 356
Parabolic segment, 356
Paradoxes, 193–222
Achilles and the tortoise, 37, 38, 57–58, 62
arrow in flight, 37, 38–39, 58–60, 62, 301, 321
Cantor's, 43, 44
Dichotomy, 37–38
Epimenides, 62–63, 221
of infinite classes, 43, 44
logical, 194, 213–219
railway train, 200
rolling circles, 195–196
rolling coin, 194–195
Russell's, 216–217
Parallel lines, 65, 134–140, 143, 154–155
Parhexagon, 14–16
Pascal, Blaise, 112, 156, 239–240
Patho-circle, 13
Pathological curves, 343–355
Peirce, Benjamin, 103
Peirce, Charles Sanders, 237–238, 256–257, 264, 367
Permutations, 242–243
Peterson, 261
π, 41–42, 49, 50, 61, 66–67, 71, 72, 74, 75–79, 89, 109, 110, 111, 246, 247
Plane, 120, 124, 147
Plato, 134
Plücker, Julius, 118
Poe, Edgar Allan, 233
Poincaré, Henri, 35, 63, 272, 298, 367
Point sets, theory of, 201–204, 219, 291
Poisson, Simeon Denis, 160, 190
Polygenic functions, 14
Possibility, mathematical, 242
Positional notation, 80, 110
Positive curvature, 147–148
Pouring problems, 161–162, 190, 191
Pretzel, 278, 283, 284, 287
Prime numbers, 64, 108, 187–188, 192
Principia, 80, 261
Principia Mathematica, 218
Principle of insufficient reason, 229–230, 263–264
Probability (223–264):
compound, 248–251
equiprobability, 241–242, 248, 251
independent events, 244–247
mutually exclusive events, 243–244
principle of insufficient reason, 229–230, 263–264
relative frequency view, 230–237
statistical interpretation, 230–237
subjective view, 227–230
truth frequency theory, 238
Problems:
of Apollonius, 12–13
Brouwer's 291–297
of coat and vest, 286–287

Problems (*Continued*):
- drawing cards from a pack, 237, 243–244
- of existence, 61–62
- of falling bodies, 316–321, 326–328, 342–343
- four-color, 287–291
- of interlocking rings, 285, 286, 287
- map-coloring, 186, 287–291
- Möbius surfaces, 118, 186, 284–286
- needle, 110, 246–247
- number theory, 186–188
- pouring, 161–162, 190, 191
- relationship, 183–185
- ring, 168–169
- river-crossing, 159, 189–190
- Russian multiplication, 167–168
- of seven bridges, 265–268, 270
- shunting, 159–160
- spider and fly, 181–182
- string, 191–192
- tower of Hanoi, 169–171

Projective geometry, 151
Prolate cycloid, 198–199
Propositional functions, 263
Pseudosphere, 140, 142–143, 146, 154–155
Ptolemy, 74, 117
Pure mathematics, 114, 116, 150–152
Puzzles, 156, 157, 158, 162, 168–169, 177–180, 189–190, 191–192, 300
Pythagoras, 28, 357, 359
Pythagoreans, 268
Pythagorean theorem, 121–122, 124

Quadratic equation, 17, 298
Quartic equations, 17
Queen of the Sciences, The, 363
Quételet, Lambert Adolphe Jacques, 260
Quintic equations, 17–18

Radical, 16–18
Radius of curvature, 328
Railway train paradox, 200
Ramsay, Frank Plumpton, 221
Rate of change, 111, 305, 316, 321–323, 326
Rational fractions, 48
Rational numbers, 49, 50
Real numbers, 49, 50, 51, 53, 54, 55
Reason and Nature, 154
Reasoning by probable inference, 224, 226
Reasoning by recurrence, 36
Recreations, mathematical, 156–192
Récréations Mathématiques, 189
Rectifying the parabola, 331, 356
Reducibility (*see* Axioms)
Relationship problems, 183–185
Relative frequency view of probability, 230–237
Rest, absolute, 24
Richter, 77
Riemann, Georg Friedrich Bernhard, 139, 140, 142–150, 181, 271
Riemannian geometry, 139–140, 142, 143, 144, 145, 146, 149, 150
Ring, 5, 279, 281
River-crossing problem, 159, 189–190
Rolling circles paradox, 195–196
Rolling coin paradox, 194–195
Rubber-sheet geometry, 265–298
Ruffini, Paolo, 17
Russell, Bertrand Arthur William, 37, 39, 56, 60, 218, 221, 222, 299, 367
Russell's paradox, 216–217
Russian multiplication, 167–168

Sanctions, 294
Sand Reckoner, The, 33–34
Sceptical Chymist, The, 257
Scheherezade, 235
Schubert, Hermann Caesar Hannibal, 176
Science and Hypothesis, 298
Seventh-Day Adventists, 148
Second derivative, 326–328

Self-consistency, principle of, 62–63, 115–116, 124
Semantics, 90
Series, infinite, 38, 64, 65, 66, 69–70, 75, 76, 77, 78, 79, 80, 87, 109, 346–347
Seven bridges' problem, 265–268, 270
Sexagesimal notation, 190
Shanks, W., 77–78
Sharp, Abraham, 77
Shunting problem, 159–160
Simple curves, 7, 276–282
Sine, 355–356
Skewes, 32
Skewes' number, 32, 33
Slope of a tangent, 324–325, 330, 356
Smith, David Eugene, 191, 367
Smith, Percey Franklyn, 365
Socrates, 228
Sophie Charlotte, Queen of Prussia, 39
Sophists, 213
Soviet Russia, 93
Space continuum, 58
Space, physical, 28, 56, 65, 112–113, 117, 119, 124, 131, 132, 133, 137, 149–150
Space-filling curve, 352–353
Space, Time, and Gravitation, 154, 365
Spencer, Herbert, 6
Sphere, 143–145, 147, 181, 203–204, 282, 283, 284, 290
Spider and fly problem, 181–182
Square root, 108, 209–210
Squaring of the circle, 12, 65, 66–69, 71–79, 109, 331
Squaring of the parabola, 330, 331
Statistical interpretation of probability, 230–237
Statistical view of nature, 254
Statistics:
  of air-raid casualties, 262–263
  in anthropology, 260–261
Steinhaus, Hugo, 367
Stereometria, 332
Story, William Edward, 191
Straight line, 146, 181–183
String problem, 191–192
Subjective view of probability, 227–230
Sue, Eugène Joseph Marie, 183
Sullivan, John William Navin, 367
Swann, William Francis Gray, 367–368
Sylvester, James Joseph, 118, 298

Tait, Peter Guthrie, 176, 191
Tangent of an angle, 355–356
Tangent to a curve, 324–325, 356
Tarski, Alfred, 205–207, 219
Tartaglia, Niccolò, 159
Terminating decimal, 64
Tesseract, 109, 125, 126
Theologiae Christianae Principia Mathematica, 261
Theorems:
  Banach and Tarski's, 205–207, 219
  binomial, 249–251, 264
  Cantor's, 43–44, 55
  Cavalieri's, 332
  Euler's, 266–268, 290–291
  Fermat's last, 68, 187–188
  geometric mean, 99–100
  Goldbach's, 187
  Hausdorff's, 204–205, 207, 219
  Jordan's, 7, 276, 278, 279
  parhexagon, 15–16
  probability of joint occurrence of two independent events, 244–245
  Pythagorean, 121–122, 124
Theory:
  of cycles, 12
  of groups, 5
  of point sets, 201–204, 219, 291
  of probability, 223–264
  of types, 63, 218, 221
Thirring, Hans, 366
Three-dimensional geometry, 120–123
Three-dimensional manifold, 120–124, 154, 282–284

Three-dimensional space, 56, 115–116, 130, 205, 206
Time continuum, 58
Todhunter, Isaac, 261
Topology, 266, 271–274, 276–297, 300, 359
Tossing coins, 225, 231, 232, 244, 245
Tower of Hanoi problem, 169–171
Tractrix, 141, 142, 143
Transcendental numbers, 6, 49, 50, 53, 64, 78, 79, 84, 111, 117
Transfinite numbers, 45, 47, 49, 53, 54, 55, 63, 194, 359
Treatise on Probability, A, 238, 263, 366
Trigonometric functions, 355–356
Trisection of an angle, 12, 68, 72, 109
Truth frequency theory of probability, 238
Turbine, 9–10
Twain, Mark, 156, 157
Two-dimensional space, 128
Two-dimensional manifold, 278, 279, 282
Two New Sciences, 365
Types, theory of, 63, 218, 221

Ultraradical, 16–18
Universe Around Us, The, 258

Variable quantities, 5, 305–306, 310–312
Variables:
dependent, 314, 315, 316, 321
independent, 314, 315, 316, 321, 326
Veblen, Oswald, 151
Velocity, 301, 342–343, 356
Veronese, 118
Vieta, Francisco, 74–75, 80–81
Vizetelly, Frank, 3
Vom Punkt zur vierten Dimension, 155
Vorstudien zur Topologie, 272

Wahlert, Howard E., 364
Waismann, Friedrich, 219
Wallis, John, 75–76, 78, 118
Weierstrass, Karl Theodor Wilhelm, 39, 43, 298, 332
Wessel, Caspar, 100
Weyl, Hermann, 221
Whitehead, Alfred North, 218, 368
Wolf, Abraham, 356

Yoneyama, 291
Young, John Wesley, 151, 154, 368

Zeno, 37–39, 57, 194, 301, 321
Zero, absolute, 24
Zeuxippus, 34

OTHER TITLES FROM TEMPUS BOOKS

THE WORLD OF MATHEMATICS
A Small Library of the Literature of Mathematics from A'h-mosé the Scribe to Albert Einstein. Presented with Commentaries and Notes by James R. Newman.
Foreword by Philip and Phylis Morrison

"...the most amazing selection of articles about mathematics yet unpublished...a delight to readers with a wide range of backgrounds."
The New York Times

"Newman writes with unfailing wit, tinted with irony and never dulled by cliché. But good writing would not be enough. He is able to make these 133 selections, chosen with great success to allow access by the nonspecialist, more accessible still by a brilliant page or two of his own introduction." Philip and Phylis Morrison, from the Foreword

First published in 1956 to widespread acclaim, THE WORLD OF MATHEMATICS is now available again. Out of print for many years, this four-volume anthology is a rich and spirited collection of 133 articles from the literature of mathematics. You'll find intriguing essays that provide windows on the history and concepts of pure mathematics, the laws of probability and statistics, puzzles and paradoxes, and the role of mathematics in economics, art, music, and literature. The selections range from hard-to-find classical pieces by Archimedes, Galileo, and Mendel to works of twentieth-century thinkers, including John Maynard Keynes, Bertrand Russell, and A. M. Turing. This new Tempus edition is completely recomposed, and reindexed.

2784 pages in four volumes, 6 x 9 boxed set
softcover, $50.00 Order Code: 86-96544
hardcover, $99.95 Order Code: 86-96593

MATHEMATICS: QUEEN & SERVANT OF SCIENCE

Eric Temple Bell

Foreword by Martin Gardner

"Bell is a lively, stimulating writer, with a good sense of historical circumstance...a sound grasp of the entire mathematical scene, and a gift for clear and orderly explanation."

James R. Newman, editor of *The World of Mathematics*

MATHEMATICS: QUEEN & SERVANT OF SCIENCE is a fascinating and lively survey of the development of pure and applied mathematics. E.T. Bell explores mathematical theories from those of Pythagoras and Euclid to those of Einstein and Gödel with unsurpassed vigor, intelligence, and wit. He covers a broad range of traditional subjects—algebra, geometry, calculus, topology, logic, rings, and groups—making even the most abtruse subject come alive. Published in cooperation with The Mathematical Association of America.

432 pages, 6 x 9, softcover, $11.95 Order Code: 86-96825

TIME: THE FAMILIAR STRANGER

J.T. Fraser

"Fraser is perhaps the leading intellectual authority in the world on the study of time. Now he has given us a book that is both stimulating and provocative." Jeremy Rifkin, author of *Time Wars*

This wide-ranging and learned book surveys the enormous variety of our understanding of time, both in the everyday world and in the sciences and humanities. From the visions of time and the timeless in religion, to those in contemporary physics and cosmology and the more common conceptions of time, J.T. Fraser offers you a fascinating history of the ideas and experience of time. The distillation of three decades of research, this volume considers such issues as the beginning and end of the universe; the biology of aging and death; human perception of time; dreaming *vs* waking reality expectation and memory; calendars and chronolgies; and the ways in which technological rhythms control our lives.

400 pages, 6 x 9, softcover, $9.95 Order Code: 86-96809

THURSDAY'S UNIVERSE

A Report from the Frontier on the Origin, Nature, and Destiny of the Universe

Marcia Bartusiak

"Marcia Bartusiak entered the universe of research at the frontiers of astronomy and cosmology and returned with a gem of a book. I recommend it." Heinz Pagels, author of *The Cosmic Code*

Cited by the *New York Times* as one of the best science books of 1987, and named a 1987 Astronomy Book of the Year by the Astronomical Society of the Pacific, THURSDAY'S UNIVERSE takes you on a tantalizing journey of discovery to the outer reaches of the cosmos. How did the universe begin? How will it end? What populates the endless expanse of space? Through THURSDAY'S UNIVERSE you will explore current ideas about the moment of creation; the birth and death of stars, quasars, and galaxies; the composition of black holes and neutrinos; and still-unanswered questions about the cosmos. Bartusiak will introduce you to the brilliant astronomers, astrophysicists, and researchers behind today's theories about the universe.

336 pages, 6 x 9, softcover, $8.95 Order Code: 86-96627

INVISIBLE FRONTIERS

The Race to Synthesize a Human Gene

Stephen S. Hall

"An important and pioneering book, dealing with events of high scientific and economic consequence...[Hall] succeeds marvelously in making the science accessible to the general reader."

New York Times Book Review

In 1976 scientists realized that synthesizing, or cloning, a human gene was imminently possible. INVISIBLE FRONTIERS takes you through the intense and dramatic race to clone a human gene and then to engineer the mass production of the life-sustaining hormone insulin. Stephen Hall recounts the developments of this high-stakes race, which resulted in the birth of the biotechnology revolution, a Nobel Prize, and the founding of the first genetic engineering company. INVISIBLE FRONTIERS, based on interviews with both major and minor participants, is an authentic and vivid account.

360 pages, 6 x 9, softcover, $8.95 Order Code: 86-96817

Simon & Schuster published the first edition of *Mathematics and the Imagination* in 1940. They reprinted it in 1967 as part of their Fireside Books series. In the twenty-two years since then, this timeless work has remained out of print. In an effort to preserve the original spirit of this classic as it first appeared nearly fifty years ago, this edition is reproduced from the original hot metal typography of the 1940 first printing. As a result, certain imperfections in the line and type may appear.

Cover illustration by Greg Hickman and Becky Geisler-Johnson
Color separations by Wescan Color Corporation
Principal production art by Mark Souder